Studies in Applied Philosophy, Epistemology and Rational Ethics

Volume 40

About this Series

Studies in Applied Philosophy, Epistemology and Rational Ethics (SAPERE) publishes new developments and advances in all the fields of philosophy, epistemology, and ethics, bringing them together with a cluster of scientific disciplines and technological outcomes: from computer science to life sciences, from economics, law, and education to engineering, logic, and mathematics, from medicine to physics, human sciences, and politics. It aims at covering all the challenging philosophical and ethical themes of contemporary society, making them appropriately applicable to contemporary theoretical, methodological, and practical problems, impasses, controversies, and conflicts. The series includes monographs, lecture notes, selected contributions from specialized conferences and workshops as well as selected Ph.D. theses.

Advisory Board

More information about this series at http://www.springer.com/series/10087

Sara Dellantonio · Luigi Pastore

Internal Perception

The Role of Bodily Information in Concepts and Word Mastery

Springer

Sara Dellantonio
Department of Psychology and Cognitive Sciences
University of Trento
Rovereto
Italy

Luigi Pastore
Department of Education Science, Psychology, Communication Science
University of Bari
Bari
Italy

ISSN 2192-6255 ISSN 2192-6263 (electronic)
Studies in Applied Philosophy, Epistemology and Rational Ethics
ISBN 978-3-662-55761-7 ISBN 978-3-662-55763-1 (eBook)
DOI 10.1007/978-3-662-55763-1

Library of Congress Control Number: 2017951182

Printed on acid-free paper

This Springer imprint is published by Springer Nature
The registered company is Springer-Verlag GmbH Germany
The registered company address is: Heidelberger Platz 3, 14197 Berlin, Germany

Acknowledgements

It has not been easy to write this book. It took a long time and considerable effort to bring together the many aspects and ideas, to ensure they were consistent with each other, to develop them into a coherent thesis, and to provide a good synthesis of the interdisciplinary fields of research under consideration. We tried to analyze the evidence available from various areas of philosophy and psychology and to develop a truly theoretical—i.e., philosophical—perspective. We leave the reader to judge whether this attempt was successful.

The very first person we would like to thank is Magda Elizabeth Altman. She not only edited this book from beginning to end, but was our first reader, a reader whose opinion also mattered. Often her comments were even more important than the editing itself: Her suggestions made this book far more clear and accurate. She not only helped us with the manuscript; her friendship and support were essential over this long endeavor.

The seminal ideas for this book date back many years and began to be developed in 2007–2008 when we were working on a project on the epistemological status of the notion of representation that was financed by CRUI, DAAD, and Ateneo Italo-Tedesco within the ambit of the 'Vigoni-Programm.' We would therefore like to thank Prof. Hans-Jörg Sandkühler for having promoted and supported this project and for his important theoretical contributions. We would also like to thank all the people who, directly or indirectly, contributed to the various initiatives we organized at the *Zentrum Philosophische Grundlage der Wissenschaften* at the University of Bremen in Germany and the Department of Bioethics at the University of Bari in Italy, by making suggestions and offering opinions on the perspectives we were proposing. Among them: Sahran Douhib, Silja Freudenberger, Lothar Knatz, Andreas Jürgens, Marco Innamorati, Furio Semerari, Francesco Saverio Trincia and Manfred Stöckler. These ideas were then further developed during a period of research we spent at the *Centre for Critical and Cultural Theory* at the University of Cardiff in the UK. Our sincere gratitude goes to Professor Ian Buchanan for sponsoring us as visiting fellows and for taking the time to discuss our work with us.

A first draft of some of the positions we argue for in this book was presented at several conferences and seminars held at the Universities of Bari, Bremen, Cologne, Guangzhou, Osnabrück, Siena, and Trento. The discussions we had at these events, with both our collaborators and the highly qualified public that attended, helped us to greatly improve our arguments. In this regard, we would especially like to thank Jagna Brudzinska, Vittorio De Palma, Remo Job, Ping Li, Dieter Lohmar, Valerio Meattini, Claudio Mulatti, Sandro Nannini, Henning Peuckert, Manfred Sommer, Achim Stephan, Michela Summa and Giuseppe Varnier. Moreover, our gratitude also goes to a number of colleagues and friends (mostly friends who also happen to be colleagues). They contributed in many ways to improving the quality of this book, through mutual discussions, collaborations, suggestions, criticisms, questions, reflections, and their own work: Saulo de Freitas Araujo, Pierluigi Barrotta, Claudia Bonfiglioli, Andrea Bosco, Roberto Cubelli, Antonio Di Oronzo, Nevia Dolcini, Gianluca Esposito, James Hartzell, Pim Haselager, Alexei Krioukov, Diego Azevedo Leite, Ferdinando Menga, Maria Paola Paladino, Alessandra Pastore, Francesco Pavani, Giuseppe Saracino, Maxine Sheets-Johnstone, Achille Varzi, and Francesco Vespignani. Some other friends—most notably Giorgio Busatto, Maria Micaela Coppola, Marta Cazzanelli, and Mario Velluso—were invaluable merely by their presence and by pursuing other projects with us that we very much care about.

Figures 1 and 2 in Chap. 3 and Fig. 1 in Chap. 4 include some hand-drawn illustrations: We are very grateful to Livia Cattani for her artwork.

Parts of sections §3, §4, §5 in Chapter 4 were taken from an article we published under the title *Sensing aliveness* in *Integrative Psychological & Behavioral Science*. We thank the editor-in-chief, Professor Jaan Valsiner, for giving us the permission to use this material.

We would also like to thank both our families. Even though they do not always understand what we do or why we do it, they are always supportive and extremely patient, tolerating our bad habit of working on Sundays and holidays.

A special thanks is due to Massimo Barale for supporting and encouraging us in so many ways. Sadly, he did not live to see this work in print: his unexpected departure was a great loss to us.

Last but not least, we would like to thank Prof. Lorenzo Magnani for believing in our project and for having the courtesy and the patience to let us work without pressure for such a long time, even though we made commitments to finish this work much earlier. Our thanks also go to our Springer Publishing Editor, Leontina di Cecco, who offered us her professional assistance from the book proposal to the submission of the manuscript.

This book was truly cooperative. Even though at the beginning we divided the chapters, thinking that each of us would be the first author for half of them, in the end we exchanged, modified, and rewrote them together so many times, moving sections back and forth between the chapters, that it is now impossible to say which one should be considered first authored by whom. So, all of them should be considered co-authored, while the order of the authors' names on the book is merely alphabetical. Of course, we both take full responsibility for any mistakes we may have made.

Contents

Introduction

The study of language is extremely important for an understanding and analysis of mental phenomena since it provides us with a window into our thought processes and grants us privileged access to the workings of the mind. In particular, investigating the semantic structure of language can shed light on the 'contents' our thinking processes rely on and help us address important questions such as what this content is, how it is constituted, and where it comes from.

Cognitive research assumes that our capacity to master (to use and to understand) natural languages can be explained on the basis of mental representations. Simply defined, representations are considered to be information units stored in the mind that convey the knowledge an organism has acquired and inform its behavior. At the heart of the cognitivist argument is the idea that the capacity to represent states of affairs in the world is a prerequisite for any kind of intelligent behavior.

Classical cognitive sciences distinguish between various kinds of representations, in particular between perceptual and conceptual representations. Perceptual representations convey information provided by a specific sensory modality, while conceptual representations are the product of categorization and are the means by which we identify and classify objects and events in the external world. Conceptual representations do not only include external information coming from various sensory channels, but are information structures that organize the incoming perceptual information into instances (tokens) and classes (types).

Conceptual representations are the precondition for acquiring and mastering the meanings of words in natural languages. To acquire a language, we need to develop particularly sophisticated ways of categorizing things in order to organize phenomena according to the criteria specified by the word meanings in that language. It is possible or even plausible that we also have concepts that do not correspond to any of the word meanings in our language; i.e., that some of the classifications we create mentally do not have any direct linguistic expression. However, in order to master the word meanings of our language, we must develop concepts that allow us:

- to identify instances referentially in the way our language does (to identify and classify as cats in the world all and only those instances denoted by the word 'cat');
- to create inferential links with other concepts (to master 'cat,' we must know that cats are animals; to master 'animal,' we need to know at least some basic things about animals; if we know cats are animals, we also know that they have the properties of animals, and so on).

Thus, if conceptual representations are the internal informational units that allow us to master word meanings, they must include all the information needed for linguistic competence; i.e., they must convey referential (perceptual) information about the instances words denote as well as inferential knowledge related to word meanings. Exactly what kind of perceptual and inferential information constitutes our conceptual representations is still an open research question, which this work aims to address.

The scope of this work is an investigation of conceptual representations. Semantic mastery—and, in particular, lexical competence—is treated as an external expression of our capacity to categorize. We consider various aspects of our semantic competence and examine some important deficits in word mastery exhibited by people suffering from specific pathological conditions. This provides indirect evidence as we seek to understand how categorization works and identify the kind of information categorization works on.

Our inquiry starts from what we consider to be a weak point of the classical studies on categorization and semantic mastery: Both the philosophical and the psychological traditions tend to explain conceptualization in relation to external perceptual experiences, i.e., starting from how we perceive the external world through the senses, placing considerable emphasis on vision, *without considering the contribution of bodily experiences*. In this work, we sketch the historical and methodological reasons that have led to neglecting the importance of internal perception (bodily experience) in the study of cognition. We maintain that—in spite of the undeniable difficulties in addressing such an evanescent phenomenon in a rigorous and scientific manner—the contribution of internal perception can no longer be disregarded. In particular, through our analysis we try to show that bodily experience is an essential informational component in many of the conceptual representations we have and use for semantic mastery.

In this book, we deal at length with the general issue of internal information—i.e., we examine why it was largely disregarded in earlier research, how it has been considered in previous theories, and how it should be conceptualized/understood. Furthermore, we discuss why/how this understanding is essential in explaining how we come to master at least some word classes, i.e., why/how it must be part of the information that composes at least some of our conceptual representations. The word classes/the kind of conceptual representations we consider in detail are those denoting *internal bodily states* (such as hunger or pain) and *emotions*. In addition, we examine *abstract words/concepts* (like freedom and friendship) and the case of *categories* meant as means of classification more general than concepts, which are distinguished from and preliminary to conceptualization and through which we partition sets of entities in ontological domains perceived to be homogeneous.

This book starts with a brief overview on the contemporary research on mind. The aim is to show that this research is characterized by a particular epistemological stance that has influenced the entire course of its development: Mind must be investigated on the basis of a scientific approach which meets the criteria of scientificity set by the natural sciences. To achieve this result, any reference to introspection and to those qualitative dimensions of mental states that are accessible only to the first-person should be excluded as much as possible from the study of mental processes. However, in spite of the methodological precepts dictated by this stance, we argue that subjective—intrinsic and phenomenal—elements may be crucial for some mental operations. Minimally, they appear to play a role in the identification and recognition—and thus categorization—of internal states.

In Chap. 2, we discuss how the issue of categorization has been addressed by the cognitive sciences and we pinpoint two main issues for this line of research. First of all, we briefly examine the well-known 'grounding problem' to show that the relationship between perception and cognition, or—to put it differently—the way in which representations in the central system (conceptual representations) are related to perceptions allowing us to referentially identify instances in the external world, has always been a challenging problem in cognitive research. We suggest that to deal with this problem we need to assume that conceptual representations have a perceptual core.

Secondly, we go back through the history of cognitive research in order to investigate the way in which it defines information processing, considered not only as one of the most basic of mental operations but also as central to understanding categorization in this tradition. We show that—as Ulric Neisser has already pointed out—the notion of information used in cognitive research is structured around the idea of information defined in the field of telecommunications. In the case of telecommunications, informational units are conveyed by neutral and unselective channels, which do not modify or influence the object of transmission. In the case of human information processing, however, information is not only transmitted, but is continuously selected and recorded. Furthermore, this information is processed by means of the body, and this process itself modifies the original stimuli in many ways as new information originating in the body itself is added. Thus, internal, bodily information is always part of the information we cognitively process. We call this kind of information propriosensitive or, in line with James Gibson's description, 'proprioceptive' (in inverted commas to distinguish our use of this notion from its current neuropsychological definition which is narrower in scope, indicating only a particular kind of internal information related to body position and body movements). We define 'proprioception,' on the basis of Charles Scott Sherrington's and James Gibson's view, as referring to all kinds of information produced by the body (including movement and postural information as well as any kind of sensations related to the general state of the body and its parts). Following Antonio Damasio and Bud Craig, we conjecture that living beings are equipped with an internal propriosensitive monitoring system, which maps all the bodily changes that constantly occur, providing us with perceptual ('proprioceptive') information on what happens inside us. In line with, e.g., Peter Goldie's and Matthew Ratcliffe's views, we argue that emotional information is also basically proprioceptive information. The working hypothesis we develop based on this analysis is that 'proprioception' is an essential component of the information living organisms rely on to develop their knowledge about the world. Our conclusion is that propriosensitive information must be as relevant for cognition as perceptual information.

We begin to address the matter of semantic competence in Chap. 3 of this work. Here, we consider the relationship between linguistic meaning and concepts (conceptual representations) and try to sketch some of the major features of our capacity to master words meanings. Relying on a distinction made by Diego Marconi, we define two aspects of our ability to master words, i.e., referential competence and inferential competence, and maintain that conceptual representations should enable both. Furthermore, we show that the most widely used and acknowledged theories of concepts do not distinguish between the referential and inferential components of our lexical competence. They rather characterize all information used for categorization indistinctly in terms of feature lists propositionally described. We suggest that this is precisely what prevents such theories from giving a satisfying account of referential competence. This argument leads us to reconsider the grounding problem discussed in the previous chapter from a different angle and to consider the position of authors such as Ray Jackendoff, Jesse Prinz, and Lawrence Barasalou, according to whom conceptual representations must include or at least be related to some kind of perceptual representations, which can ensure the referential relationship. However, we also argue that conceptual representations cannot be identified with perceptual representations *tout court*; otherwise, we would not be able to account for inferential competence. In the view we put forward, in order to explain how both referential competence and inferential competence work, conceptual representations should be viewed as information unities characterized by both inferential information and a perceptual core. To describe this core, we rely on Jesse Prinz's notion of 'proxytype' and on Mark Johnson, George Lakoff, and Jean Mandler's idea of a 'schematic image.' We integrate and amend these views. As a result, we define the notion of 'perceptual schema' and suggest that conceptual representations are characterized by a core consisting of perceptual schemata: They include nonlinguistic, structured experiential gestalts (patterns or maps) made up of information taken from all sensory modalities, including propriosensitive information. These accomplish a quasi-conceptual function: They allow us to recognize and classify different instances.

In Chap. 4 of the book, a particular aspect of categorization is considered which is preliminary for any further analysis of the human conceptual apparatus and for understanding how propriosensitive information contributes to its constitution, i.e., the difference between the notions of 'category' and 'concept,' which in the contemporary cognitive literature is often treated as synonyms. The chapter starts with a brief discussion on how the classical philosophical tradition (Aristotle, Kant, and Husserl) defines 'categories' and 'concepts' and how it distinguishes between them.

The view on categories we propose in this chapter is developed starting from an Husserlian perspective and references the work of Ray Jackendoff, Jean Mandler, George Lakoff, and Mark Johnson. We suggest that our knowledge is organized first of all categorically; this categorical organization is prelinguistic, works on the basis of universal structuring principles, and precedes the development of a conceptual system. This allows us to form homogeneous sets of entities partitioned into regions that describe the ontological boundaries of the objects that humans perceive. Starting from this early structuring of our experience, humans can then develop more specific classifications (concepts) which are the basis for language learning. One of the main issues we consider in relation to categories is whether they are innate or are rather formed before concepts on the basis of information and organizational structures that become available at a very early developmental stage. This point is addressed by considering a specific example of categorical partition, i.e., ANIMATE/INANIMATE, and we conjecture that at least with respect to this example there is no need to assume that categories are innately present in the human mind. Their formation can instead be explained starting from basic mechanisms that work on bodily information, i.e., on 'proprioception' as it was previously defined. Other general partitions like living/non-living or biological/non-biological can then be built on the basis of the ANIMATE/INANIMATE partition with the addition of cultural factors.

In Chap. 5, we consider the mastery of words denoting bodily experiences (e.g., 'pain,' 'hunger,' 'thirst') and emotions. The hypothesis we argue for is that—even though these words do not have any observable referent in the external world—our understanding of them depends on a specifically referential competence. Indeed, we identify and classify our bodily experiences and our emotions on the basis of how they feel, i.e., using 'proprioceptive' information provided by our internal, propriosensitive monitoring system. In this section, we defend a perceptual theory of emotions of the kind put forward originally by William James and Karl George Lange according to which emotions are something we experience in a direct and qualitative manner on the basis of the bodily sensations they give rise to. Moving from this theory, we try to show that there is continuity between bodily experiences such as hunger and pain and emotions. Following authors such as Keith Oatley, Philip Johnson-Laird, and Nico Frijda, we distinguish between basic and complex emotions and argue that basic emotions can be identified and classified uniquely relying on how they feel, while the identification of complex emotions needs to be additionally supported by cognitive factors. We propose that the identification and classification of both bodily experiences and basic emotions occur in the same way, more specifically based on four characteristics of the sensations we experience, i.e., the specific quality of the sensation (how it feels); its somatic localization; its intensity; and the presence or absence of specific concomitant sensations.

Emotional feelings are described starting from Anthony Marcel and John Lambie's view according to which—even though emotions correspond to specific bodily changes (we experience an emotion through certain bodily changes, e.g., changes in heartbeat, respiration, body temperature, muscle tension, facial expression)—these bodily changes *per se* are not identical with our emotional experience. In fact, to have an emotional experience, these bodily states need be integrated and reach our awareness not as a mere sum of physiological states but as a unitary, integrated, and conscious emotional experience. We show how this distinction between two stages in the processing of emotions is compatible with our criteria for how we identify and classify emotions. Furthermore, to support our view, we bring some evidence related to the semantic competence exhibited by people suffering from Alexithymia and Autism Spectrum Disorder.

Finally, in the last chapter, we consider the issue of how we master abstract words, i.e., of how we identify and categorize abstract entities or conditions. The main aim of this chapter is to challenge the standard picture of language according to which words in natural languages can be positioned along a spectrum that goes from concrete terms (whose meaning is determined primarily referentially on the basis of observable properties) to abstract terms (whose meaning is fixed inferentially by means of linguistic definitions). In the view we put forward, this opposition is misleading since it presupposes that all the words that do not rely on observable properties of the external word are *ipso facto* abstract. This would automatically include in the class of abstract words all the terms that are grounded in 'proprioception' instead of external perception. In this way, words such as those considered in the previous chapter denoting bodily experiences (e.g., 'pain,' hunger,' and emotions) should be viewed as abstract, whereas we concluded that they are deeply rooted in a form of propriosensitive experience that we described as largely analogous to perceptual experience. This is however not the only problem with this kind of approach to language. Indeed, here we try to show that there are a number of words which are in some respects definitional, but still rely heavily on propriosensitive information such that competence in their use also depends on this information. We follow authors such as George Lakoff and Lawrence Barsalou in maintaining that the knowledge we have of (at least some) abstract words depends on internal information that is collected through body monitoring. To sustain this position, we present two different lines of evidence coming on the one hand from psycholinguistic studies and on the other hand from research on the deficits of semantic competence exhibited by people with Asperger's syndrome. On the basis of our analysis, we suggest a classification which allows us to distinguish between different kinds of concreteness and different degrees of abstraction: Concepts/words referring to bodily experiences and basic emotions are described as analogous to concrete concepts/words because they are grounded in perceptual (i.e., propriosensitive) experience, while abstract concepts/words are considered more or less 'abstract' depending on how heavily they rely on inferential information or—inverting the problem—how heavily they rely on perceptual or propriosensitive information. In the last section of the chapter, we consider two scales that have been used in psycholinguistic research to measure the degree of concreteness versus abstractness of words and we show that—used jointly—they can give a measure of the internal versus external grounding of specific words.

Chapter 1
First Person Access to Mental States

Mind Science and Subjective Qualities

Abstract The philosophy of mind as we know it today starts with Ryle. What defines and at the same time differentiates it from the previous tradition of study on mind is the persuasion that any rigorous approach to mental phenomena must conform to the criteria of scientificity applied by the natural sciences, i.e. its investigations and results must be intersubjectively and publicly controllable. In Ryle's view, philosophy of mind needs to adopt an antimentalist stance to achieve this aim. Antimentalism not only definitively rejects the idea that mind is a substance separated from the body, it also denies that mental phenomena radically differ from physical phenomena by virtue of several unique features. Most problematically, mental phenomena have a conscious character (mental states are related to specific qualitative feelings) and are accessible only to the first-person (only the subject knows directly what s/he is experiencing inside his/her mind). Ryle takes a strong stance on antimentalism going so far as to maintain that an approach to mind which aims to meet the criteria of scientificity set by the natural sciences must avoid any reference to internal, unobservable mental states. In his view (which is considered a specifically philosophical version of psychological behaviorism and also addresses questions put forward in psychological research), mental states can be redescribed in terms of behavioral dispositions. In this chapter, we address the historical roots of the antimentalist view and analyze its relation to the later tradition of research on mind. We show that, compared to the antimentalist stance, functionalism and cognitivism take a step back when they maintain that direct reference to mental states is necessary since mental states cause and therefore explain human behavior. This step backwards is often interpreted as a return to mentalism. However, this is only partially true. Indeed, we suggest that these later traditions retain one important element of Ryle's antimentalism, i.e. the idea that mental states must be uniquely identified using external and publicly observable criteria, while excluding any reference to introspection and those qualitative dimensions of a mental state, which are accessible only to the first-person. According to the perspective we put forward, this epistemological stance has continued to influence contemporary research on mind and current philosophical and psychological theories which both tend to exclude the subjective qualities of human experience from their accounts of how the mind works. The issue we raise here is whether this is

S. Dellantonio and L. Pastore, *Internal Perception*, Studies in Applied Philosophy, Epistemology and Rational Ethics 40, DOI 10.1007/978-3-662-55763-1_1

legitimate or whether subjective qualities do play a role with respect to the way our mind works. The conclusion of this chapter anticipates the argument the book makes in favor of this latter position, starting from a particular angle, i.e. the problem of how we categorize concepts related to our internal states.

1 Mentalism and Contemporary Philosophy of Mind

Mental phenomena have always been an object of interest for philosophical reflection; the gnoseological tradition and the theory of knowledge are two paradigmatic examples in this regard. However, philosophy of mind, as the autonomous and specific research field that we know today, arose only in the second half of the 20th century. A possible birthdate could be 1949, when Gilbert Ryle published his best known work, *The Concept of Mind* (Ryle [1949] 2009).

With the publication of Ryle's book, there was a trend reversal in the way the philosophical tradition dealt with mental phenomena. Indeed, this volume lead philosophy to definitively give up *mentalism*,[1] i.e. the approach to mind and to mental phenomena embraced by most of the previous philosophical tradition. According to mentalism, the mind exists as an autonomous entity that can be studied separately and independently from all other phenomena in the world and the mind is causally responsible for human actions and behaviors as well as for the human capacity to develop knowledge. To investigate mental phenomena and to describe their properties and functions, mentalism developed a specific vocabulary gleaned in part from the previous philosophical tradition. The approach to mind initiated by the publication of Ryle's book challenged both this perspective and this vocabulary.[2] *In this sense, we can say that philosophy of mind as we know it today was born from an antimentalistic perspective.*

[1]The word *mentalism* was first used as a neologism by the English philosopher H. Sidgwick in 1901 to indicate a number of phenomena that by their nature cannot be explained materialistically (Sidgwick 1901). Sidgwick tried to find some form of alignment between idealistic vocabulary and the explanatory needs of the emerging field of experimental psychology. In this context the word mentalism was used to indicate all the knowledge and mental states that could be accessed from the inside (the interior life) e.g. perceptions, conscious states, emotions, etc. which had traditionally been the object of study of philosophy and had fallen within the focus of interest of psychology. As the French psychologist Burloud (1948: 9) usage of the word demonstrates, at a later stage *mentalism* was used to indicate all those views according to which subjective states are accessible through introspection. In the work of psychologists from the positivistic or behavioristic traditions who denied the utility of introspection and its epistemic legitimacy as a method to acquire knowledge on mind, this specific sense of the word assumed a negative value. For an overview of the various uses of this word made by the philosophical tradition and the related problems, see Marres (1989). To be thorough, it should be mentioned that the term mentalism is also used in the field of parapsychology to indicate a magical form of insight. However, this use of the term is completely extraneous to this work.

[2]Ryle's book represents an important systematic change in the way mental phenomena were approached since it addresses them from a linguistic point of view. Historically, this is not the first

Ryle considers mental phenomena as something asymptotically converging with the observable dimension of actions and behaviors. He suggests that research on mind, including philosophical research, be conducted similarly to the process of knowledge acquisition carried out in the natural sciences, i.e. its investigations and results must be intersubjectively and publicly controllable. In this vein and in contrast to the previous tradition of research, Ryle argues in favor of an *anti-speculative* and *anti-metaphysical* approach to mind and to mental phenomena. In the traditional philosophical lexicon, words like *speculation* and *metaphysics* had assumed different meanings, which were both positive and negative.[3] While e.g. the

(Footnote 2 continued)

attempt to interpret mental phenomena from a linguistic perspective. Another theoretically relevant, even if less influential proposal in this direction is Bertand Russell's volume *The Analysis of Mind* (1921). This work has the merit, among others, of bringing the behavioristic approach developed by Watson for psychology in the US to the attention of logical empiricists.

[3]Originally the word *speculation* indicated an essentially theoretical investigation and specifically a form of disinterested reflection, distinct from the empirical domain which is by nature linked to action and production. In speculation, one seeks truth relying on the strength of his/her thought alone. In this sense, 'speculation' (which is meant to capture the sense of the Greek term *theoria*) alludes to the strength of thought to grasp the truth: to see it with 'the mind's eye' as something like an intuitive, direct and immediate knowledge. In fact, both the word root *or*-occurring in *theoria* and the word root *spec*-occurring in speculation are semantically related to the act of seeing. Speculation retains this uniquely positive meaning through the idealistic period when the term is used (e.g. by Hegel) to indicate true knowledge par excellence because the highest form of knowledge is ideal, and not empirical, since empirical knowledge is at best based on partial or contingent realizations. In this regard, see Pastore (2010a, b). As for the word 'metaphysics', this was introduced almost casually in the philosophical discussion, but then it was universally accepted in the philosophical dictionary as one of the defining notions of this discipline. Indeed, the origin of the word can be indirectly traced back to Aristotle: when Andronicus of Rhodes made the first edition of Aristotle's writings, he classified some of them under the headword *metà ta physikà* (what comes after physics, i.e. after the study of the natural world which is always changing). The matter of these writings was qualified by Aristotle as *prote philosophia* (first philosophy) and it concerned the study of being qua being, i.e. the study of the fundamental principles which are shared by all forms of knowledge. From a methodological point of view, Aristotle's first philosophy has a purely rational, deductive and speculative character; its object is ultimate reality which is unchanging and immobile, insensible to time and therefore eternal. Thus, in the later philosophical lexicon the word 'metaphysics' that described the studies on first philosophy came to denote an inquiry into those issues considered to be the most fundamental and which are therefore prerequisite for the study of any particular matter. Because of its purely logical-deductive character, metaphysics became a discipline devoted to the study of "the pure concepts of reason" (to use an expression taken from Kant). Its aim is to describe the objects and principles which can be known by the human mind by means of reason alone; thus, its area of concern is something that transcends perceptual experience, is a world that remains hidden from the senses and from our everyday experience (see Barone 1997). By virtue of these characteristics, metaphysics is closely linked to both ontology on the one hand and religion on the other (especially with theological and spiritualistic reflections). This is the reason why in the contemporary philosophical research of the 19th and 20th century the term metaphysics and the adjective metaphysical are often saddled with negative associations often sharing the same negative connotations as speculation and speculative (on this see e.g. Stekeler-Weithofer 2010; Crane and Farkas 2004).

Aristotelian and Idealistic traditions had promoted a uniquely positive use of these words to indicate a direct and intuitive form of knowledge (*speculation*) and a thorough investigation of the fundamentals of knowledge (*metaphysics*), in the context of the rising philosophy of mind they assumed a definitively negative connotation. Both terms were used to describe self-referring views and research practices, put forward mainly by individual thinkers in an abstract manner, without taking into account the results obtained in other disciplines using empirical methods.

Investigations described as speculative were accused of systematically overlooking empirical grounding. For this reason, they were considered to be of no value for the progress of science or—as a consequence—for understanding the dynamics of everyday life. Thus, the speculative character of philosophical investigation also came to represent the specific methodological and epistemological limits of philosophical research and in this context, 'speculation' and 'speculative' became derogatory definitions of philosophical knowledge stigmatizing the weakness and arbitrariness of philosophical methods.

Ryle considers the speculative character of philosophy as the effect of the dominant, traditional philosophical view of mind—i.e. ontological dualism. According to dualistic positions, mind and body are considered to be two ontologically separate substances, which cannot be reduced to each other or compared. Dualism offers a *meta-physical* perspective on mind in the sense that it considers the mind as something other than/beyond the physical world; as something that can exist independently of matter and more specifically independently of the body. For this reason, 'metaphysics' came to be considered as a word used to denote an obsolete and mythical conception, i.e. the doctrine of the soul and its legacy. Philosophy of mind is considered anti-metaphysical because it definitively rejects the idea that the mind is separate from the body, and can therefore be studied and described independently of the body, even though it inhabits and guides it, directly causing actions, volitions, decisions, etc.

Classical dualistic views postulate the impossibility of reducing the mind and its functions to matter and thus also to that particular form of living matter which is the living body. As a consequence, investigative strategies applied to acquire knowledge about matter and the body are methodologically inadequate for the mind and vice versa. Since the mind, its contents and its properties radically differ from the material world, those mathematical instruments and measures considered essential for investigating natural phenomena since Galilei's 'revolution' could not be applied to the analysis of mental phenomena.

First and foremost, dualistic views maintain that, unlike material phenomena, mental phenomena cannot be observed objectively from an intersubjective, third person perspective. According to these views the knowledge we acquire about the mind and its functions relies uniquely on what the mind 'knows' about itself and this involves something like 'self-perception', 'self-intuition' or 'self-observation'. Empirical investigations cannot help us to understand the mind since these investigations can only address aspects related to the materiality of the body and would therefore 'degrade the ontological rank of the mind'. For this reason, dualism is an anti-materialistic perspective. And for Ryle this is considered by definition an obstacle to the progress of psychological research, above all that of experimental research.

In the context of this contraposition between the old substantialist and spiritualist philosophical stances and a new empirically grounded approach to mind, the whole language of the philosophical tradition was considered suspicious, since it provided descriptions that were not based on observations and that were therefore potentially mistaken. Above all, the new philosophy of mind engaged in a battle again the "official doctrine" of traditional mentalism "which hails chiefly from Descartes": "'the dogma of the Ghost in the Machine" (Ryle [1949] 2009: 1, 5), i.e. the myth of a material body "guided", "directed", and "governed" by an unobservable and immaterial entity. As Ryle describes it, the official doctrine of the philosophical tradition is related to Descartes' dualistic view[4] and maintains that: "With the doubtful exceptions of idiots and infants in arms every human being has both a body and a mind. Some would prefer to say that every human being is both a body and a mind. His body and his mind are ordinarily harnessed together, but after the death of the body his mind may continue to exist and function" (Ryle [1949] 2009: 1).

In opposition to the Cartesian tradition, the new philosophy of mind aims to show that the mind *cannot* exist independently from the body. Moreover, it also challenges the idea that mental phenomena are characterized by properties of a different kind than those exhibited by physical phenomena and that the mind is the cause of behavior (that it is something like a "helmsman" who steers the body and its actions).

This position stems from an epistemological argument: if the only possible model of scientificity is that of the natural sciences and the natural sciences rely on observation and experimentation to provide reliable and true knowledge about the world, then everything that cannot be observed or verified experimentally is impossible to know and we cannot even be certain that it exists. Thus, the new philosophy of mind denies that mental phenomena are absolutely different in kind from physical phenomena and cannot be traced back to them or explained on the basis of them. In the debate, this latter position was called *antimentalism*.

In fact, antimentalism is a complex view that challenges a specific aspect of the traditional conception of mental phenomena, namely the idea that they are radically different from physical phenomena. The criticism developed by the antimentalist stance specifically addresses the view that mental states are:

- *conscious* ("What has physical existence is composed of matter, or else is a function of matter; what has mental existence consists of consciousness, or else is a function of consciousness." Ryle [1949] 2009: 5);
- *introspectively accessible* ("Besides being currently supplied with these alleged immediate data of consciousness, a person is also generally supposed to be able to exercise from time to time a special kind of perception, namely inner

[4]For an overview of Cartesian dualism, see Rozemond (1998).

perception, or introspection. He can take a (non-optical) 'look' at what is passing in his mind." Ryle [1949] 2009: 4);
- *certain* ("In consciousness, self-consciousness and introspection he is directly and authentically apprised of the present states and operations of his mind. He may have great or small uncertainties about concurrent and adjacent episodes in the physical world, but he can have none about at least part of what is momentarily occupying his mind." Ryle [1949] 2009: 2);
- *non-spatial and therefore both private and not subject to mechanical laws* ("Human bodies are in space and are subject to the mechanical laws which govern all other bodies in space. Bodily processes and states can be inspected by external observers. So a man's bodily life is as much a public affair as are the lives of animals and reptiles and even as the careers of trees, crystals and planets. But minds are not in space, nor are their operations subject to mechanical laws. The workings of one mind are not witnessable by other observers; its career is private." Ryle [1949] 2009: 2);
- *intentional* (this term is meant in the sense introduced by Franz Brentano: the mind is characterized by possession of mental states like beliefs, desires, intentions etc. which have a content, i.e. they are/refer to something external).[5]

The mentalist tradition assumes the existence of the mind as an autonomous and independent substantial entity whose fundamental characteristic is the activity of thinking. To think means always *to be conscious of yourself and of your own states*, an idea that traces back in Ryle's opinion to a fundamental tenet of Descartes's view (Descartes [1641] 1984). More specifically, this idea of thinking as consciousness includes at least two aspects.[6] On the one hand, from a mentalist perspective and in line with the Cartesian tradition, to say that a mental state is conscious means that the person who is in a state knows that s/he is in that state (i.e. is aware of being in that state). This awareness is due to the fact that the subject has immediate access to his/her own state and this allows her/him to also know what this state is about.[7] On the other hand, according to Cartesianism, mental states are conscious because they exhibit characteristics that in more contemporary language

[5]The thesis of the intentionality of mind originated in the context of so-called rational psychology in the scholastic period, which also in part influenced the Cartesian tradition. It was then further developed by Franz Brentano at the end of 1800, who defined it as follows: "Every mental phenomenon is characterized by what the Scholastics of the Middle Ages called the intentional (or mental) inexistence of an object, and what we might call, though not wholly unambiguously, reference to a content, direction toward an object (which is not to be understood here as meaning a thing), or immanent objectivity. Every mental phenomenon includes something as object within itself […]" (Brentano [1874] 1995: 88). Intentionality is considered a particular character of mental phenomena that unequivocally distinguishes mental from physical phenomena because only mental phenomena have the property of being intentional, i.e. of having a content referring to something outside in the world; of developing beliefs about the world.

[6]For an overall perspective on the properties that Descartes attributes to the mind, see Clarke (2003).

[7]This resulted from the Cartesian conceptions of ideas: see Boyle (2009).

can be described as 'phenomenal' (in the sense of so-called phenomenal consciousness[8]): the subject who is in this state experiences directly what it is like to be in that state, i.e. how it feels to be in that specific state (on this see below Sect. 5). Thus, to be conscious of e.g. a state of pain does not mean primarily to have an abstract knowledge of the fact that you are experiencing pain and maybe of the reasons why you are experiencing pain (to know e.g. that you are injured), but it means to have a conscious sensation of pain; it means to have a *specific, qualitative sensation* which you classify as 'painful' and which is intrinsically connected with this state and inseparable from it. In this sense, thought is always conscious because to have mental states means to be directly aware of the sensations one experiences when s/he is in that state.

The view that consciousness is an essential feature of thought has many implications for both the nature of mental states and for the methods of accessing them. If mental states are characterized by consciousness, then people have a direct knowledge of *their own* mental states. Thus, mental states—the objects and the contents of thoughts—are *private* states that *exist internally to the subject* and *can be known directly only by the first-person* who has a *privileged access* to them. The idea that features like internal (i.e. introspective) access to the mind and *first-person authority* over our own mental states (i.e. the certainty of the knowledge we have of our own mental states) are fundamental properties of the mental is indeed implicit to the structure of Descartes philosophy.

In fact, Descartes describes the mind as a *res cogitans*, as "thing that thinks [...] a thing that doubts, understands, affirms, denies, understands a few things, is ignorant of many things, is willing, is unwilling, and also [...] imagines and has sensory perceptions" (Descartes [1641] 1984: 24). According to Descartes the *res cogitans* has the capacity to know immediately (i.e. to have immediate access to) itself, without needing any instrument beyond itself: "I know plainly [*cognosco*] that I can achieve an easier and more evident perception of my own mind than of anything else" (Descartes [1641] 1984: 22–23). As Descartes illustrates with the example of him imagining something material, the content of our own thoughts "can be perceived by the mind alone" (Descartes [1641] 1984: 21); this is known as the principle of the *solius inspectio mentis*.

Because of the influence Descartes exercised on the later philosophical tradition, the principle of the *solius inspectio mentis* gave rise to a specific method for inspecting mental content as an object directly present to the mind and to a corresponding lexicon to describe these contents. Both the method and this lexicon were inherited by pre-scientific experimental psychology[9] whose approach is

[8]On the notion of *phenomenal consciousness* see first of all Block [1995] 1997 and Crane 2001: 70–101.

[9]We consider those schools of thought prior to behaviorism as 'prescientific experimental psychology' while we define only the movements subsequent to behaviorism as 'scientific psychology'. The reason for this is that at the end of the 17th century psychology was still consolidating its identity after its separation from philosophy. Although the foundations for the future development of an experimental discipline were already laid in this period (the first laboratories date back to

characterized by a systematic and controlled use of *introspection* as a strategy to access mental phenomena (and first of all consciousness) and to describe them. Moreover, the idea that the first-person has privileged knowledge of his/her own mental contents since s/he can access them directly 'from the inside' has its origin in this principle and its interpretation by the later Cartesian tradition. In this view, no other approach or instrument can provide a comparably reliable perspective. Take for example the case of pain. According to the mentalist tradition, the content of a state of pain is transparent only for the person who experiences it directly. In consequence, the truth of this state (e.g. the fact that it occurs here and now) can be known only by the subject who experiences it. Nobody else can confirm or deny that a person is actually experiencing pain in a certain moment.

This idea—which in the modern and contemporary philosophical tradition takes the name of first-person authority—is fundamental to and constitutes a necessary presupposition for the development of mentalism. Because of first-person authority, the subject has an exclusive, indubitable and indisputable knowledge of his/her states and thus also of the content of his/her thoughts, while introspection—understood as the capacity of the mind to self-inspect itself—represents the only instrument that allows access to mental states and their knowledge. For this reason, first-person authority has an epistemological consequence of crucial importance: mental contents cannot be intersubjectively verified and therefore studied using a scientific approach. This conclusion is clearly unacceptable for the new philosophy of mind which aims to develop an observational and empirically grounded investigation of mental phenomena and to make the philosophical approach compliant with the scientific method. For this reason, mentalism and especially this particular aspect of mentalism is the main object of criticism of this new philosophical trend; the subjectivity of mind is something any scientifically oriented research on mind must avoid and refute.

2 The Influence of Logical Positivism and of Psychological Behaviorism on the Study of Mind

From a methodological and epistemological point of view, the strategy for analysis of mind initiated by Ryle and incorporated in the roots of the contemporary philosophy of mind complies with the approach put forward by two important traditions of research which were developed independently from each other starting

(Footnote 9 continued)

1870), experimental praxis was still centered only on conscious activity and conscious content and didn't fulfill the criteria of scientificity introduced by Watson's behaviorism which aimed to render the study of psychology analogous to that of natural science. The adoption of these criteria led psychological research to depart from the study of conscious activity and to focus on variables that are directly quantifiable, observable and reproducible as is prescribed by good laboratory practice, i.e. on stimuli and behavioral responses.

from the beginning of the 20th century: the philosophical movements called logical positivism and the psychological movement initiated by Watson called behaviorism. What Ryle takes from logical positivism and psychological behaviorism is not primarily their ontological hypotheses, but rather the strategy he uses to describe the entities that are the object of investigation.[10]

Indeed, logical positivism and first-generation behaviorism exhibit from both an epistemological and a methodological point of view relevant analogies that find their point of convergence in the same ideal of scientificity. Both these lines of research give serious attention to linguistic factors which are the vehicle by which the fundamental concepts and methods of a discipline are to be conveyed and communicated. The language of a discipline must ensure the stability of the reference of the concepts it relies on as well as the stability of the criteria used to individuate its objects, since the intersubjectivity of the conceptual apparatus of a discipline is a precondition for considering it a science. To qualify as a science, a discipline must rely on linguistic procedures that guarantee that its methods and results will achieve the maximum degree of control and reproducibility and therefore be as objective as possible. The degree of objectivity increases when the level of subjectivity decreases and approaches zero. Thus, for both logical positivism and psychological behaviorism using strict and precise language is essential in ensuring that a discipline is objective and can therefore be considered to be a science.

In line with this principle, the epistemological project of logical positivism is to show how the language of science must be purified in order to fit the criteria of objectivity and what rules must be followed to achieve this aim. Among the most important rules is the need to link in a strictly controlled manner the language (the conceptual apparatus) of science to its observations. In this way, scientific language becomes anchored to empirical evidence and its function is just to describe reality as it is, without illegitimately introducing any description of objects that is not intersubjectively verifiable. This movement follows Russell in maintaining that to make an investigation truly scientific we need to adopt a Galilean perspective carrying out "the substitution of piecemeal, detailed, and verifiable results for large untested generalities recommended only by a certain appeal to imagination" (Russell 1914: 4). Intersubjective verifiability is a precondition for science and this requires, once again, a language which is directly linked to experience and highly standardized. As stated in *The Scientific World-Conception,* a work considered the *manifesto* of logical positivism, to reach scientific objectivity we need a "neutral system of formulas, a symbolism freed from the dross of historical languages" (Verein Ernst Mach [1929] 2012: 81) that allows us to identify and to reject from the field of science all the statements that seem to speak of reality but that are

[10]For an overview of the relationship between logical positivism and behaviorism, see Smith (1986) and Thyer (1999). For a discussion of the differences between a mentalist and a behavioristic view, see Uttal (2000).

actually de facto "meaningless, because unverifiable and without content" (Verein Ernst Mach [1929] 2012: 83).

According to logical positivism reality must coincide with experience: "Something is 'real' by virtue of being incorporated into the total structure of experience" (Verein Ernst Mach [1929] 2012: 83). Thus, we can say that a scientific statement describes something real only if we can trace it back to experience: "it must be possible to give the meaning of every statement of science by reduction to a statement about the given [and] it must also be possible to give the meaning of any concept, whatever branch of science it may belong to, by a stepwise reduction to other concepts, down to the concepts of the lowest level which refer directly to the given" (Verein Ernst Mach [1929] 2012: 84). A language that abides by this structure of reference has the highest degree of commonality and reproducibility. This allows us to individuate and describe the objects, phenomena and events of the world starting from their empirical structure, independently of factors that might depend on the observer and his historical and cultural perspective. Subjective or cultural interpretations of the phenomena which are not directly empirically grounded must be excluded from science. Through this precept, logical positivism expresses its opposition to speculation and metaphysics, i.e. to any kind of knowledge that goes beyond experience.

Logical positivists consider it problematic that the language used in the sciences is still influenced by the (mentalist) philosophical tradition from which each particular science originally arose putting them at risk of metaphysical and speculative deviations. This contamination by the philosophical lexicon is problematic for at least two main reasons. On the one hand, philosophical concepts were formed on the basis of non-standardized observations which were therefore highly dependent on individual and qualitative factors. On the other hand, philosophical vocabulary gives rise to linguistic confusion due to an imprecise use of grammatical categories which in turn leads to treating different kinds of objects (things, qualities, relationships etc.) in the same manner, as if they had the same ontological status. For logical positivists, this lack of precision in scientific language is due to a "too narrow tie to the form of traditional languages […]. Ordinary language […] uses the same part of speech, the substantive, for things (such as 'apple') as well as for qualities ('hardness'), relations ('friendship'), and processes ('sleep'); therefore, it leads one to astray into a hypostatized, substantialized, or thing-like conception of functional concepts" (Verein Ernst Mach [1929] 2012: 83).

The only way to avoid a misleading conceptual apparatus, which might be fatal to science (and to philosophy) is to anchor scientific language as strictly and precisely as possible in the structural characteristics of reality through codified and standardized procedures that rely on observations. These procedures guarantee that science expresses knowledge and not just subjective experiences. The only aspects that science can take into consideration are those that *can be described in quantitative terms, avoiding any reference to qualitative elements*: "subjectively experienced qualities – redness, pleasure – are as such only experience, not knowledge" (Verein Ernst Mach [1929] 2012: 84). This warning against qualities raises an important epistemological issue for scientific research. Indeed, it requires excluding all the 'human' factors which are by nature subjective from scientific research since these cannot be described in a standardized manner, independently of a specific observer, and his/her culture and personal experience.

Logical positivists point out that some sciences have reached a higher degree of development and greater explanatory success than others; in their view, this is due to the fact that these disciplines were able to exactly and rigorously define their objects of study; this allowed them to increase their capacities to penetrate and to explain empirical evidence. Physics exemplifies this achievement:

> epistemological analysis of the most important concepts of natural science has freed them more and more from metaphysical admixtures which had clung to them since ancient times […] the concepts of space, time, substance, causality, and probability have been purified […] the doctrines of absolute space and absolute time have been made obsolete by the theory of relativity; space and time are no longer absolute containers but only ordering manifolds for elementary processes. Material substance has been dissolved by atomic theory and field theory. Causality was divested of the anthropomorphic character of 'influence' or 'necessary connection' and reduced to a relation among conditions, a functional coordination. Furthermore, statistical laws have replaced some of the laws of nature which had been thought to be strictly valid […] the concept of probability is reduced to the concept of relative frequency which can be expressed in empirical terms (Verein Ernst Mach [1929] 2012: 86).

No one can question that physics is a science. This discipline achieved its indisputable status as a science by progressively purifying the language used to express its methods, observations and theories and by linking it in a fixed and rigorous manner to observations. The objects of observations and their relations were then conveyed mathematically making the descriptions even more impersonal and a-subjective: anyone could reproduce them using the same criteria and instruments. For logical positivism, all disciplines should adopt the same approach in order to become proper sciences.

Given this idea of science, the case of psychology turns out to be particularly problematic because the objects and properties considered by this discipline seem to be irremediably compromised by the private and subjective dimension of personal experience. Thus, to become a science in the proper sense, psychology must give up any kind of subjective method for individuating and describing phenomena. Subjective feelings and introspection have to be banned, while a conceptual apparatus in which all descriptions are linked to observations in a stable and standardized manner has to be adopted.

> the linguistic forms which we still use in the field of the psychological today have their origin in certain ancient metaphysical notions of the soul. The formation of concepts in psychology is made particularly difficult by these defects of language […] consequently, most of the concepts used in psychology are inadequately defined so far […] in this field nearly everything in the way of epistemological analysis still remains to be done […] analysis here is more difficult than in physics (Verein Ernst Mach [1929] 2012: 88).

To become a science psychology must redefine most of the notions that compose its lexicon which—like spirit, mind, intelligence etc.—are not directly linked to empirical data. These notions must be rendered into others which directly rely on empirical data concerning the publically observable, and therefore more objective behaviors of people (see Carnap [1932] 1959, 1935; Feigl 1963).

For these reasons, logical positivism shows appreciation for an independent movement that started in the US at the beginning of the 20th century, i.e. the behaviorism of John Watson. Indeed, Watson tried to impose scientific criteria analogous to those used in the natural sciences, criteria that fully comply with the neopositivist ideal on psychological research: "the attempt of behaviorist psychology to grasp everything psychological through the behavior of bodies, which is at a level accessible to perception, is, in its principled attitude, close to the scientific world-conception" (Verein Ernst Mach [1929] 2012: 88). Because of the compliance of the epistemological principles adopted by behaviorism with those put forward by logical positivism, in the 1930s many of the exponents of this movement suggested giving up the term 'psychology' or 'science of mind' to describe research on the human psyche as *Behavioristics* (in German: *Behavioristik*) (on this see e.g. Neurath [1932] 1987; Carnap 1938; Stanowski 1971; Sorell 1991). As e.g. Carl Gustav Hempel enthusiastically remarks: "We find in Behaviorism […] an attempt to construct a scientific psychology which would show by its successes that even in psychology we have to do with purely physical processes, and that therefore there can be no impassable barrier between psychology and physics" (Hempel [1935] 1980: 16). In his view, this shows that "all the branches of science are in principle of one and the same nature: they are branches of the unitary science, physics" (Hempel [1935] 1980: 21).

Behaviorism—and especially the so-called 'first-behaviorism' (we will say something more on the difference between first- and second-behaviorism later in this chapter)—seeks to align psychology with the natural sciences through a radical change in both its methods and its lexicon. In this way, psychology can become a natural science. To achieve this aim, psychology must follow the same development that was achieved by the natural sciences: qualitative methods must be excluded in favor of quantitative ones. Thus, any reference to objects, events and properties that are not directly observable and measurable in the laboratory must be avoided. Terms referring to supposed mental states must be purged: the behaviorist "'ignores' [mental states] in the same sense that chemistry ignores alchemy, astronomy, horoscopy, and psychology telepathy and psychic manifestations. The behaviorist does not concern himself with them because as the stream of his science broadens and deepens such older concepts are sucked under, never to reappear" (Watson 1920: 94).

All of these epistemological principles developed by behaviorism in the field of psychology and by logical positivism in the field of philosophy of science penetrated the roots of the rising philosophy of mind whose aim was to bring to philosophical research on mind the same rigor and the same adherence to empirical evidence that characterize scientific approaches to phenomena. The price for this was the adoption of an antimentalist perspective which did not only gives up Cartesian dualism but also radically transformed the traditional approach to mind. The new approach to mind abandoned all attempts to address phenomena that could only be directly accessed from the inside, such as, e.g. visions, conscious states (i.e.

qualitative experiences), or the perception humans have of their internal feelings (introspection), or even their capacity to represent the external world in their mind (intentionality in Brentano's sense), etc. which were traditionally the privileged object of study in philosophy.

With Ryle the focus of antimentalism changed from ontology to epistemology. Originally (e.g. in the case of logical positivism) this was conceived in contraposition to the idea that the mind is an immaterial substance which can be known only separately from the body and cannot be reduced to anything that can be described in a vocabulary compatible with the lexicon of physics. With Ryle antimentalism became above all a methodological stance, whose aim was to bring to the philosophical research on mind the same rigor and the same adherence to empirical evidence that characterize scientific approaches to phenomena. To achieve this, any mental state or mental property must be described in terms of observable behaviors. Beliefs, desires, volitions must be defined in relation to the actions that are actually carried out by people. In this sense, the antimentalist stance consists primarily in the idea that a rigorous philosophy of mind cannot rely on a first-person perspective of internal states, but must be based on a third-person description of them.

3 Third Person Approaches to Mental States as a Strategy to Develop a Scientific Psychology

In the previous sections, we argued that philosophy of mind as we know it today starts with Ryle in the first half of the 20th century and arises as an antimentalist perspective. There we discussed the relationship between classically conceived mentalism and Cartesian dualism and we described the methodological rationale underpinning the antimentalist stance. In particular, we called attention to the idea that mentalism conceives of mental states as private phenomena that cannot be verified intersubjectively in order to be studied using a scientific approach. In fact, the idea that subjective phenomena cannot be an object of scientific investigation is a basic principle of both the tradition of logical positivism in philosophy and that of behaviorism in psychology. Ryle's view is influenced especially by logical positivism, but shares important overlaps with psychological behaviorism as well. In this section, we focus on this latter aspect of his work. In considering the epistemological foundations of behavioristic psychology more closely, it is first of all useful to give an idea of the cultural importance and spread of antimentalism in the first part of the 20th century and second to explain why and how later philosophical and psychological traditions of research on mind remained in many respects tied to this perspective.

Psychological behaviorism aims to develop a *scientific* psychology. It assumes that rejecting the idea that the first person has a privileged access to his/her own mental states is essential in achieving this aim. In fact, the first-person perspective must be rejected and all psychological descriptions must be given from the point of view of the third person, i.e. from an external point of view rather than from the internal point of view of the subject.[11] Any reference to conscious aspects of mind and to introspective access to our mental states and feelings must be avoided. As pointed out by Watson in the very first part of his programmatic essay *Psychology as the Behaviorist Views it* (1913a), commonly considered the *Manifesto* of Behaviorism:

> Psychology as the behaviorist views it is a purely objective experimental branch of natural science. Its theoretical goal is the prediction and control of behavior. Introspection forms no essential part of its methods, nor is the scientific value of its data dependent upon the readiness with which they lend themselves to interpretation in terms of consciousness (Watson 1913a: 158).

According to behaviorism, psychology must become part of the natural sciences. In order to achieve this aim, its research subject must be modified with respect to the previous tradition of study to make sure that all methods and results have the maximum degree of objectivity, i.e. that they can be reproduced and verified intersubjectively. For this reason, all aspects of psychological research must be expressed in quantitative rather than in qualitative terms.

Already before behaviorism, psychology had made important steps towards the adoption of quantitative means (in particular, thanks to the traditions of research started by Wilhelm Wundt and Gustav Fechner). These aimed at standardizing the observation procedures in laboratories and at precisely defining from a linguistic point of view the lexicon of the discipline in order to express the outcomes of experiments in the most precise and stable way possible. In spite of such efforts, from the point of view of behaviorism their achievements were still characterized by relevant and unacceptable qualitative components which represented the main obstacle to psychology becoming a natural science. These components were characterized by the great importance given to the subjectivity of the observer and experimental participants as well as the qualitative aspects of their experiences (of their sensations and perceptions).

[11]The discussion on whether we should describe phenomena starting from an internal or from an external point of view results in an important epistemological debate in the field of the philosophy of science of 1900 which involves two opposite fronts: externalism and internalism. This debate concerns the methodological issues related to the adoption of an internal versus an external prospective and the consequences of the adoption of one or the other of these perspectives for the way in which the references of the words that form the lexicon of a discipline are fixed, since these need to be univocal and publicly verifiable. For an overview of this debate from an epistemological point of view see e.g. Kornblith (2001), McGrew and McGrew (2006) and Farkas (2008). The effects of this dispute in the field of psychology can be recognized in the theoretical developments occurring after the so-called *crisis of introspection*: see e.g. Lyons (1986) and Smithies and Stoljar (2012). The problem of the adoption of an internal point of view is especially relevant with respect to the problem of Self-Consciousness and with how we access to our Self: on this see Gertler (2011), Michel (2014), Carruthers (2005) and Carruthers (2010). On the developments of this debate in the field of philosophy of language see e.g. Nimtz (2002), Dellantonio (2007), Goldberg (2007, 2015) and Kallestrup (2011).

> Take the case of sensation. A sensation is defined in terms of its attributes. One psychologist will state with readiness that the attributes of a visual sensation are quality, extension, duration, and intensity. Another will add clearness. Still another that of order. I doubt if any one psychologist can draw up a set of statements describing what he means by sensation which will be agreed to by three other psychologists of different training. Turn for a moment to the question of the number of isolable sensations. Is there an extremely large number of color sensations – or only four, red, green, yellow and blue? Again, yellow, while psychologically simple, can be obtained by superimposing red and green spectral rays upon the same diffusing surface! If, on the other hand, we say that every just noticeable difference in the spectrum is a simple sensation, and that every just noticeable increase in the white value of a given color gives simple sensations, we are forced to admit that the number is so large and the conditions for obtaining them so complex that the concept of sensation is unusable, either for the purpose of analysis or that of synthesis" (Watson 1913a: 164).

Not only are sensations highly variable and unstable per se, but they can be reported only verbally. In fact, psychologists need not only quantify and measure the stimuli they present to participants, but must also make sure that the experimental outcomes are properly (i.e. objectively and exactly) identified and described. However, these outcomes consist mainly in verbal reports on the kind and intensity of the sensation the subject has experienced. Even though introspectionist research tried to define the language of these reports in a way that was as stable as possible and subjects were trained in the use of the lexicon, reporting remained an uncontrollable factor influencing the reliability of experiments. Thus, if on the one hand the physical characteristics of the stimuli could be measured and quantified, on the other hand the responses were irremediably qualitative and subjective.

In Watson's view these qualitative elements had to be purged. All aspects of psychological research had to be quantifiable, measurable, directly observable and reproducible. This required rejecting not only introspection as a method of psychological research but also internal states as objects of research. The central object of psychology was therefore redefined and changed from mind to behavior.[12] The word 'behavior' is not here intended in its usual sense to indicate the various ways in which people act and interact with each other to achieve particular goals. It is

[12]One of the reasons why Watson believes in the possibility of making psychology a natural science is that he is deeply persuaded that the same quantitative standards used in animal psychology can be applied to human psychology. In the case of animal psychology, the object of research is uniquely the behavior of the animals in its observable dimension, i.e. in relation to the stimulus that give rise to it. Because of this rigorous definition of the object of research, animal psychology achieved important results which were considered objective and results could be interpreted and explained without resorting to unobservable entities or processes. Watson suggests that, if we could apply the same principles we use in animal psychology to human psychology, we would thereby free human psychology from any reference to unobservable entities and processes which threaten its scientificity. Moreover, the use of the same principles in both disciplines would be compliant with the idea that nature is a continuum in which animals and humans differ only in degree not in nature. On this see Watson (1914).

rather defined more precisely in terms of any observable response that humans exhibit as a consequence of specific stimuli.[13] As Watson specifies:

> The psychology which I should attempt to build up would take as a starting point, first, the observable fact that organisms, man and animal alike, do adjust themselves to their environment by means of hereditary and habit equipments [...] secondly, that certain stimuli lead the organisms to make the responses. In a system of psychology completely worked out, given the response the stimuli can be predicted; given the stimuli the response can be predicted. Such a set of statements is crass and raw in the extreme, as all such generalizations must be (Watson 1913a: 167).

In this context, the notion of behavior refers uniquely to the link between specific environmental stimuli and the specific responses they elicit. Any reference to events, states and processes which cannot be directly observed is rejected; notions like 'consciousness', 'affect', 'mental image' and the like, which had been at the center of previous introspectionist psychological research, should be deleted from the psychological lexicon (see Watson 1913b).

> I do not wish unduly to criticize psychology. It has failed signally, I believe, during the fifty-odd years of its existence as an experimental discipline to make its place in the world as an undisputed natural science. Psychology, as it is generally thought of, has something esoteric in its methods. [...] The time seems to have come when psychology must discard all reference to consciousness; when it need no longer delude itself into thinking that it is making mental states the object of observation. We have become so enmeshed in speculative questions concerning the elements of mind, the nature of conscious content [...] that I, as an experimental student, feel that something is wrong with our premises and the types of problems which develop from them (Watson 1913a: 163).

The need to purge psychological language of any reference to unobservable entities and processes would imply a deep redefinition of its lexicon in order to link all notions to empirical evidence. Emotions provide a useful example of this process of redefinition that helps us understand in what way qualities can be transformed into quantities and how subjective experiences can be described from a third person point of view. Indeed, emotions as internal states that are experienced only by the first person cannot be included in the field of investigation of psychology. To become part of a psychological science, these need to be described in an objective way, in terms of characteristics that are intersubjectively verifiable.

In *Behaviorism* Watson redefines emotional states in terms of specific reactions of the organism elicited by particular environmental stimuli. These reactions consist of precise (observable and measurable) bodily modifications (mainly visceral and glandular reactions) or bodily movements (e.g. escaping, smiling, jumping, etc.).

> [...] notwithstanding the fact that in all emotional responses there are overt factors such as the movement of the eyes and the arms and the legs and the trunk, *visceral and glandular factors predominate*. The 'cold sweat' of fear, the 'bursting heart,' the 'bowed head' in

[13]It is worth mentioning that speaking is also considered to be a form of behavior. As Watson specifies: "we can observe behavior—what the organism does or says. And let me make this fundamental point at once: that saying is doing, that is, behaving. Speaking overtly or to ourselves (thinking) is just as objective a type of behavior as baseball" (Watson 1930: 6).

> apathy and grief, the 'exuberance of youth,' the palpitating heart' of the swain or maiden, are more than mere literary expressions; they are bits of genuine observations (Watson 1930: 165).

In line with this view, Watson describes e.g. the emotion that it is usually called "fear" in terms of the reaction that people have as a response to stimuli like "loud sounds" and "loss of support" which manifests itself in behaviors like "checking the breathing, 'jump' or start of whole body, crying, often defecation and urination (and many others not worked out experimentally. Probably the largest group of fear reactions are visceral)" (Watson 1930: 156). Other examples Watson considers are the emotions usually called 'love' and 'jealousy'. 'Love' is described as the reaction caused by stimuli like "Stroking skin and sex organs, rocking, riding on foot, etc." which reveals itself in behaviors like "Gurgling, cooing and many others not determined. That visceral factors predominate is shown by changes in circulation and in respiration, erection of penis, etc." (Watson 1930: 157). As for 'jealousy', this is described as the effect of stimuli like "Sight (or sound) of loved object being tampered or interfered with" which shows itself in behaviors like "Stiffening of whole body, clenching of hands, reddening and then blackening of face – pronounced breathing, fighting, verbal recrimination, etc." (Watson 1930: 194)[14]. Through these re-descriptions, the object of the psychological investigation can be approached while avoiding any qualitative characterization related to consciousness and the first person perspective. The direct access to mental states is fully avoided through the assumption of the equivalence between internal states (e.g. emotions) and their manifestations in behavior.

The kind of psychology that emerges from this redescription is very different from the traditional formulation. In order to become a science, psychology modified its object in such a radical manner that all connections with the previous mentalistic tradition of research were lost. As Watson acknowledges:

[14]In this perspective, emotions are no longer lived experiences which have a specific intensity and quality, but they are rather behavioral and externally observable reactions. Emotions are considered as behavioral reactions that lead to a change in the normal course of action, i.e. in the behavioral schemes that people usually follow to adequately respond to their environment (Watson 1930: 113 ff). Emotions bring about reactions that are described as accessory, slowed, negative, or atypical/inappropriate with respect to the normal course of action settled on the basis of habituation, i.e. of the adaptation of the organism to the environment. Since they lead to the suspension of the execution of a settled behavioral scheme, they are considered regressive in the sense that they activate primordial physiological reflexes towards unexpected (out of the ordinary) situations in which the support of past experience goes missing. For some examples see Watson (1930: 114): "Accessory reactions: The subject does the task quickly and correctly but he becomes Pale, he may even cry, urinate or defecate, his mouth glands may become inactive. He reacts steadily and correctly in spite of his emotional state. Other examples of accessory reactions arc whistling, talking, singing, while at work […] Slowed reactions: He does the act but his reaction time is increased he may fumble and drop his work, or react with too much or too little energy. Response to questions comes slowly or very rapidly […] Negative reactions: He may show fear at food push it away or run away from it himself. Instead of the ordinary reactions to dog or horse, the subject may walk away from them. Phobias belong in this group".

> Will there be left over in psychology a world of pure psychics, to use Yerkes' term? I confess I do not know. The plans which I most favor for psychology lead practically to the ignoring of consciousness in the sense that term is used by psychologists today. I have virtually denied that this realm of psychics is open to experimental investigation. I don't wish to go further into the problem at present because it leads inevitably over into metaphysics. If you will grant the behaviorist the right to use consciousness in the same way that other natural scientists employ it – that is, without making consciousness a special object of observation – you have granted all that my thesis requires (Watson 1913a: 175).

This redefinition of emotions by behaviorism is a glaring example of what anti-mentalism actually means: of what it is motivated by and what its effects are. Anti-mentalism constitutes the main trait of the philosophy of mind at its origin: is a view aimed at removing from the scientists' tool chest (and from his/her vocabulary) all those entities whose existence cannot be specifically demonstrated in controlled experiments and whose features cannot be uniquely determined in an objective (i.e. intersubjective and repeatable) manner using quantitative measures. According to the approach initiated by Watson and developed by Edward C. Tolman, Clark L. Hull and Burrhus F. Skinner (see Smith 1986: 69–297), the very same notion of mind defined by both the philosophical and the introspectionist psychological traditions cannot be investigated scientifically. This notion is compared to a 'black box' that we cannot inspect from the inside. If we want to know and to explain the human being in a rigorous manner—i.e. linking theoretical descriptions to observable, verifiable and quantitatively measurable data—it is not possible or even useful to appeal to its content.[15]

Indeed, Ryle's view is considered to be a particular, specifically philosophical version of behaviorism which creates an original synthesis of elements taken from the approaches suggested by both logical positivism and psychological behaviorism as introduced by Watson.[16] In line with behaviorism, Ryle argues that words indicating mental states or properties need to be redefined in terms of behavioral reactions and more specifically in terms of *behavioral dispositions*, roughly, in terms of the acts that typically accompany these states or properties. This idea that any property can be accounted for in dispositional terms is taken from logical

[15]The notion of 'black box' is one of the elements of the behaviorist lexicon that most captured people's imagination in the 20th century. This metaphor indicates that the behaviorists are not interested in the notion of mind as an entity (or as a set of processes), that it cannot be the object of direct observation and therefore cannot be assigned any explanatory function. On this see e.g. Zuriff (1985), Mills (1998) and Catania and Harnad (1988).

[16]On Ryle's model as a specific philosophical version of behaviorism see e.g. Byrne (1994), Bechtel (1988a, b), Goldman (1993) and Nannini (2006). In the literature, the impact of the behaviorist tradition on philosophy is usually described as twofold. On the one hand it gives rise to so-called *logical behaviorism*, i.e. a family of theories influenced by epistemology and by the ontology of logical positivism which inherits that particular form of monistic reductionism called physicalism (Carnap, Neurath, Hempel, Brunswick). On the other hand, from this it develops into a form of *methodological behaviorism* that does not share the ontological reductionism of logical behaviorism and that takes over another component from this legacy, i.e. the idea that scientific language must be linked to elements in the external world that are publicly and intersubjectively observable (Russell, Wittgenstein). Gilbert Ryle belongs to this latter tradition.

positivism and from the attempts made by this line of research to show how sciences which—like psychology—do not directly address observable phenomena can uniquely anchor their statements to an empirical basis. In Carnap's view, for example, all statements concerning the psychological states (mental states or sensations) of individuals should be described by analogy to statements that describe properties of material objects by referring to behavioral dispositions. As he specifies, psychological states can be "clarified by means of analogy with a sentence about a physical property, defined as a disposition to behave (or respond) in a specific manner under specific circumstances (or stimuli). To take an example: a substance is called 'plastic' if, under the influence of deforming stresses of a specific sort and a specific magnitude, it undergoes a permanent change of shape, but remains intact" (Carnap [1932] 1959: 170).

While internal states are subjective and not verifiable from a third-person point of view, behavioral dispositions are publicly accessible; thus, they allow us to account for unobservable properties in terms of other phenomena that are instead directly observable from a third person point of view and thus compliant with the physical sciences (on this see also Hempel [1935] 1980). In physics, we accept that a perceivable property like 'hardness' can be described as a manifest phenomenon that depends on the degree of cohesion of the atoms that compose the object; according to Carnap, an analogous explanatory strategy can be adopted in psychology to account for psychical states like anxiety, excitement, anger etc. They can be described as manifestations depending on a physical reaction (of the brain or of the nervous system) to solicitations (stimuli) coming from the external environment that might be considered as the real and direct causes of these states.

If we accept this way of describing psychic states, we implicitly agree that the manifest qualities of material objects can be treated as analogous to qualities we attribute to our internal states. The qualities of material objects can be considered as changes of state which are directly observable, are caused by external factors and result from measurable processes. In this vein, a qualitative aspect of e.g. an object made of glass such as *being fragile* can be treated as a specific physical property consisting in the disposition or tendency of this object to shatter in the presence of specific external causal factors (e.g. when the object collides with another object with a certain force or when the pressure on it exceeds a certain threshold). Analogously, *solubility* can be considered as the manifest effect of atomic structure and be described as the disposition or the tendency of a substance to dissolve in the presence of the right causal factors (i.e. of the right solvent) (Carnap 1936). This explanatory strategy brings to light an observable conditional link and an observable regularity that allows us to explain qualitative aspects as extrinsic properties of objects. If we apply it to mental properties, we can get rid of the old mentalistic vocabulary and describe mental qualitative phenomena in the terms of what brings them about and what they give rise to.

The fundamental traits of this descriptive strategy are adopted by Ryle (Ryle [1949] 2009: 100–135; Alvarez 2014: 74–96), because they allow us to trace phenomena that are traditionally considered as not observable back to others which can be directly observed in the external world. In Ryle's philosophy, Carnap's idea of dispositional explanation takes a particular form which relies specifically on *behavioral* dispositions: a mental state or quality is explained in terms of the dispositions or tendency individuals exhibit to act (to behave) in a certain way every time they are under a specific environmental pressure (they are exposed to specific stimuli). The behavior and the stimuli are described in the terms and at the level of our everyday language: they are described in the way we naturally observe them in the external world.

Dispositional descriptions constitute an attempt to emulate the descriptions made by natural sciences, which individuate specific regularities and conditional links even though they diverge from them in their nature as tendencies. They are tendencies in the sense that—even though they express an actual regularity—this is not as exact and certain as the relationships between cause and effect described by natural sciences. However, like descriptions in the natural sciences they approach phenomena from an extrinsic, third person perspective. This is one of the characteristics of Ryle's view that permeates later research traditions and survives unchanged in the philosophy of mind even after the crisis and decline of logical positivism and behaviorism.[17]

4 After and Beyond Ryle: Antimentalist Heritage Within Mentalistic Renaissance

According to the most widespread position in the literature, the denial of a mentalistic perspective is something that starts and ends with (philosophical and psychological) behaviorism,[18] while the views on mind that are developed later—especially philosophical functionalism and psychological cognitivism—go back to

[17]Ryle's approach survives the crisis of both logical positivism and the more radical behaviorist models in psychology. The reason for this is that Ryle's approach keeps a distance from all the ontological aspects of the positions developed by logical empiricism and by psychological behaviorism (above all in Watson's version). Indeed, Ryle takes up neither the physicalism and the reductionism of logical positivism nor the naturalistic implications of psychological behaviorism: see Nannini (2006) and Marres (1989).

[18]This clear-cut distinction between the antimentalistic view of the behaviorist movement and the mentalistic recovery carried out by cognitivism in psychology and by functionalism in philosophy is a simplification which is useful for the aims of this work, but leads us to overlook an important family of theories that is temporally and theoretically in between these two views, i.e. the so-called identity-theories. These are developed in the field of philosophy of mind and represent a first step —even though 'in disguise'—in direction of the renaissance of mentalism. Identity theories argue for a monistic view as far as the ontology of mind is concerned and they maintain that mental states are de facto brain states. However, they acknowledge that the linguistic descriptions of the sensations we are aware of is relatively autonomous from those of the brain states they are supposed to be identical to. Moreover, identity theories go back to the view that behaviors have an internal cause, which is not observable from the outside, even though this cause is purported to be of a physical kind, i.e. the brain. For an overview of identity theories see Feigl (1958), Smart (1959), Place (1956) and Heil (1998).

mentalism, since they argue that mental states are the *internal causes* of *observable behaviors*. While it is certainly true that these lines of research again assign a central role to mental states as causes of behavior, on the other hand it is also undeniable that they maintain a fundamental trait of behaviorism and of the anti-mentalism that characterizes it. Indeed, according to these new perspectives on mind, mental states must still be described according to an extrinsic, third person perspective.

By 'functionalism' we mean the multiform perspective developed within the philosophy of mind and the philosophy of psychology starting from the 1960s on the basis of the mind/computer analogy which interprets the mind as software implemented by the brain as the hardware of the system (for an overview see e.g. Bechtel 1988a; Heil 1998). Since the software can be investigated in a way that is partially independent of the hardware that implements it, this metaphor legitimizes an approach to mind which is relatively autonomous from the study of its physical basis, i.e. the brain, in which mental states are seen as internal, functional states of a computing machine. To understand what an internal state of a program consists of, we need to determine first of all how it 'behaves': how it interacts with other states to produce an output. From this point of view, mental states are states of a system that can be characterized on the basis of their *causal role* inside the system; as states that—given certain inputs and interactions with specific mental states—lead to a particular output.

Even though behaviorism avoids any direct reference to internal states while functionalism makes use of them in its explanatory apparatus, the two lines of thought share important features since both describe mental states and psychological processes in *dispositional terms*. According to both behaviorism and functionalism, mental states such as e.g. 'pain' are caused by specific inputs such as injuries or diseases and cause in turn certain dispositions such as the disposition to complain and to take certain drugs. The kernel of the difference between behaviorism and functionalism is that the latter includes in its explanation one element that is not allowed from a behavioristic perspective, i.e. the fact that mental states are not just dispositions, but also have an independent existence, at least in the sense that they are related to each other and that the final outcome (the behavior caused by these mental states) depends on the causal relationships that each state entertains with others.

This additional element allows us to deal with the traditional objection made against behaviorism, i.e. that being in a certain mental state—e.g. being in 'pain'—does not necessarily lead to a specific, reliable behavioral reaction, such as e.g. complaining and/or taking a drug (Chisholm 1957; Geach 1957; Putnam [1963] 1975). If one is in a context in which it is not appropriate to show that s/he is in pain, or if one believes that for some reason s/he should not take drugs, then pain will not lead to the expected behavioral disposition (complaining and/or taking drugs). This is the reason why functionalism changes the behaviorist perspective on this aspect and includes mental states in its explanatory model, defining them in a relational manner: a mental state (such as e.g. *pain*) is something that is caused by specific inputs (such as e.g. injuries or diseases) and that—*if it is related with specific other mental states* (like the belief that one is allowed to freely express his/her pain or that drugs can help and there is no reason not to take them)—*then* causes a particular disposition (e.g., the disposition to complain and take some drugs).[19] However, in this view if this same state—e.g. this very same pain—is causally related with different other states (e.g. the belief that one needs to hide his/her feelings or that all drugs are toxic), then this will cause different behavioral reactions. As e.g. Block, speaking of the characterization of mental states on functional terms, specifies: "what makes a pain (and, generally, what makes any mental state the state it is) is its having a certain causal role" (Block 1980: 173).

Even though the functionalist definition of mental states differs from the behavioristic definition in acknowledging that mental states have causal power, this does not change the earlier behavioristic presupposition that mental states must be identified uniquely starting from external and publicly observable criteria, excluding any reference to introspection and to the qualitative dimension of the state itself, i.e. to the way it is perceived by the subject who is in it. Going back to the example of *pain*, the account of a state like *pain* given by functionalism is in the third person only: *pain* is not considered to be something with specific qualities that a person experiences in a certain way, but it is viewed as a state that has certain causes (inputs) and leads to certain effects (outputs), depending on what other states it is causally related to. Indeed, functionalism shares with behaviorism and with logical positivism the methodological presupposition that mental states need to be identified on the basis of extrinsic criteria and that the first person perspective—i.e. the subjective and qualitative dimension of mind—must be excluded from a rigorous, scientific account of mind itself. From this point of view, the functionalist perspective also remains antimentalist by rejecting any form of introspective access and first person authority over one's own mental states and thus excluding any approach to the conscious aspects of mind that starts from the first person perspective.[20]

This kind of antimentalism also characterizes the multifaceted line of thought that goes under the name of *cognitivism* which is in many ways related to functionalism. This connection is primarily due to two factors. On the one hand, both approaches share the same interest in computers, computation and artificial intelligence as possible models for mental processes and mental architecture. On the other hand, cognitivism borrows many elements of its view of mental representations and mental structures from the theory developed by Jerry Fodor, who is one of the founding fathers of functionalism.[21]

---

[19]On this see e.g. Block (1978); on the issue of the definition and identification of mental states according to a functionalist perspective, see also Lewis (1966, 1972).

[20]These general remarks apply to all kinds of functionalism—to metaphysical functionalism, to computational functionalism, as well as to common sense functionalism. On this see e.g. Brandt and Kim (1963).

[21]For a history of the development of cognitivism see e.g. Gardner (1987), Chaps. 3, 4 and 5. For a recent overview of cognitive psychology see e.g. Best (1998) and Sternberg and Sternberg (2011).

To maintain that cognitivism embraces—in some way—an antimentalist perspective seems to be paradoxical, since the innovative character of this line of research with respect to behaviorism is usually considered to be the recovery of the mental. It is for example Chomsky who, in his famous review on Skinner's book *Verbal Behavior*—considered by many to be the 'coupe de grace' of behaviorism and the opening act of cognitive science—points out that we need to go back to a *mentalistic psychology* since the behavioristic approach has proved inadequate to explain verbal behavior: "It is clear from such examples, which abound, that any talk of 'stimulus control' simply disguises a complete retreat to mentalistic psychology. We cannot predict verbal behavior in terms of the stimuli in the speaker's environment, since we do not know what the current stimuli are until he responds" (Chomsky 1959: 32, see also 58). The mentalistic perspective to which Chomsky suggests we return is based mainly on the idea that mind is innately characterized by internal structures which contribute to the organization of stimuli in a way that allow us to explain how people can acquire and master a language. From this point of view, to account for our linguistic capacities we need to define the organization and the functioning of these innate structures and this is, in Chomsky's view, a mentalistic project. As he specifies in his *Aspects of the Theory of Syntax*: "Hence, in the technical sense, linguistic theory is mentalistic, since it is concerned with discovering a mental reality underlying behavior" (Chomsky 1965: 4).

The notion of mentalism plays a central role also in Fodor's view. Indeed, in *The Language of Thought* Fodor (1975) criticizes Ryle for having erroneously identified Cartesianism (in the sense of dualism) *toto modo* with mentalism. "Ryle assumes (as most psychologists who take a Realistic view of the *designata* of mental terms in psychological theories would not) that a mentalist would be a dualist" (Fodor 1975: 4). Fodor thinks that this identification is wrong since—through the metaphor of the mind as software—functionalism showed that it is possible to argue that mental states really exist without resorting to a dualistic metaphysics. Moreover, in his view assuming that mental states really exist and that they entertain causal relationships with each other as well as with inputs and outputs is the only possible way to explain behavior (on this see also Fodor 1968).

Thus, the point of contention between Fodor and Ryle is not dualism and the idea that mind is a separate, immaterial substance, since they both deny such a metaphysical position. Their disagreement concerns rather the model we should adopt to explain behavior and especially the issue of whether (both practical and verbal) behavior should be explained as a product of mental causes.

> It is frequently remarked that contemporary approaches to language are 'mentalistic'. What is usually meant by this is just that items in the theoretical vocabulary of linguistics and psycholinguistics are presumed to designate nonbehavioral states and processes. Any psychologist who is not a behaviorist is *ipso facto* a mentalist in this sense, and I should have thought that it was no longer possible seriously to doubt that useful theorizing about language will have to be in this sense mentalistic (Fodor 1975: 108).

Generally speaking, the specificity of the cognitive approach which characterizes all disciplines involved in the project of the cognitive sciences (which in the main stems from cognitive psychology and functionalism) is to explain the behaviors and the capacities exhibited by human beings as products of mental processes. These are mostly understood in terms of computational processes carried out on entities—i.e. representations—internal to the human mind. As e.g. Barbara Von Eckardt points out: "Cognitive science seeks to explain the human capacities by reconceptualizing them as a system of computational and representational capacities" (Von Eckardt 1992: 303). The same idea is also expressed by Ulrich Neisser in his seminal work *Cognitive Psychology*: "Whatever we know about reality has been mediated, not only by the organs of sense, but by complex systems which interpret and reinterpret sensory information" (Neisser 1966: 3). And further: "'cognition' refers to all the processes by which the sensory input is transformed, reduced, elaborated, stored, recovered, and used" (Neisser 1966: 4).

From this perspective, cognition consists in information processing carried out on the medium of representations, which convey information about the external world and constitute the mediating element between the external world, on the one hand, and the behaviors and capacities exhibited by human beings on the other. As was also acknowledged by Skinner—one of the leading figures of so-called second-generation behaviorism—this notion of representation as an internal element of the mind which mediates between world and behavior is one of the main theoretical aspects, possibly even the most important one, that differentiates behaviorism from cognitivism or more generally from the cognitive perspective:

> The difference between the cognitive and behavioral approaches is perhaps best seen in the field of perception. For cognitive science the direction of action is from organism to environment. The perceiver acts upon the world, perceives it in the sense of taking it in (Perceive comes from a Latin word meaning 'capture'). [...] There is some ambiguity in calling perception the processing of information. Are sensory data the information to be processed, or is information extracted from them? In either case, processing must have a product, and for cognitive science that product is, as it was for the Greek philosopher, a representation. We do not see the world, we see copies of it (Skinner 1987: 95).

The general idea in cognitivism and more generally in the cognitive sciences that cognition works on representations and processes internal to the mind is most certainly coherent with mentalism as this was described by Fodor and Chomsky and makes the cognitive perspective—as well as the functionalist view—definitively different from behaviorism. However, there is also an important element of continuity between this line of thought and behaviorism that accounts for the conclusion that functionalism and cognitivism—and thus also cognitive sciences in their first and classical version which relies on these views—continue to endorse an antimentalist approach to mind. They continue to be antimentalist because they posit that mental states can be identified and described uniquely from a third person perspective, on the basis of their causal role and without any reference to their qualitative and 'conscious' dimension. The behaviorist notion of stimuli is replaced by the notion of information which is qualified on the basis of the analogy between mind and computer and is thus conceived as something whose content can be precisely defined by its link to external evidence. In particular, information is defined from a third-person point of view on the basis of its causal role with respect to the external situation and to behavior. This becomes *the only* internal (specifically mental) factor that causally determines people's behavior.[22] Thus, the structure of the explanation of mental processes given by cognitivism (and by philosophical functionalism) remains unchanged with respect to Ryle insofar as both rely uniquely on a third person perspective and are based on external, observable factors only.

Unlike classical (Cartesian) mentalism, the cognitive sciences neither qualify consciousness as one of the main aspects of our mental life nor consider introspection as a privileged way to gain access to any aspect of our mental states. Congruent with this principle, they focus primarily on the comprehension of the perceptual and rational aspects of the human mind, which can be best explained as computations in which information taken from the external world is processed in different ways, while consciousness is considered to be a hard research problem which should be left on the fringes of psychological investigation. To be more precise, only some aspects of consciousness are legitimate topics in the field of cognitive research, while all others are basically excluded.

The aspects of consciousness that are comprised in the cognitive perspective on mind are those which fall under the notion of *access-consciousness* (shortened: *A-consciousness*) as defined by Ned Block. This describes the cognitive contents that are explicitly possessed by the subject who can access them and which are under rational control through action and language. As Block specifies: "A perceptual state is access-conscious, roughly speaking, if its content – what is represented by the perceptual state – is processed via that information processing function, that is, if its content gets to the Executive System, whereby it can be used to control reasoning and behavior" (Block [1995] 1997: 379). The aspects of consciousness that are, on the contrary, basically excluded from this analysis are those captured by the notion of *phenomenal consciousne*ss (shortened: '*P-consciousness*') which describes the truly conscious aspects (strictly speaking) of mental states, i.e. the 'feeling' we experience by being in these mental states and more generally the feelings we experience because we are creatures with consciousness. As Block clarifies, "phenomenal consciousness is experience; what makes a state phenomenally conscious is that there is something 'it is like' to be in that state" (Block [1995] 1997: 377). P-consciousness denotes the opaque, subjective and truly qualitative aspects of mental life, i.e. it denotes what people feel when they experience something—e.g. the qualitative feeling one experiences when s/he is in pain or sees red or tastes the flavor of something (we will discuss the issue of qualities and of Nagel's notion of *what-it-is-like* more specifically in the next section).

The position taken by classical, computational cognitive science on this point suggests that the qualitative component of our experience described by the notion of P-consciousness and the cognitive component characterized by the notion of

---

[22]On the change of the notion of stimulus to that of information and on the epistemological implications of this change see e.g. Toccafondi (2002).

A-consciousness must be kept separate with respect to the functions they accomplish in our mental life. While A-consciousness accomplishes a central function in our cognition, since it describes the contents—i.e. the knowledge—that people acquire/can access and on the basis of which they think, reason, act etc., P-consciousness is considered to be irrelevant with respect to cognition. In this view, phenomenal qualities do not play any causal role in the computational processes by which we reason and act.[23] As e.g. Ray Jackendoff—one of the leading representatives of this line of thought—explicitly states: "consciousness is *causally inert*. This may seem harmless enough until we realize its ugly consequence. *Consciousness is not good for anything*. The only way it can be good for anything is for it to have effects, and such possibility has just been denied" (Jackendoff [1987] 1994: 26).

As was pointed out e.g. also by Francisco Varela, Evan Thompson and Eleanor Rosch, the fact that within cognitive science consciousness is relegated to a minor role is a consequence of the explanatory strategy of the theory which relies on computationalism and the analogy between mind and computer: "cognitivism challenges our conviction that consciousness and the mind either amount to the same thing or there is an essential or necessary connection between them" (Varela et al. 1991: 49). From a computational viewpoint, the mind works on representations using operations that do not themselves need to be conscious, and that rather occur mostly below the threshold of consciousness. Thus, while cognitivism needs to assume the existence of representations that convey information about the external world (that are intentional in Brentano's sense—see Sect. 1 of this chapter), and these representations are an essential part of the way in which cognition is explained, the same does not apply to consciousness: cognitivism does not need to assume that representations have conscious qualities, or that consciousness is relevant with respect to cognitive processes; "Thus for cognitivists, cognition and intentionality (representation) are the inseparable pair, not cognition and consciousness" (Varela et al. 1991: 50).

From this point of view, it is clear that the cognitive sciences do not go back to all aspects of mentalism as they were defined by the classical philosophical tradition. The view they argue for not only denies dualism, but also rejects from its explanatory apparatus the conscious (qualitative and introspective) aspects of mind. Indeed, their inclusion would be an obstacle to the project of pursuing a scientific approach to mind, compliant with the criteria of scientificity devised by logical positivism and by behaviorism, which require studying the mind uniquely from a third person perspective and without any recourse to introspection. However, the compliance with these criteria is obtained by paying a high cost: a third person

[23]In regard to this point, it might be relevant to specify that Block ([1995] 1997) do not agree with this view. According to Block P-consciousness does play a role in the knowledge of external objects. Even though it does not contribute to categorization processes or analogous functions, it enriches our experience which would not be the same without P-consciousness.

approach to cognition can be held in its most rigorous form—i.e. without any concession to introspection—only if one accepts the view that qualitative properties have no causal role to play with respect to cognition.

5 Qualitative Properties and First Person Experience

In his book *Il Saggiatore* (*The Assayer* [1623] 2008, §48) Galileo Galilei distinguishes two kinds of qualities: the ones that exist in objects themselves and depend on their configuration and structure (e.g. their shape, their weight etc.) and the ones that are determined by our perception of the objects and correspond to the sensations that they evoke (e.g. whether they are soft or hard; smooth or rough, whether they test sweet or sour, whether the sound they emit is acute or deep, etc.). A similar distinction is also suggested by Locke ([1690] 1999) in *An Essay Concerning Human Understanding* where he differentiates between primary and secondary qualities.

In Locke's view primary qualities are those that inhere in bodies, are utterly inseparable from them and are independent of the perception we have of them:

> First, the bulk, figure, number, situation, and motion or rest of their solid parts. Those are in them, whether we perceive them or not; and when they are of that size that we can discover them, we have by these an idea of the thing as it is in itself; as is plain in artificial things. These I call primary qualities (Locke [1690] 1999, II, VIII, §15).

On the contrary, the secondary qualities are not in the objects themselves, but are produced in us by the objects through their primary qualities and are therefore determined by the fact that the objects affect our senses in a particular manner:

> […] such qualities which in truth are nothing in the objects themselves but power to produce various sensations in us by their primary qualities, i.e. by the bulk, figure, texture, and motion of their insensible parts, as colors, sounds, tastes, etc. These I call secondary qualities (Locke [1690] 1999, II, VIII, §10).

In this view, secondary qualities are the effect of the interaction between the primary qualities of the objects and our senses: they consist in the sensations that the external objects produce in us by virtue of their primary qualities. Hence, the difference between primary and secondary qualities lies in the fact that only the primary qualities truly reveal how the objects are and tell us something about the external world, while the secondary qualities are subjective, in the sense that they describe the sensations evoked in humans by virtue of contact (i.e. by the perception) of objects with specific properties (i.e. the primary qualities).[24]

> Ideas of primary qualities are resemblances; of secondary, not. From whence I think it easy to draw this observation, that the ideas of primary qualities of bodies are resemblances of them, and their patterns do really exist in the bodies themselves, but the ideas produced in us by these secondary qualities have no resemblance of them at all (Locke [1690] 1999, II, VIII, §15).

[24]For a recent overview of the debate on primary and secondary qualities, see e.g. Ross (2015).

Already in Galilei's opinion as well as in Locke's one and in the later philosophical tradition, the distinction between qualities (properties) that depend on objects themselves and qualities (properties) that depend on the subject is of great relevance with respect to scientific research since it allows us to identify the domain of what can be investigated scientifically, delimiting it from other phenomena that cannot be addressed on the basis of a scientific method. Indeed, in Galilei's view only the qualities that exist in the object itself can be measured and described in an objective manner using—among others—mathematical instruments, while the qualities that depend on perception have an intrinsically subjective character and cannot therefore be described in an objective way. This is the reason why only the former can be investigated scientifically, while the latter must be excluded from the domain of scientific research.

As we described in the previous sections, this view which Galilei put forward in relation to physics was extended by logical empiricists to all kinds of sciences. In particular, the behaviorist movement tried to comply with this restriction by excluding secondary qualities from the domain of psychological research. Subsequent traditions—functionalism and cognitivism—also inherited the same skepticism toward secondary qualities and continued therefore to marginalize them and to consider them irrelevant with respect to our cognitive processes. However, even though secondary qualities have a subjective character and it may be problematic to address them using a scientific approach, they are nevertheless also a kind of experience that exists, that is accessible to the first person and that, at least potentially, can provide information to be used for cognitive purposes. As e.g. Varela, Thompson and Rosch maintain, if we neglect this kind of experience, our theory remains at least incomplete:

> Our view is that the current style of investigation is limited and unsatisfactory, both theoretically and empirically, because there remains no direct, hands-on, pragmatic approach to experience with which to complement science. As a result, both the spontaneous and more reflective dimensions of human experience receive little more than a cursory, matter-of-fact treatment, one that is no match for the depth and sophistication of scientific analysis (Varela et al. 1991: xiii).

And further:

> At this point, the tension between science and experience should be obvious and tangible. [...] We cannot simply dismiss this experience without explanation (Varela et al. 1991: 51).

In the field of the contemporary philosophy of mind, the properties that describe this kind of experience and which accompany or characterize the way in which humans perceive the world are generally called '*qualia*'. As e.g. Jagwood Kim states, qualitative states are all those

> that involve sensations: pains; itches; tickles; afterimages; seeing a round, green patch; hearing screeching car tires against pavement; feeling nausea and so on. These mental states are thought to have 'phenomenal' or 'qualitative' aspects – the way they *feel* or the way things *look* or *appear*; thus, pains are thought to have a special qualitative feel that is distinctive of pains – they hurt. When you look at a green patch, there is a distinctive way the patch looks to you: it *looks green* and your visual experience involves this green look

> [...]. Each such sensation has its own distinctive feel [...] The expressions 'raw feel' and 'qualia' are also used to refer to these qualitative mental states (Kim 1996: 13).

Qualia characterize both external perceptions as well as internal states: each perception has its own distinctive feel. One of the most intuitive and brilliant characterizations of the notion of *qualia* that has been offered by the literature can be found in Thomas Nagel's paper '*What is it like to be a bat?*' which discusses the issue of the perceptual experience of bats (Nagel 1974). Nagel is interested in bats essentially because of two factors. On the one hand, they are at a fairly high level of the evolutionary scale; thus, it is almost undeniable that bats have subjective experiences. On the other hand, the sensory apparatus of bats is radically different from that of humans:

> bats [...] perceive the external world primarily by sonar, or echolocation, detecting the reflections, from objects within range, of their own rapid, subtly modulated, high-frequency shrieks. Their brains are designed to correlate the outgoing impulses with the subsequent echoes, and the information thus acquired enables bats to make precise discriminations of distance, size, shape, motion, and texture comparable to those we make by vision. But bat sonar, though clearly a form of perception, is not similar in its operation to any sense that we possess, and there is no reason to suppose that it is subjectively like anything we can experience or imagine (Nagel 1974: 438).

Echolocation allows bats to perceive almost all the characteristics of the objects that human beings can capture through vision. However, since perception in bats radically differs from human perception, it is plausible to assume that the subjective experience of perceiving will nevertheless be extremely dissimilar. Exactly because of this extreme diversity, the subjective experience of bats and the 'world view' that bats form on the basis of their experience seems to be inaccessible to humans:

> I want to know what it is like for a bat to be a bat. Yet if I try to imagine this, I am restricted to the resources of my own mind, and those resources are inadequate to the task. I cannot perform it either by imagining additions to my present experience, or by imagining segments gradually subtracted from it, or by imagining some combination of additions, subtractions, and modifications (Nagel 1974: 439)

The notion of *quale* describes the qualitative and ineffable character of this subjective experience; it describes its *what-it-is-like*; i.e. the phenomenal and introspectively accessible aspects of our mental lives that makes the conscious experience people have of themselves in the first person radically different from that of bats. As Nagel points out:

> So if extrapolation from our own case is involved in the idea of what it is like to be a bat, the extrapolation must be incompletable. We cannot form more than a schematic conception of what it is like. For example, we may ascribe general types of experience on the basis of the animal's structure and behavior. Thus we describe bat sonar as a form of three-dimensional forward perception; we believe that bats feel some versions of pain, fear, hunger, and lust, and that they have other, more familiar types of perception besides sonar. But we believe that these experiences also have in each case a specific subjective character, which it is beyond our ability to conceive (Nagel 1974: 439).

A similar argument in support of the view that experience is characterized by a qualitative dimension is given also by Frank Jackson (1986) in his famous article *What Mary Didn't Know*. Jackson's discussion is centered on the issue of the knowledge of colors. Jackson imagines the case of a scientist called Mary: Mary has acquired a complete physical, chemical and neurological knowledge of colors—of what causes them and of how we come to perceive them—but since she was born, she has been confined in a black-and-white room, and educated through black-and-white books and television. Thus, she has never had a direct experience of colors. Jackson argues that her complete knowledge of all the physical, chemical and neurological facts related to colors and with their perception is not per se sufficient to have a complete knowledge of colors *tout court*. In Jackson's view, the day that Mary is let out of the black and white room and sees colors in the world, she acquires something new, she gains a knowledge of a phenomenal kind that she did not have before because it was not included or implicit in her physical knowledge of the world: i.e. she learns what colors look like. The day she finally comes to see colors, she learns what it is like to see a color; and in this way she also understands what it is like for other people to see colors:

> [...] after Mary sees her first ripe tomato, she will realize how impoverished her conception of the mental life of others has been all along. She will realize that there was, all the time she was carrying out her laborious investigations into the neurophysiologies of others and into the functional roles of their internal states, something about these people she was quite unaware of (Jackson 1986: 292).

Both Jackson's and Nagel's papers have theoretical goals that go beyond the interests of this analysis. Jackson's aim is mainly to challenge the position of physicalism according to which the world is entirely physical and therefore complete physical knowledge is complete knowledge *sic et simplicitier* (Mary is supposed to have a complete knowledge of colors from a physical and neurological point of view, but she still misses an essential piece of information on colors since she does not know what it is like to see red and she learns this only after leaving her room. Therefore, there is a piece of information about her and other people's knowledge of the world that go beyond physicalism—see Jackson 1986: 293). On the other hand, the main goal of Nagel's example is to challenge reductionism, i.e. the thesis that mental processes are indeed physical processes (he thinks that reductionist views cannot account for the subjective character of experience—see Nagel 1974: 444ff). However, regardless of the specific positions these authors aim to defend, both raise the same issue: can we really assume that an objective, third person explanation of perception that describes it in a physical and functional way gives a complete description of the experience and of its features and thus also of how the mind processes information, or shall we acknowledge that—if we disregard the first person perspective and qualitative experience—our analysis of perceptual and cognitive processes will be incomplete or lacking?[25]

[25]This issue has been discussed in the literature on the basis of various arguments. For some recent perspectives, see e.g. Bayne (2009), Kriegel (2002, 2003) and Robinson (2004).

This issue can be addressed starting from different angles. One of these concerns categorization, a topic which is central for cognitive sciences (for an overview on this see e.g. Cohen and Lefebre 2005): can we study categorization—i.e. the way in which people identify and classify what they perceive through concepts—without considering the qualitative dimension of perception?

In the case of the perception of external objects we have the impression that perpetual quale can be separated from the perceived object or even that the two have nothing in common. The vision e.g. of a green tomato implies two things at the same time: the awareness of the perceived object and the awareness of the what-it-is-like of the sensation of seeing green which is part of the perception of the tomato. In spite of the fact that this perception includes two factors—the awareness of qualia and that of the object—it is possible to maintain that the object awareness is clearly distinguished from the qualitative sensation and that it is the only component that plays a role in the categorization process. When we are seeing the green tomato, we do have a qualitative feeling, but what we identify and classify through concepts is not the feeling but the green tomato. It is on the basis of observations like these that many authors belonging to the classical tradition of cognitive science developed the hypothesis that qualia are irrelevant with respect to categorization and to cognitive processes and that they are just *epiphenomena* which accompany our cognitive activities without actively participating in them and without playing any causal role with respect to them. Such a position is actually supported also by Jackson himself who argues that—even though qualia certainly exist—they are useless from a cognitive point of view:

> [...] they are an excrescence. They do nothing, they explain nothing, they serve merely to soothe the intuitions of dualists, and it is left a total mystery how they fit into the world view of science. In short we do not and cannot understand the how and why of them (Jackson 1982: 135).

The theory that phenomenal consciousness has no causal power and thus it does not exhibit any effect in the world has already been mentioned in the previous section and is called "*consciousness epiphenomenalism*" (on this see also e.g. Jackendoff [1987] 1994: 26; Chalmers 1996: 152; Kirk 2005: 42[26]). According to consciousness epiphenomenalism, the qualitative dimension of our knowledge does not have any cognitive function.[27]

Applied to the problem of categorization, this view suggests that qualia do not play any function as for the identification and classification of instances in a conceptual form. However, while this position might seem plausible when we consider the identification and classification of external objects (such as e.g. the green tomato we mentioned before), it becomes very unlikely when it comes to the categorization

[26]The discussion on consciousness epiphenomenalism is closely related to the debate concerning the possibility of the existence of 'zombies', i.e. of creatures who are identical to us as in terms of their cognitive architecture in a functional sense, but lack a phenomenal consciousness. See, i.e., Braddon-Mitchell and Jackson (1996: 122–143), Lyons (2009), Güzeldere (1997) and Kirk (2005).

[27]For a critical overview of this debate see e.g. Kriegel (2006) and Flanagan (1997).

of bodily states like pains, itches, tickles, feeling nausea etc. because in these cases the qualitative feelings *are* the objects we need to classify and to conceptualize. As a matter of fact, when we perceive something like a pain or an itch and identify/classify this perception as 'pain' or as 'itch', the information we rely on can only be how this feels. Indeed, we are able to identify the specific kind of pain we are experiencing now (e.g. as a stinging pain in the stomach) and to distinguish it from the pain of a different kind we experienced some days ago (e.g. as a burning pain in the stomach). This means that we have the capability to conceptually classify our bodily states. And the classification of these states does not seem to depend on the features of some external referents, as in the case of the perception of a green tomato, but is much more likely to depend on the feeling itself: on what we are feeling or on how what we are feeling feels. More specifically, this appears to depend on the characteristics of these feelings in the sense of its specific 'what-it-is-like'. Thus, when it comes to the classification of bodily states, it seems extremely plausible to assume that qualia do have a cognitive function insofar as they are—i.e. they coincide with—the information that we use to identify and to classify our internal feelings.

Actually, this coincidence between the qualitative feeling on the one hand and the objects we need to identify and classify on the other also occurs in the case of external perception when it comes to the 'secondary qualities' defined by Galilei and Locke such as e.g. colors, flavor, sound, odors etc. Let's consider for example our capacity to discriminate different tonalities of colors, e.g. of red. This is the precondition for learning any system of classification of colors, from the simplest to the most complicated. In this case, we could argue that our perceptual and conceptual capacity to differentiate between different tonalities of colors and to distinguish e.g. between 'purple' and 'violet' relies on the color itself, not on the qualitative feel of the color, i.e. on the what-it-is-like of color perception. However, in this case it seems difficult to distinguish the color itself from the what-it-is-like of the color perception, while it appears more plausible to maintain that the *quale* and the color are one and the same thing. An analogous conclusion can be drawn in the case of an odor (think e.g. of the smell or cinnamon), of a flavor (the sour taste of a lemon) or of a sound (the acute noise of the dentist's drill). In all these cases, it seems that the qualitative aspect of the sensation is the element we use for its identification and classification: i.e. to identify a particular smell or to classify it as the smell of cinnamon; to identify a particular flavor or to classify it as the flavor of a lemon; to identify a particular acute sound or to classify it as the typical noise of a dentist's drill. In all these cases, qualia seem to have a cognitive function because they are the information we use to identify and to classify the sensory experience we have. In such cases, taking qualia into account seems of essential relevance in understanding how cognitive processes work. Here, phenomenal character and representational content seem to be practically indistinguishable.

As pointed out also by Varela, Thompson and Rosch in the book which defines the main research lines of the position called embodied cognition:

> It is important to appreciate just how problematic the relation between the computational mind and the phenomenological mind is. The problem centers on how intentionality and consciousness are related. We have seen that cognitivism draws a sharp and fundamental distinction between these two aspects of cognition. Our cognition, however, seems to be directed toward the world in a way that intimately involves consciousness. Thus notice that our cognition is directed toward the world in a certain way: it is directed toward the world as we experience it. For example, we perceive the world to be three-dimensional, macroscopic, colored, etc.; we do not perceive it as composed of subatomic particles. Thus our cognition is directed toward an experiential world, or in the terms of phenomenology, toward a lived world. How, then, if intentionality and consciousness are fundamentally distinct, does cognition come to be about the world as we consciously experience it? (Varela et al. 1991: 52)

As it is shown by Nagel's argument, and at least in part also by Jackson's example, the experience that a person has depends on his/her bodily constitution and on the way in which his/her sense organs work: the human way of seeing—i.e. tridimensional, characterized by specific color distinctions etc.—depends on the bodily constitution of human beings and it makes our perception and our way of experiencing the world very different from that e.g. of a bat. As a matter of fact, the kind of experience of—i.e. of 'access' to—the world any organism has depends on its bodily constitution. For this reason, it would be difficult to argue that the qualitative and perspectival aspects of the mapping of the external environment made by living organisms do *not* influence their 'knowledge' (their 'cognition') about the world. The kind of world view an organism has and the kind of information processing it carries out seem to depend indeed on the kind of information it can access and on the kind of experiences it has. This makes the position that qualitative properties are irrelevant with respect to the way in which humans or creatures in general conceive (cognitively process) information highly unlikely.[28]

Even if it is subjective and accessible only to the first person, the qualitative component of our experience has its own reality status, in the sense that it is an undeniable part of the way in which every creature provided with consciousness has access to the world. For this reason, it must be considered empirically relevant, i.e. as a piece of evidence that needs to be included in the explanation of how the mind works. As also pointed out by Nagel, it is improbable that neglecting this aspect will help us gain a better understanding of the human experience: "Certainly it appears unlikely that we will get closer to the real nature of human experience by leaving behind the particularity of our human point of view and striving for a description in terms accessible to beings that could not imagine what it was like to be us." (Nagel 1974: 444). To obtain a full understanding of human experience, our investigations must consider all the aspects that characterize it, including the qualitative—phenomenal—aspects which can be addressed starting from the first person perspective only.

From this point of view, the precept of Galilean origin that secondary qualities must be excluded from scientific investigation led to an impoverishment of the

[28]For a meaningful overview of this see e.g. Thompson (2007) and Stewert et al. (2010).

study of mind, giving rise to a paradox of difficult solution, that is excellently described by Nagel in the following terms:

> What has made modern physical science possible is the method of investigating the observable physical world not with respect to the way it appears to our senses [...] but rather as an objective realm existing independently of our minds. [...] In order to do this, it was necessary to find ways of detecting and measuring and describing features of the physical world which were not inextricably tied to the ways things looked, sounded and felt to us; and this resulted in the discovery of objective, spatiotemporal properties of the physical world which could be mathematically described and related by general laws of extraordinary power and universality, thus enabling us to transcend the rough and particular associations available at the level of merely human appearances. The result is an understanding of objective physical reality almost unrecognizably different from the familiar world of our theoretically unaided experience. But it was a condition of this remarkable advance that the subjective appearances of things be excluded from what had to be explained and described by our physical theories. [...] The whole idea of objective physical reality depends on excluding the subjective appearances from the external world and consigning them to the mind instead. But it follows inexorably from this strategy that the same methods of objective physical understanding cannot be successfully applied to the subjective contents of the mind themselves. The method can be used on the body, including its central nervous system, and on the relation of neural activity to the observable behavioral functioning, because they are all aspects of objective physical reality. But for the subjective qualities of experience themselves, we need a different form of understanding. We cannot hope to understand them completely as an aspect of objective physical reality, because the concept of objective physical reality depends on excluding them from what has to be understood. [...] So when science turns to the effort to explain the subjective quality of experience, there is no further place for these features to escape to. (Nagel 1994: 65–66)

In the light of these considerations, we can say that Nagel's and Jackson's arguments point to a theoretical deficiency of psychological theories that concerns not only behaviorism but also post-behavioristic theories. In fact, since the latter also agree to the methodological principle of excluding any kind of phenomenon which is not publicly observable and objectively measurable from their investigations, they support an approach to mind which is intrinsically limited, because it relies exclusively on an extrinsic, third person identification of mental states, avoiding any reference to qualitative, subjective experience. To overcome this limitation, some philosophers argue that we need to go back to a deeper form of mentalism, because it is impossible to develop a full explanation of mental phenomena without considering also the intrinsic, qualitative and subjective dimension of mind.

6 First Person Experience and Embodiment

The explanatory apparatus used in cognitive science to describe mental processes is built on the so-called mind/computer analogy which suggests that cognition consists of computations on representations (for a more detailed analysis of this aspect see Chap. 2) On the basis of this analogy, the cognitive system is viewed as an *immaterial, quasi-cartesian device* that makes calculations "according to rules on data which take the form of atomic facts" (Dreyfus 1972: 143). This analogy appears to rule out the idea that our knowledge has a specific, first personal and qualitative dimension and above all that the phenomenal character of our experience somehow enters into and determines our cognitive processes and their outcomes. The apparent incompatibility of the idea that cognitive processes are computations with the need to account for the first personal, qualitative character of experience raises the issue of whether the relation between the computational mind and the phenomenological mind can somehow be accommodated.

One possible angle to start addressing this issue is to focus on those aspects of human cognition that differ from the information processing carried out by a computer and that are obscured by the mind/computer analogy. Ontologically, a computer and a human being are certainly not one and the same thing. As biological organisms—as living systems—humans are equipped with a body that cannot be compared *tout court* with a mechanical system of information transmission. If cognition consists of information processing, and yet this information travels through a medium, first of all we must investigate the differences between information transmission that is transmitted through inert versus living matter. Classical studies in the field of cognitive science ignore the function of the body in cognitive processes (see e.g. Putnam 1960; Fodor 1975; Pylyshyn 1984; Churchland 1986; Edelman 1992). As e.g. Lawrence Barsalou's has also pointed out:

> According to the functionalist perspective that has dominated modern cognitive science, the symbol system underlying human intelligence can be disembodied. Once we characterize the computational properties of this system successfully, we can implement it in other physical systems such as computers. Thus, functionalism implies that the computational system underlying human intelligence can be understood independently of the human body. [...] Just as computer software can be characterized independently of the particular hardware that implements it, the human symbol system can be characterized independently of the biological system that embodies it (Barsalou 1999: 598).

Unlike machines, humans are equipped with a body and human cognitive processing takes place in and through the body. Even if we think that we can explain cognition in terms of computational processes consisting of operations on representations whose content is information, we must investigate the consequences of the fact that human cognition is embodied, i.e. that it travels through the medium of the body.

This issue is certainly not new in the history of cognitive research and it goes back at least to Norbert Wiener, one of the founding fathers of cybernetics. According to Wiener, any empirically based approach to mind that can give rise to a logically and formally well-grounded account of how mind works must rely on the notion of information. Yet, information cannot be interpreted in terms of "matter or energy": "Information is information, not matter or energy. No materialism which does not admit this can survive at the present day" (Wiener [1948] 1961: 132). This view is completely congruent with the abstract notion of information which is fostered by functionalism. Indeed, in spite of this so to speak 'immaterialist' idea of information according to which it is something 'ideal' that we extract from the external world and use in cognition, Wiener opens the door to a very different interpretation of how information processing works that takes into consideration where it is processed and how this processing affects it.

> the central nervous system no longer appears as a self-contained organ, receiving inputs from the senses and discharging into the muscles. On the contrary, some of its most characteristic activities are explicable only as circular processes, emerging from the nervous system into muscles, and re-entering the nervous system through the sense organs, whether they be *proprioceptors* or organs of the special senses. This seemed to us to mark a new step in the study of that part of the neurophysiology which concerns not solely the elementary processes of nerves and synapses but the performance of the nervous system as an integrated whole (Wiener [1948] 1961: 8, italics added).

Weiner's emphasis on perceptual and especially proprioceptive factors as crucial elements in describing how knowledge is processed opens up new territory for cognitive research—an angle that wasn't explored by subsequent investigators, who instead focused only on the computational aspects of Wiener's theory.[29] Indeed, as Wiener points out, the body is not just a neutral means for extracting information from the external environment and transmitting it to the inside, but is itself a source of (proprioceptive) information: while information "circulates" through the body, it is contaminated by it and by the information the body produces through its own activity. As Clancey suggests, cognition is a *concrete form* of *activity developing within the activity itself* (Clancey 1997: 4).

Wiener's point has, however, been taken up by contemporary theorists who posit the cognition is essentially situated. *Situatedness* can be interpreted in many ways,[30] but essentially means that the context in which a process takes place is not only the container of that process, but becomes an essential part of the process itself, since context and process depend on each other and are constituted jointly (Clark 2008a, b; Prinz 2009).[31] Theories that consider the body as the most important context where cognition takes place are called embodied (see e.g. Varela et al. 1991).[32] One of the main ideas of embodied cognition is that "the presence of a humanlike mind depends quite directly upon the possession of a humanlike body" (Clark 2008a: 43). According to this embodied perspective, our mental capacities "[depend] on the possession and use of a body, not just a brain" (Prinz 2009: 420). Indeed, according to this perspective, no form of knowledge can be considered as

[29]This becomes particularly clear if we consider how much the later tradition of research relies on the idea of multiple realizability *a là* Putnam according to which psychological states or properties are not bound to any specific physical support, but the same state or property can be realized by many different physical systems (for classical arguments for multiple realizability see Putnam 1967; Fodor 1974).

[30]For an overview of the theories of situated cognition see e.g. Robbins and Aydede (2009).

[31]The context in which cognition takes place can be interpreted in various ways; these different interpretations give rise to more or less diverging views such us the so-called '*4-e-conceptions*' of cognition, *embodied cognition, embedded cognition, extended cognition and enactive cognition*. For a critical discussion see Walter (2014) and Walter and Kästner (2012).

[32]The manifesto of this line of thought is usually considered to be Varela et al. (1991).

separate from and as insensible to the system that produces it because it is always a function of this biological organism (Varela et al. 1991; Thompson 2007).

The reason why the embodied tradition considers possession of a body to be an essential ingredient in any account of cognition is, in the first place, the same as that mentioned by Wiener: our cognitive system not only draws information from the external world, but also from the body itself which influences and contaminates all other sources of information and information processing (Varela et al. 1991). Thus, according to this view, neither perception nor cognition can be studied in abstract as general functions and purely symbolic manipulations, without taking into account the embodiment of the cognitive system.

Prima facie this characterization of embodied cognition and its aims does not really address the question we started with in this section concerning how it is possible to reconcile the phenomenal character of our experience with a classical cognitive account of cognition based on information processing. However, if we look more closely, an answer is already at hand. Firstly, and more generally, the phenomenal character of experience can be plausibly interpreted as an effect of the fact that information is processed by a biological system. Thus, the study of biological systems is a necessary precondition to understanding qualitative experience. Secondly and more specifically, bodily information is a particularly interesting aspect to investigate if one is interested in understanding whether first person, qualitative experience is essential for understanding cognition or whether we can consider it as an 'excrescence'. As we already mentioned in the previous section, in some cases—take e.g. vision—the content of the perceptual act can be distinguished by the phenomenal quality that accompanies this act: when I see an apple I can distinguish the phenomenal quality of my perception from its content (the apple). Granted that this distinction is possible, an analysis of our cognitive capacities might focus only on the content—on the 'matter'—of the perceptual act and ignore the first person experience. However, in the case of bodily information this distinction is awkward because the experience we have of the body and its activities is uniquely first personal and qualitative. In cases of bodily states for example, the content of our perception cannot be separated from how it feels to us —by the phenomenal character of the experience we have.

Take the case of pain. The information we have about our pain as well as the way we represent this pain is constituted by conscious qualitative information which we access from a first personal point of view (i.e. introspectively). What we know about pain is due to subjective and qualitative information acquired through proprioception. No one else can experience my pain and to know what pain is, a person must have experienced it in the first person. Pain is not the only direct example of internal qualitative information we have access to that supports our knowledge of the world and of ourselves. Many bodily sensations allow us e.g. to acquire knowledge of the internal states of the body: think for example of hunger, thirst, itch, relaxation, tension, etc. We know we are in these states because we have a phenomenal (qualitative) feel that informs us about them. Moreover, many other kinds of knowledge we have depend most plausibly on first-personal bodily experience. How do we know, for example that the thought or the perception we have right now is *our* thought and *our* perception and not somebody else's?[33] How do we know that we are in love or angry?

One hypothesis that certainly deserves to be explored is that our knowledge of all kinds of internal states including emotions depends primarily not on information we get from the external world but on bodily, qualitative information we can access only from a first personal point of view. On the other hand, internal feelings might be involved in many other forms of knowledge we develop. Various types of abstract knowledge rely (directly or indirectly) on bodily information (on bodily states and emotions). We conceive of a political situation as *tense* because we have a concept built on bodily experience of what 'tense' means and implies. We understand that a person is a *friend* not only because s/he behaves in a certain way toward us but also or primarily because we have certain emotions (feelings) toward him/her. So, the general question of embodied cognition as we would like to practice it in this work becomes: *how do the bodily feelings that we consciously experience from a first-person perspective contribute to constituting (some of) the knowledge we have about ourselves and about the world?*

Indeed, those lines of study that follow a truly embodied perspective maintain that the cognitive system is an embodied system, i.e. a biological, living system that gives rise to multiple forms of information of a qualitative character which are available to first person access only and which cannot be traced back to a third person perspective. In their view, this qualitative, first-person information generated by the body must be included in our explanation of cognitive processes (see e.g. Varela et al. 1991; Thompson 2007; Berthoz and Petit [2006] 2008; Borghi and Cimatti 2010). In this work, we share this basic principle and we try to show that it has important implications and consequences for our study of cognitive processes.

We address this issue starting from a specific angle, i.e. from the angle of categorization processes: we consider specific cases of information processing related to our categorization capacities and we try to show in a concrete way that—if we assume that phenomenal (qualitative) experience does play a role with respect to cognition—we can shed new light on some unresolved questions concerning the way in which we form our conceptual representations and develop some of our capacities. The line we will follow can be summarized by a formulation of Lawrence Shapiro:

> [...] we are going to consider research in embodied cognition that purports to show a relationship between the kind of body an organism possesses and the kinds of concepts an organism can acquire. In its most dramatic form, the hypothesis under review is this: to conceive of the world as a human being does requires having a body like a human being's (Shapiro 2011: 71).

---

[33]This idea is implicitly suggested by Goldman when he speaks of a proprietary code of representations: see Goldman (2006: 261ff). Goldman also suggests that the knowledge of our mental states is introspective and bodily-related (Goldman 2012), but we prefer not to enter this territory for now and to focus our analysis on specific concepts directly and indirectly related to internal bodily states.

More specifically, we address the issue of how people develop the capacity to categorize internal states and to form corresponding concepts that allow them to speak about what they feel. The hypothesis we will put forward and argue for here is that our bodily feelings determine or at least constrain the concepts we develop, since our concepts and the categorizations we carry out depend intrinsically (also) on *proprioceptive* information that originates in our bodies.

The general line of this work starts from a suggestion briefly developed by Alvin Goldman on "first person attributions". As part of his criticism of anti-mentalistic views—including post-behaviorist ones—according to which mental states are and must be identified from a third person perspective, without any reference to a first personal perspective, Goldman raises the question of how people come to classify (i.e. to identify and to conceptualize) states/conditions that do not have an external reference, but which denote some internal experience.

> When a person classifies a perceived object as a chair, presumably he has some active perceptual information about that object that matches his stored concept of a chair, i.e., the representation in long-term memory associated with the word 'chair' […] Let us assume that the same sort of thing transpires in the case of mental classifications. If a person classifies himself as currently having a headache, or an itch, his cognitive system must have received some sort of information that matches his concept of headache, or his concept of itch (Goldman 1993: 84).

Thus, this is the issue brought up by Goldman: *what is the information we use to carry out these kinds of categorizations?* Shall we assume that we identify our own internal states starting from extrinsic information, e.g. of a functional kind (e.g. what caused my pain or how it affects my behavior), or shall we rather suppose that the identification and the classification of our internal states is carried out on the basis of some phenomenal information of a qualitative kind which we have access to primarily through some introspective channel? This leads us on to the issue of whether the concepts we develop to categorize internal states which support our semantic competence in using words denoting these states, consist of external information only, or whether they are composed (at least also) of qualitative information.

Goldman's answer to this issue is that the subjective—intrinsic and phenomenal—element is crucial for the identification and for the recognition—and thus for the categorization—of mental states: "Certainly it is highly plausible that one classifies such sensations as headaches or itches on the basis of qualitative feel" (Goldman 1993, 87—for a more detailed discussion of Goldman's position see Chap. 5, Sect. 2). In this work we further develop Goldman's intuition and try to show that information concerning first person experience—experience that is accessible uniquely through introspection and that has to do with *what-it-is-like*—is indispensable for developing (at least) some conceptual classes closely related to internal states and is therefore also essential for mastering the corresponding words.

The problem we are addressing is analogous to—and could, at least in part, be traced back to—the (more classic) issue of phenomenal concepts. In the philosophical literature, the concepts called 'phenomenal' are those that refer to phenomenal experiences, i.e. to the qualitative character of experience (for an overview see e.g. Balog 2009). However, our approach in this work does not comply with the philosophical discussion of phenomenal concepts, diverging in several respects.

First of all, we do not address the problem of phenomenal concepts in general but consider it only with reference to specific conceptual classes. Indeed, the class of phenomenal concepts is too fuzzy: starting from specific conceptual classes, we aim to address the issue of how qualitative information contributes to categorization in a more specific and concrete manner. Our goal is also to determine what concepts rely on qualitative information and must therefore be included in or related to this class. Indeed, at the end of this research we will show that there are many more concepts than one might imagine and that categories like animate/inanimate, or concepts referring to emotions as well as a number of abstract concepts are also built starting from qualitative information.

Secondly, our inquiry is based on the view that concepts support our semantic mastery of the corresponding words (on this see Chap. 3, Sects. 1 and 2) and that therefore the issue of whether our direct, first-personal phenomenal experience does contribute to our cognitive processing and especially to categorization can be addressed starting from the analysis of our semantic competence. In our view, it is more promising to address the issue in this way since it allows us to answer this question using empirical evidence on semantic mastery under normal circumstances as well as in the case of particular speech impairments.

Finally, we do not consider qualitative experience in relation to external perception, but focus primarily if not uniquely on qualitative experience related to the internal states of the body and to our bodily feelings. As we specified before, bodily feelings are an ideal starting point to discuss whether the qualitative component of our experience does play a role in categorization because—unlike the case of external perception where we can possibly distinguish the content of the perceptual act from the sensation—in the case of internal feelings this distinction does not apply. Thus, the case of internal feelings is more extreme and unequivocal. In addition, as we will show in the next chapter, internal feelings are the element in our experience that has been most overlooked and neglected by cognitive research in spite of the fact that—if we are right—these feelings play a central role in categorization processes. The thesis we will try to argue for is that the development of concepts directly or indirectly related to internal states—which support our mastery of the words denoting such states—is based on something like an *internal ostension*, i.e. on the identification of specific, qualitatively characterized features of these states which serve as a perceptual basis for categorizing them and as referents for the words describing those states.

References

Alvarez, M. (2014). Ryle on motives and dispositions. In D. Dolby (Ed.), *Ryle on mind and language* (pp. 74–96). London: Palgrave MacMillan.

Balog, K. (2009). Phenomenal concepts. In A. Beckermann, B. P. McLaughlin, & S. Walter (Eds.), *The Oxford handbook of philosophy of mind* (pp. 292–312). Oxford: Oxford University Press.

Barone, F. (1997). *Metafisica: Il mondo nascosto*. Rome: Laterza.

Barsalou, L. W. (1999). Perceptual symbol systems. *Behavioral and Brain Sciences, 22*(4), 577–660.

Bayne, T. (2009). Perception and the reach of phenomenal content. *The Philosophical Quarterly, 59*(236), 385–404.

Bechtel, W. (1988a). *Philosophy of mind. An overview for cognitive science*. Hillsdale, NJ: Erlbaum.

Bechtel, W. (1988b). *Philosophy of science. An overview for cognitive science*. Hillsdale, NJ: Erlbaum.

Berthoz, A. & Petit, J.-L. ([2006] 2008). *Physiology and phenomenology of action*. (C. Macann, Trans.). Oxford: Oxford University Press.

Best, J. B. (1998). *Cognitive psychology* (5th ed.). New York, London: Wiley.

Block, N. ([1995] 1997). On a confusion about a function of consciousness. In N. Block, O. J. Flanagan & G. Güzeldere (Eds.), *The nature of consciousness. Philosophical debates* (pp. 375–415). Cambridge, MA: MIT Press.

Block, N. (1978). Troubles with functionalism. In C. W. Savage (Ed.), *Minnesota studies in the philosophy of science. Volume 9: Perception and cognition* (pp. 261–325). Minneapolis: Minnesota University Press.

Block, N. (1980). What is functionalism? In N. Block (Ed.), *Readings in the philosophy of psychology* (Vol. 1, pp. 171–184). Cambridge, MA: Harvard University Press.

Borghi, A. M., & Cimatti, F. (2010). Embodied cognition and beyond: Acting and sensing the body. *Neuropsychologia, 48*(3), 763–773.

Boyle, D. A. (2009). *Descartes on innate ideas*. London, New York: Continuum.

Braddon-Mitchell, D., & Jackson, F. (1996). *Philosophy of mind and cognition*. London: Blackwell.

Brandt, R., & Kim, J. (1963). Wants as explanation of actions. *The Journal of Philosophy, 60*(15), 425–435.

Brentano, F. ([1874] 1995). In T. Crane & J. Wolff (Eds.), *Psychologie vom empirischen Standpunkt*. English edition: *Psychology from an empirical standpoint* (2nd ed.). London: Routledge.

Burloud, A. (1948). *Psychologie*. Paris: Hachette.

Byrne, A. (1994). Behaviourism. In S. D. Guttenplan (Ed.), *A companion to philosophy of mind* (pp. 132–140). Oxford: Blackwell.

Carnap, R. ([1932] 1959). Psychologie in physikalistischer Sprache. English edition: *Psychology in physical language*. In A. J. Ayer (Ed.), *Logical positivism* (pp. 165–198). New York: Free Press.

Carnap, R. (1935). Les concepts psychologiques et le concepts physiques sont-ils foncièrement différents? *Revue de Synthèse, 10*(1), 43–53.

Carnap, R. (1936). Testability and meaning. *Philosophy of Science, 3*(4), 419–471.

Carnap, R. (1938). Logical foundation of the unity of science. In O. Neurath, R. Carnap, & C. Morris (Eds.), *International encyclopedia of unified science* (Vol. 1, pp. 42–62). Chicago: Chicago University Press.

Carruthers, P. (2005). *Consciousness: Essays from a high-order perspective*. Oxford: Oxford University Press.

Carruthers, P. (2010). *The opacity of mind. An integrative theory of self-knowledge*. Oxford: Oxford University Press.

Catania, A. C., & Harnad, S. (Eds.). (1988). *The selection of behavior. The operant behaviorism of B.F. Skinner: Comments and consequences*. Cambridge, New York: Cambridge University Press.

Chalmers, D. J. (1996). *The conscious mind: In search of a fundamental theory*. New York, Oxford: Oxford University Press.

Chisholm, R. (1957). *Perceiving*. Ithaca: Cornell University Press.

Chomsky, N. (1959). A review of B.F. Skinner's verbal behavior. *Language, 35*(1), 26–58.

Chomsky, N. (1965). *Aspects of the theory of syntax*. Cambridge, MA: MIT Press.

Churchland, P. S. (1986). *Neurophilosophy: Toward a unified science of the mind/brain*. Cambridge, MA: MIT Press.

Clancey, W. J. (1997). *Situated cognition. On human knowledge and computer representations*. Oxford: Clarendon Press.

Clark, A. (2008a). Pressing the flesh: A tension in the study of the embodied, embedded mind? *Philosophy and Phenomenological Research, 76*(1), 37–59.

Clark, A. (2008b). *Supersizing the mind*. Oxford: Oxford University Press.

Clarke, D. (2003). *Descartes' theory of mind*. Oxford: Oxford University Press.

Cohen, H., & Lefebre, C. (Eds.). (2005). *Handbook of categorization in cognitive science*. Amsterdam: Elsevier.

Crane, T. (2001). *Elements of the mind. An introduction to the philosophy of mind*. Oxford: Oxford University Press.

Crane, T., & Farkas, K. (Eds.). (2004). *Metaphysics. A guide and an anthology*. Cambridge: Cambridge University Press.

Dellantonio, S. (2007). *Die interne Dimension der Bedeutung. Externalismus, Internalismus und semantische Kompetenz*. Peter Lang: Frankfurt a.M.

Descartes, R. ([1641] 1984). *Meditationes de prima philosophia*. English edition: *Meditations on first philosophy*. In J. Cottingham, R. Stoothoff & D. Murdoch (Eds.), *The philosophical writings of Descartes* (Vol. 2, pp. 3–62). New York, Cambridge: Cambridge University Press.

Dreyfus, H. L. (1972). *What computers can't do: A critique of artificial reason*. New York: Harper & Row.

Edelman, G. M. (1992). *Bright air, brilliant fire: On the matter of the mind*. New York: Basic Books.

Farkas, K. (2008). *The subject's point of view*. Oxford: Oxford University Press.

Feigl, H. (1958). The 'mental' and the 'physical'. In H. Feigl, M. Scriven, & G. Maxwell (Eds.), *Minnesota studies in philosophy of science. Volume 2: Concepts, theories, and the mind-body problem* (pp. 370–397). Minneapolis: Minnesota University Press.

Feigl, H. (1963). Physicalism, unity of science and foundations of psychology. In P. A. Schilpp (Ed.), *The philosophy of Rudolf Carnap* (pp. 227–267). La Salle: Open Court.

Flanagan, O. J. (1997). Conscious inessentialism and the epiphenomenalist suspicion. In N. Block, O. J. Flanagan, & G. Güzeldere (Eds.), *The nature of consciousness. Philosophical debates* (pp. 357–373). Cambridge, MA: MIT Press.

Fodor, J. A. (1968). *Psychological explanation: An introduction to the philosophy of psychology*. New York: Random House.

Fodor, J. A. (1974). Special sciences: Or the disunity of science as a working hypothesis. *Synthese, 28*(2), 97–115.

Fodor, J. A. (1975). *The language of thought*. New York: Thomas Y. Crowell Company.

Galilei, G. ([1623] 2008). Il Saggiatore. Partial English edition: *The assayer*. In M. A. Finocchiaro (Ed.), *The essential Galileo* (pp. 179–189). Indianapolis, Cambridge: Hackett Publishing.

Gardner, H. E. (1987). *The mind's new science: A history of the cognitive revolution*. New York: Basic Books.

Geach, P. (1957). *Mental acts. Their content and their objects*. London: Routledge & Kegan Paul.

Gertler, B. (2011). *Self-knowledge*. London: Routledge.

Goldberg, S. C. (Ed.). (2007). *Internalism and externalism in semantics and epistemology*. New York: Clarendon Press.

Goldberg, S. C. (2015). *Externalism, self-knowledge and skepticism: New essays*. Cambridge: Cambridge University Press.
Goldman, A. I. (1993). *Philosophical applications of cognitive science*. Boulder: Westview Press.
Goldman, A. I. (2006). *Simulating minds: The philosophy, psychology, and neuroscience of mindreading*. New York, Oxford: Oxford University Press.
Goldman, A. I. (2012). A moderate approach to embodied cognitive science. *Review of Philosophy and Psychology, 3*(1), 71–88.
Güzeldere, G. (1997). The many faces of consciousness: A field guide. In N. Block, O. J. Flanagan, & G. Güzeldere (Eds.), *The nature of consciousness. Philosophical debates* (pp. 1–67). Cambridge, MA: MIT Press.
Heil, J. (1998). *Philosophy of mind: A contemporary introduction*. London, New York: Routledge.
Hempel, C. G. ([1935] 1980). The logical analysis of psychology. In N. Block (Ed.), *Readings in philosophy of psychology, Volume 1* (pp. 14–23). Cambridge, MA: Harvard University Press.
Jackendoff, R. ([1987] 1994). *Consciousness and the computational mind* (4th ed.). Cambridge, MA: MIT Press.
Jackson, F. (1982). Epiphenomenal qualia. *The Philosophical Quarterly, 32*(127), 127–136.
Jackson, F. (1986). What Mary didn't know. *The Journal of Philosophy, 83*(5), 291–295.
Kallestrup, J. (2011). *Semantic externalism*. London: Routledge.
Kim, J. (1996). *Philosophy of mind*. Boulder: Westview.
Kirk, R. (2005). *Zombies and consciousness*. Oxford: Clarendon Press.
Kornblith, H. (Ed.). (2001). *Epistemology: Internalism and externalism*. London: Wiley-Blackwell.
Kriegel, U. (2002). Phenomenal content. *Erkenntnis, 57*(2), 175–198.
Kriegel, U. (2003). Consciousness as sensory quality and as implicit self-awareness. *Phenomenology and Cognitive Science, 2*(1), 1–26.
Kriegel, U. (2006). Consciousness: Phenomenal consciousness, access consciousness, and scientific practice. In P. Thagard (Ed.), *Handbook of philosophy of psychology and cognitive science* (Vol. 4, pp. 195–217). Amsterdam: North Holland.
Lewis, D. K. (1966). An argument for identity theory. *The Journal of Philosophy, 63*(2), 17–25.
Lewis, D. K. (1972). Psychophysical and theoretical identifications. *Australasian Journal of Philosophy, 50*(3), 249–258.
Locke, J. ([1690] 1999). *An essay concerning human understanding*. Electronic Classic Series, The Pennsylvania State University. ftp://dca.fee.unicamp.br/pub/docs/ia005/humanund.pdf.
Lyons, J. C. (2009). *Perception and basic beliefs: Zombies, modules, and the problem of external world*. New York: Oxford University Press.
Lyons, W. E. (1986). *The disappearence of introspection*. Cambridge, MA: MIT Press.
Marres, B. (1989). *In defence of mentalism. A critical review of the philosophy of mind*. Amsterdam: Rodopi.
McGrew, T., & McGrew, L. (2006). *Internalism and epistemology: The architecture of reason*. London: Routledge.
Michel, C. (2014). *Self-knowledge and self-deception. The role of transparency in first personal knowledge*. Paderborn: Mentis.
Mills, J. A. (1998). *Control: A history of behavioral psychology*. New York, London: New York University Press.
Nagel, T. (1974). What it is like to be a bat? *The Philosophical Review, 83*(4), 435–450.
Nagel, T. (1994). Consciousness and objective reality. In R. Warner & T. Szubka (Eds.), *The Mind-body problem: A guide to the current debate* (pp. 63–86). Oxford: Blackwell.
Nannini, S. (2006). *Seele, Geist und Körper. Historische Wurzeln und philosophische Grundlagen der Kognitionswissenschaften*. Frankfurt a.m: Peter Lang.
Neisser, U. (1966). *Cognitive psychology*. New York: Appleton Century Crofts.
Neurath, O. ([1932] 1987). Unified science and psychology. In B. McGuinness (Ed.), *Unified science* (pp. 1–23). Dordrecht: Riedel.
Nimtz, C. (2002). *Wörter, Dinge, Stellvertrer*. Paderborn: Mentis.

Pastore, L. (2010a). Intuition. In H.-J. Sandkühler (Ed.), *Enzyklopädie philosophie* (pp. 1160–1168). Hamburg: Felix Meiner.

Pastore, L. (2010b). Intelligenz. In H.-J. Sandkühler (Ed.), *Enzyklopädie philosophie* (pp. 1120–1126). Hamburg: Felix Meiner.

Place, U. T. (1956). Is consciousness a brain processes? *The British Journal of Psychology, 47*(1), 44–50.

Prinz, J. J. (2009). Is consciousness embodied? In P. Robbins & M. Aydede (Eds.), *The Cambridge handbook of situated cognition* (pp. 419–437). Oxford: Oxford University Press.

Putnam, H. (1960). Minds and machines. In S. Hook (Ed.), *Dimensions of mind* (pp. 138–164). London: Collier-MacMillan.

Putnam, H. ([1963] 1975). Brains and behavior. In H. Putnam (Ed.), *Philosophical papers. Volume 2: Mind, language and reality* (pp. 325–341). London: Cambridge University Press.

Putnam, H. (1967). Psychological predicates. In W. H. Capitan & D. D. Merrill (Eds.), *Art, mind, and religion* (pp. 37–48). Pittsburgh: University of Pittsburgh Press.

Pylyshyn, Z. W. (1984). *Computation and cognition*. Cambridge, MA: MIT Press.

Robbins, P., & Aydede, M. (Eds.). (2009). *The Cambridge handbook of situated cognition*. Cambridge: Cambridge University Press.

Robinson, W. S. (2004). *Understanding phenomenal consciousness*. Cambridge: Cambridge University Press.

Ross, P. (2015). Primary and secondary qualities. In M. Matthen (Ed.), *The Oxford handbook of philosophy of perception* (pp. 405–421). Oxford: Oxford University Press.

Rozemond, M. (1998). *Descartes' dualism*. Cambridge, MA: Harvard University Press.

Russell, B. (1914). *Our knowledge of the external world*. London: Allen & Unwin.

Russell, B. (1921). *The analysis of mind*. London: G. Allen & Unwin.

Ryle, G. ([1949] 2009). *The concept of mind* (60th Anniversary ed.). London, New York: Routledge.

Shapiro, L. (2011). *Embodied cognition*. New York: Routledge.

Sidgwick, H. (1901). The philosophy of T.H. Green. *Mind, 10*(1), 18–29.

Skinner, B. F. (1987). *Upon further reflection*. Englewood Cliffs, NJ: Prentice-Hall.

Smart, J. J. C. (1959). Sensations and brain processes. *The Philosophical Review, 68*(2), 141–156.

Smith, L. D. (1986). *Behaviorism and logical positivism. A reassessment of the alliance*. Stanford, CA: Stanford University Press.

Smithies, D., & Stoljar, D. (Eds.). (2012). *Introspection and consciousness*. Oxford: Oxford University Press.

Sorell, T. (1991). *Scientism: Philosophy and infatuation with science*. London, New York: Routledge.

Stanowski, A. (1971). Postulates of behavioural terminological empiricism. *Quality and Quantity, 5*(2), 297–310.

Stekeler-Weithofer, P. (2010). Metaphysik/Metaphysikkritik. In H.-J. Sandkühler (Ed.), *Enzyklopädie philosophie* (pp. 1584–1591). Hamburg: Felix Meiner.

Sternberg, R. J., & Sternberg, K. (2011). *Cognitive psychology* (6th ed.). Wadsworth: Cengage Learning.

Stewert, J., Gapenne, O., & Di Paolo, E. A. (Eds.). (2010). *Enaction: Toward a new paradigm for cognitive science*. Cambridge, MA: MIT Press.

Thompson, E. (2007). *Mind in life: Biology, phenomenology, and the sciences of mind*. Cambridge, MA: Harvard University Press.

Thyer, B. A. (Ed.). (1999). *The philosophical legacy of behaviorism*. Berlin: Springer.

Toccafondi, F. (2002). Empirismo logico e psicologia. Temi, problemi e visioni consolidate. In S. Poggi (Ed.), *Le leggi del pensiero tra logica, ontologia e psicologia. Il dibattito austro-tedesco* (pp. 203–273). Milan: Unicopli.

Uttal, W. R. (2000). *The war between mentalism and behaviorism: On the accessibility of mental processes*. Hillsdale, NJ: Erlbaum.

Varela, F. J., Thompson, E., & Rosch, E. (1991). *The embodied mind: Cognitive science and human experience*. Cambridge, MA: MIT Press.

Verein Ernst Mach. ([1929] 2012). In F. Stadtler & T. Uebel (Eds.), *Wissenschaftliche Weltauffassung. Der Wiener Kreis*. Herausgegeben vom Verein Ernst Mach. Reprint on the first edition on behalf of the Institute Vienna Circle on the Occasion of its 20th Anniversary, with translations into English, French, Spanish and Italian. Wien, New York: Springer.
Von Eckardt, B. (1992). *What is cognitive science?* Cambridge, MA: MIT Press.
Walter, S. (2014). Cognition: A field guide to some open conceptual and ontological issues. *Review of Philosophy and Psychology, 5*(2), 241–263.
Walter, S., & Kästner, L. (2012). The where and what of cognition: The untenability of cognitive agnosticism. *Cognitive System Research, 13*(1), 12–23.
Watson, J. B. (1913a). Psychology as the behaviorist views it. *Psychological Review, 20*(2), 158–177.
Watson, J. B. (1913b). Image and affection in behavior. *The Journal of Philosophy, Psychology and Scientific Methods, 10*(16), 421–428.
Watson, J. B. (1914). *Behavior: An introduction to comparative psychology*. New York: Holt & Co.
Watson, J. B. (1920). Is thinking merely the action of language mechanisms? *British Journal of Psychology, 11*(1), 87–104.
Watson, J. B. (1930). *Behaviorism*. London: Kegan Paul Trench Trubner & Co.
Wiener, N. ([1948] 1961). *Cybernetics, or control and communication in the animal and the machine* (2nd ed.). Cambridge, MA: MIT Press.
Zuriff, G. (1985). *Behaviorism: A conceptual reconstruction*. New York: Columbia University Press.

Chapter 2
The Misleading Aspects of the Mind/Computer Analogy

The Grounding Problem and the Thorny Issue of Propriosensitive Information

Abstract After the crisis of behaviorism, cognitivism and functionalism became the predominant models in the field of psychology and of philosophy, respectively. Their success is mainly due to the new key they use for interpreting mental processes: the *mind/computer analogy*. On the basis of this analogy, mental operations are seen as cognitive processes based on computations, i.e. on manipulations of abstract symbols which are in turn understood as informational unities (representations). This chapter identifies two main problems with this model. The first is how these symbols can relate to and communicate with perception and thus allow us to identify and classify what we perceive through the senses. Here we limit ourselves to presenting this issue in relation to the classical symbol grounding problem originally put forward by Harnad on the basis of Searle's Chinese room argument. An attempt to address the problem raised here will be made in Chap. 3. The second point we discuss in relation to the *mind/computer analogy* concerns the idea of information it fosters. Indeed, following this analogy, information is something available in the external world which can be captured by the senses and transmitted to the central system without being influenced or modified by the procedures of transmission. This perspective does not take into account that—unlike computers—in living beings information is acquired by means of the body. As Ulric Neisser has already pointed out, the body is itself an informational source that provides us with additional sensory experience that influences (modifies or complements) the information extracted from the external world by the senses. To develop this line of analysis and to determine exactly what information is provided by the body and how this might influence cognition, we examine Sherrington's and Gibson's positions. Moving on from their views, we qualify bodily information in terms of 'proprioception'. We use 'proprioception' in a broad sense to describe any kind of experience we have of our internal states (including postural information as well as sensations related to the general state of the body and its parts). Following Damasio's and Craig's studies, we further elaborate this position, arguing that living beings are equipped with an internal propriosensitive monitoring system which maps all the changes that constantly occur in our body and that give us perceptual ('proprioceptive' or propriosensitive) information about what happens

S. Dellantonio and L. Pastore, *Internal Perception*, Studies in Applied Philosophy, Epistemology and Rational Ethics 40, DOI 10.1007/978-3-662-55763-1_2

inside us. Moreover, relying on Goldie's and Ratcliffe's view, we show that emotional information can also be considered as a form of 'proprioception' which contributes to determining everything we perceive. This analysis leads us to the second main thesis of this book: *'proprioception' is a form of internal perception and it is an essential component of the sensory information we can access and use for all cognitive purposes.*

1 The Idea of a Symbol System and the Grounding Problem

Representation is one of the foundational notions in classical cognitive science. Representations are most simply defined as any kind of internal information, organized in a discrete form, that conveys the knowledge an organism has acquired and which serves to orient that organism's subsequent behavior. At the heart of the cognitivist argument is the idea that a capacity to represent the state of affairs of the world is necessary in performing any kind of *intelligent* behavior (see e.g. Varela et al. 1991: 40). As e.g. Jean Mandler suggests:

> Representation is defined most simply as stored information (The terms representation and knowledge can be considered synonymous, but the term representation emphasizes the format in which knowledge is stored). Any organism that takes in information from the world in such a fashion that it influences its later behavior is storing information and so can be said to represent that information. All learning requires storage of information and so requires representation in some form or other (Mandler 1998: 257).

Classical cognitive science focuses mainly on a particular kind of mental representations, i.e. conceptual representations, which are the means humans use to know (to categorize) the external world and thus to organize the information taken from it. These representations allow us to carry out all the high-level cognitive processes we are capable of performing starting with thinking and speaking (i.e. acquiring natural languages).

This notion of representation is understood in relation to the so called *mind/computer analogy*, which both cognitivism and functionalism—i.e. the models of mind that became predominant after the crisis of behaviorism (see Chap. 1)—use to interpret the very concept of mind. As this analogy turns out to offer a powerful explanatory framework, it has become the reference model for many of the investigations dealing with the problem of individuating mental capacities and explaining their functioning. In this perspective, the mind is metaphorically understood as software that works independently of the kind of material (the *hardware*) on which it is implemented, e.g. in the case of human beings, the brain. Mental operations are seen as cognitive processes based on computations, i.e. they are interpreted in terms of manipulations of symbols which are in turn understood as informational unities (representations).

These informational unities or representations are available to the cognitive system thanks to the sensory systems which detect and extract them from stimuli in the external environment. According to the classical cognitive model, after being captured this sensory information is transformed and transmitted to the cognitive system and gives rise to the information the system uses. By virtue of the modularity thesis embraced by the computational view of mind (Fodor 1983; see also Fodor 2000 for an update of the original position and Carruthers 2006 for an extension), the first transformations this information undergoes are rigid: i.e. they are domain-specific (only a limited range of inputs can be computed) and encapsulated (computation is restricted: it cannot be influenced by other modules or by the central system). It is only once this information has been finally transformed into the format of the central system that the modality specific pieces of information can be blended into new amodal informational units whose format is not specific for domain.

On the basis of the metaphor of the mind as a data processor, what serves as input for mental computations can be described in terms of packets of information that are captured by the peripheral, modular systems (one might think this means these information packets are captured by the senses, but this is not accurate because according to the modularity thesis modules are subunits of different sensory modalities—see Fodor 1983, Chap. 3). These inputs then travel through modular channels towards a hypothetical general CPU (i.e. the central system, the true kernel of the computational mind). Thus, the cognitive system works on the one hand with modality specific information units at the level of sensory experience and on the other with amodal information units at the level of the central system where—among other things—semantic processing takes place.

These amodal information units are interpreted in analogy with language as pieces of information that do not depend on any particular sensory modality but can communicate with all of these modalities and that can be identified with a symbol. Just as the word 'dog' is not related to dogs through any specific modality, but just refers to dogs, so too the representations in the central system are not related to any specific sensory modality, but codify information about something in the world (e.g. about dogs) in a neutral manner. Moreover, just as the meaning of dog (the informational unit 'dog') in natural language is indicated by symbols (D-O-G), so too the informational units in the central system are hypothesized to be identified by symbols. As the form of the symbols that compose the word 'dog' does not in any way resemble the form of real dogs (i.e. is not iconic), in the same way mental symbols are hypothesized to be unrelated with the physical and functional features of their referents. They are not bound to the perceptual system: their relation to referents is arbitrary and they do not have any modality-specific properties.

This kind of model was considered to have many advantages: first of all—and this applies in general to all theories of high-level thought—it meets the basic demand of explaining how people can organize the perceptual information they get in an economical and functional manner. In fact, if the information received through the senses were retained in its raw form (i.e. in an interpreted manner), since all perceptual episodes are unique, it would be impossible to match a new perception with an old one in order to recognize that they were of the same kind and we would need to have an unlimited memory capacity to store every single episode we perceive (Pylyshyn 1973). Secondly, if conceptual representations are amodal symbols, they are not directly related to any kind of (modality specific) perceptual experience and they can therefore neutrally convey and express perceptual information acquired through different sensory modalities. In this way, having the conceptual representation (symbol) of 'chair' does not necessitate having a set of non-linguistic perceptual experiences that can be traced back to the way we experience chairs with the different sensory modalities we are equipped with. It means rather that we have an amodal representation in the mind ('CHAIR') that—much like words in natural language—is arbitrarily linked to the chairs in the external world and whose format does not depend on the perceptible features of these chairs. This view also seems to be particularly parsimonious since concepts can be described as nodes in the central system (or in the semantic memory, as the part of the central system where conceptual representations are stored) which "can be processed simply by accessing the information stored at the type level" (Paivio 1990: 12). Last but not least, being language-like these representations are also compositional: a finite number of symbols can be recombined indefinitely according to specific syntactic rules (see e.g. Fodor and Pylyshyn 1988; Newell and Simon 1972). According to this view, "knowledge is organized propositionally, with the meaning of words emerging from their relations to internal symbols. Determining the meaning of a symbol is like looking up in the dictionary in order to find which definition is given by its relation to other symbols" (Wilson and Foglia 2011; for the discussion of other positive aspects of an amodal code see the next section).

This way of interpreting thought as intrinsically linguistic is, in fact, fairly adequate for explaining adult thought, especially as far as reasoning and higher level cognitive processes are concerned. As e.g. Mandler points out:

> As adults we are so imbued with language that it just seems natural to us that the mind should consist of symbolic representations and their manipulation. There are symbols (roughly words) for each of the objects we can think about. These may or may not be decomposable into parts (e.g. semantic features), but vis-á-vis thinking or reasoning, they are the basic units that are put together to form complex propositions. In this view, thinking consists of manipulating symbols in sentence-like expressions (Mandler 1998: 258).

The identification of thinking with language has sometimes been traced back to a form of Cartesianism (Anderson 2003: 93), since Descartes maintained that only creatures equipped with language—i.e. humans—can 'think' in the proper sense of the word, while other animals are mere physical mechanisms that only have sensations.

> This denial that sensing and acting in the world require thinking, and the concomitant identification of thinking with the higher-order reasoning and abstraction paradigmatically displayed in language use is perhaps the true heart of the Cartesian attitude. Indeed, I believe that it is primarily from this inheritance that the central attitudes and approach of cognitivism can be derived. Simply put, cognitivism is the hypothesis that the central

> functions of mind – of thinking – can be accounted for in terms of the manipulation of symbols according to explicit rules (Anderson 2003: 93).

This idea of thinking as manipulation of symbols has also other advantages in an explanatory framework for cognitivism related to the possibility of offering both a material and mechanical model of thought—i.e. the computer—and a naturalistic explanation of the relationship between brain and mind. The cognitivist thesis is that behavior is caused by representational, *semantically characterized* states like propositional attitudes (beliefs, desires, intentions, and so on). But how is it possible that semantic states can cause the physical changes that are needed to carry out a behavior? Symbolic computation can help solve this issue since symbols are physical, while computations are manipulations of symbols that follow semantic rules. The analogy between thought and symbolic computations can therefore offer a model of how physical causality and semantics can be put together (Pylyshyn 1980). At the same time the relationship between software and hardware offers a naturalistic model of how thought—i.e. the symbol manipulation carried out by a software—can be physically realized in the brain—i.e. in the hardware—without any need to resort to some mysterious form of dualism (for a more detailed explanation of this argument see e.g. Varela et al. 1991: 40ff).

The strict parallelism between mind and computer and the language-like interpretation of conceptual representations that goes with it have many explanatory advantages. This explains why this model has lasted so long and why it continues to be the dominant theoretical model in cognitive research. However, despite its undeniable strength, this view has been criticized from many angles and in recent times it has been *attacked* to an even greater extent. The arguments put forward to challenge this view are numerous. In general, they argue that thought and language cannot be interpreted in a mechanical and formal manner since this does not correspond with the way in which people actually think and speak. More specifically, there are two lines of criticism that we consider to be fundamental and that we will address in this chapter. The first one concerns the link between perception and cognition; we will discuss it is the following part of this section and go back to it in Chap. 3, where we suggest a possible way to overcome the issue. The second one will be discussed in the next sections of this chapter and addresses the abstract and disembodied notion of information derived from the mind/computer analogy that is used to describe the content of representations.

The reason why the link between perception and cognition is often seen as problematic is that we lack a satisfying explanation of how amodal conceptual representations might be formed from perceptual, modal-specific experiences. Indeed, it is quite mysterious how, during its transmission to the central system, information conveyed by the input systems (that provide perceptual, modal information) might be transformed into abstract amodal symbols which lack any specific perceptual characterization and are arbitrarily related to their references. In addition, it is quite mysterious also how amodal symbols might remain connected to perceptual, modal information so that we can always put them in relation to what we perceive through the senses. As e.g. Lawrence Barsalou points out:

> In traditional theories, knowledge consists of amodal symbols that redescribe sensory, motor, and introspective states. [...] Traditional theories assume that [...] a symbolic system redescribes these states, producing amodal descriptions that reside separately from sensory, motor and introspective systems and that operate according to different principles. For example, sensory, motor, and introspective states could be redescribed as feature lists, networks of propositions, fired sets of productions, instantiated schemata, statistical vectors, and so forth. In all cases, knowledge of the original experience is a redescription in an amodal representation language (Barsalou et al. 2003: 44).

How this redescription works and above all how the redescribed symbols are linked to the perceptual experience they supposedly originate from are unclear. This issue can be made more concrete in at least two main ways. The first one is related to how the models of mind that take inspiration from the computer metaphor explain linguistic learning. As has been pointed out e.g. by Jean Mandler:

> Although it is easy to add new facts via language, there is no obvious way for most such systems to learn information on their own or to generalize their experience. This might be why the topic of learning and organizing knowledge through experience is rarely considered in most symbolic treatments of representations [...]. For developmental psychologists, this is a major failing. How can we take seriously any notion of representation in which the learning done by humans from birth is so difficult to specify? (Mandler 1998: 258)

While in the case of the computer, data are explicitly inserted in the appropriate format, in the case of the human being we need to explain how these data can be acquired starting with perceptual experience and what the (genetic and referential) relationship between the symbols and the experience may be (since the symbols must be formed starting from the experience and the experience needs to bestow the symbols with meaning).

The second and better known criticism against the symbolic view of mind is closely related to the first one—it even possibly includes it—and is known as the *Symbol Grounding Problem*. The kernel of this position is the idea that "abstract, arbitrary symbols, such as words, need to be grounded in something other than relations to more abstract arbitrary symbols if any of those symbols are to be meaningful" (Glenberg and Robertson 2000: 381). In its classical version—which is usually traced back to Harnad's 1990 paper—the *Symbol Grounding Problem* is presented on the basis of arguments which related more or less directly back to Searle's Chinese room argument (Searle 1980).[1] Searle's argument is aimed to challenge the principle that—if a computer passes the so-called Turing test (Turing 1950), i.e. can respond with correct symbols strings in a natural language to other symbols strings in the same language so that its answers are indistinguishable from those that a human being might have given—than the computer understands that natural language.

Very briefly, Searle's argument relies on a thought experiment involving an English speaking person who doesn't have a clue about Chinese. This person is

[1]For another well-known interpretation of Searle's argument in the light of the symbol grounding problem which however suggests—differently from Harnad—that we completely abandon a formal model of cognition: Thompson (1997).

closed into a room and has an instruction manual (written in English) that explains how to respond to certain Chinese symbols with other Chinese symbols. Using these instructions, the English speaker is able to write on a piece of paper the correct answer to the symbol strings that Chinese people outside the room pass him through a slot. Since his answer appears reasonable to them and they don't know about the instruction manual, they assume the person inside the room understands Chinese, even though all s/he does (and all s/he knows how to do) is copy some symbols strings associated with other symbols strings according to specific instructions. According to Searle, the person closed in the room is—exactly like a computer—only manipulating symbols according to some syntactic rule, which itself only relies on the shape of the symbol; however, both the English speaker in the example and the computer do not have any understanding of the meaning of the symbols they are manipulating. The symbols are meaningful only for the people *outside* the room and *outside* the computer who interpret the answer both of the person closed on the room and of the computer. Therefore—this is Harnad's conclusion—*the interpretation is not intrinsic to the symbol, but depends on the fact that symbols are intrinsically meaningful for people who know the language* (Harnad 1990: 338–339).[2]

Starting from this conclusion, Harnad defines two versions of the *Symbol Grounding Problem*:

> one difficult, and one, I think, impossible. The difficult version is: Suppose you had to learn Chinese as a second language and the only source of information you had was a Chinese/Chinese dictionary. The trip through the dictionary would amount to a merry-go-round, passing endlessly from one meaningless symbol or symbol-string (the definiens) to another (the definiendum), never coming to a halt on what anything meant [...]. The only reason cryptologists of ancient languages and secret codes seem to be able to successfully accomplish something very like this is that their efforts are grounded in a first language and in real world experience and knowledge. The second variant of the dictionary-go-round, however, goes far beyond the conceivable resources of cryptology: Suppose you had to learn Chinese as a first language and the only source of information you had was a Chinese/Chinese dictionary! This is more like the actual task faced by a purely symbolic model of the mind: How can you ever get off the symbol/symbol merry-go round? How is symbol meaning to be grounded in something other than just more meaningless symbols? This is the symbol grounding problem (Harnad 1990: 339–340).

In Harnad's view, the solution that symbolists usually offer for the problem that the central system seems to be merely "hooked up to peripheral devices" (Harnad 1990: 340) only trivializes the nature of the problem itself. He maintains that—in order to deal with the grounding problem—one must give an explanation of the relationship between representations and external world that allow us to account for the human capacity to discriminate among the instances of the external world and to identify them. And to achieve this aim, we need to assume that *some elementary, primary representations* have a nonarbitrary shape, are iconic, i.e. they are sensory and nonsymbolic, while higher-order symbols are composed out of them.

---

[2]For a recent review of Searle's and Harnad's arguments see e.g. Rodríguez et al. (2012).

> According to the model being proposed here, our ability to discriminate inputs depends on our forming iconic representations of them (Harnad 1987). These are internal analog transforms of the projections of distal objects on our sensory surfaces (Shepard and Cooper 1982). [...] For identification, icons must be selectively reduced to those invariant features of the sensory projection that will reliably distinguish a member of a category from any nonmembers with which it could be confused. Let us call the output of this category-specific feature detector the categorical representation. [...] both iconic and categorical representations are [...] sensory and nonsymbolic (Harnad 1990: 342).

In Harnad's classical version the symbol grounding problem takes a 'compositional form' in the sense that not all the representations must be conceived as perceptually grounded, but only an elementary subset of these which serve as basis to build all the others. Independently of whether only some basic concepts need to be grounded in experience (as Harnad's compositional model suggests) or whether all concepts need to be somehow linked to experience, the question that needs to be addressed is what does this 'grounding' exactly consist of: i.e. under what condition can one say that the knowledge of a concept is grounded in experience (for an overview on this discussion see e.g. Shapiro 2011, Chap. 4; Gibbs 2005, Chap. 4). As e.g. Michael Anderson clarifies:

> Grounding the symbol for 'chair', for instance, involves both the reliable detection of chairs, and also the appropriate reactions to them. These are not unrelated; 'chair' is not a concept definable in terms of a set of objective features, but denotes a certain kind of thing for sitting. Thus is it possible for someone to ask, presenting a tree stump in a favorite part of the woods, "Do you like my reading chair?" and be understood. An agent who has grounded the concept 'chair' can *see* that the stump is a thing for sitting, and is therefore (despite the dearth of objective similarities to the barcalounger in the living room, and despite also being a tree stump) a chair. Simply having stored the fact that a chair is for sitting is surely not sufficient ground for this latter capacity. The agent must know what sitting is and be able to systematically relate that knowledge to the perceived scene, and thereby see what things (even if non-standardly) afford sitting (Anderson 2003: 101–102).

The descriptions of the concepts—this is the point made by Anderson—are usually given linguistically in the terms of the features of the categorized objects, most often of their form and function (when we need e.g. to give a description of the characteristics that an object needs to have to be classified as a chair, we usually mention features like legs, seatback, seat, it is used to sit on, etc.). However, knowing a concept (mastering e.g. the concept of 'chair') requires a set of perceptual experiences concerning the objects and our interactions with them that go far beyond these features. To put this in different terms, to list a set of relevant features and to assign to each of them a specific weight in the categorization judgment, as many quasi-computational approaches to categorization do, is far from being sufficient to explain how categorization is carried out. To explain categorization, we need first of all to clarify how people come to individuate and to select these features on the basis of their perceptual experience; i.e. how these features relate to their perceptual experience.

We will address this problem in Chap. 3. Here we will argue that representations in the central system—i.e. conceptual representations—cannot be *entirely* neutral (amodal) and featural (we will say 'inferential'); and that they must rely also on a perceptual component, which can be more or less predominant depending on the kind of concepts we are considering. In Chap. 3, we will suggest a hypothesis as to how this perceptual information might be organized so that it can offer a link and constitute a bridge between cognition and perceptual episodes.

2 Representations and Bodily Interferences in Information

As we have already mentioned several times, the notion of representation is one of the fundamental elements in understanding how cognitivism and functionalism view the nature and the acquisition of knowledge. Representations are conceived as symbolic entities of *virtual and abstract character* which convey a *content*. This content is the basic material for the mental operations humans use to develop all kinds of knowledge and the material they use to organize their behavior. This content is essentially viewed as consisting of *information* (Gardner 1985: 38–39). In a cognitivist and functionalist perspective, information is the 'matter' representations are made of. Its function is to get the organism acquainted with the features of its environment and to mediate its behavioral responses to this environment. In this sense, clarifying exactly how this information is interpreted, what it conveys and how it is generated, constitutes an indispensable step for the understanding of both the notion of representation developed in these lines of research and the nature of the mental processes they hypothesize.

The notion of information has its roots in the classical, Greek and Latin philosophical tradition. In a nutshell, in order to explain how the mind could produce thoughts whose content was about objects or entities in the external world, the tradition hypothesized the existence of something that could cross the boundaries of the mind, pass from the outside to the inside and allow the mind to somehow relate (physically and ideally) with the world. This something was called an *intelligible form* (i.e. a form that can be grasped by the mind) (von Borman et al. 1972). *Intelligible forms* were the specific elements through which human beings could think of the objects since they were the only component of these objects that could be *mentalized*, i.e. captured and introjected by the mind. Forms could penetrate the mind and modify its states giving rise to others (perceptions, images, thoughts, etc.) (van Steenberghen 1946); functionally, they were signs internal to the mind about something external to it. The cognitivist and functionalist notion of information is analogous to these forms in many respects. As forms, information can also have a direct causal role on the internal states of the mind, determine a change in their internal organization and therefore influence behavior. Information is what conveys the structure of an entity or a property of it to the cognitive system, ensuring the possibility of a relationship between mind and world.

From the perspective of functionalism and cognitivism, the notion of information takes the place of that of the (distal) *stimulus* as this is described by behavioristic psychology. One of the main reasons why the former substituted the latter is that the distal stimulus—the things and events we perceive in the external world—is something complex that can hardly be decomposed. On the contrary, the notion of information allows us to 'unpack' the aspects included in the distal stimulus and to treat them separately and independently from each other. In this way it is easier to explain how the different aspects or properties are recorded by e.g. different sensory systems and processed through various operations depending on their configuration to obtain as a final result some form of knowledge that is independent from the specific channel it was originally gained by. Think for example of perception: on the basis of the notion on information it is possible to hypothesize that the structural dimension of a complex of modality specific proximal stimulations can be interiorized and processed in a way that is basically independent from the sensory channel through which it was acquired. Information can be captured, transmitted and transformed because it is translated into a unitary code which is common for the whole system. In this respect, this notion of information diverges from that of form, because the form is extracted from the objects and reproduces some of the characteristics they have in a modal specific way, i.e. in a way that depends on the sensory channel through which it was extracted.

From this perspective, the idea of a unitary code of information has various advantages: all we know about the world is conceived as being codified in an ordinate and limited sequence of simple signals and the contents we learn lose their specificity in the sense that the signs that describe them do not have isomorphic properties and do not 'resemble' the contents they are about. Furthermore, the unitary code accounts for the fact that our knowledge is built in a homogenous manner: all the parts of the system work on the basis of the same code which can be shared by them and transmitted among them without barriers. The cognitive system can therefore be described as a system of transmission, processing and storing of 'matter' that is amodal, i.e. independent of the modalities through which it was acquired and neutral with respect to all of them. From this perspective, it is not relevant through which channel we learn about an object in the external world. Indeed, all the information gained through the sensory channels is codified in this amodal and neutral manner and made available for storing and further processing. This clarifies the operational sense of the mind/software analogy which cognitive psychology and philosophical functionalism rely on. The mind is conceived as the program of an electronic calculator; the mind contents are interpreted by analogy with the inputs of the programs; thought is viewed in terms of computations on symbols while the various operations thoughts consist of are parts of these computations; finally, behavior is seen as the output of the program, i.e. as the result of the calculator processing.

The idea that thought and its operations can be described on the basis of a purely formal model derives from some seminal ideas suggested in the first half of the 20th century by the English logician and mathematician Alan Turing (Gardner 1985: 17ff; Pratt 1987: 177ff). Turing hypothesized the possibility of codifying in a very simple way—through a binary code—the instructions that allow a (virtual) machine to carry out an indefinite number of operations in a finite number of steps (Turing

1936). Turing realized that it was possible to translate tasks of any kind, even complex ones, into sequences of calculations on ideal objects, where the rules for the calculations and the entities that were calculated could be expressed though sequences of signals belonging to a unitary, very simple code. According to Turing, any task could be realized in this way once the operations this consisted of were expressed in an ordinate, clear and finite sequence. It was Turing himself who thought that these principles could be applied to psychology (Hodges 1983: 290), giving rise to an operational model capable of simulating human thought as a process made up of a sequence of rules for computationally operating on symbols and of the symbols themselves which also consist of limited series of signals written in the same code as the rules. Intelligence and more generally thought could be entirely described as a series of virtual operations of this kind (Turing 1950).

Applying Turing's theses on psychology appeared reasonable due to developments in biology and physiology between the end of the 19th and the beginning of the 20th century in particular the idea that living things are entities consisting of very simple mechanisms governed by an imminent teleology, i.e. directed at achieving the aim of survival. Organisms are systemic organizations characterized by performances of growing complexity which are capable of generating and maintaining a certain degree of internal balance, even though they interact with an external environment that is continually changing (Pratt 1987: 179–180). This view of living organisms raised the question of how the execution of operations that are widely different from each other in their nature and level of complexity might lead to behaviors that appear to be simple and congruent. The coherence of behavioral reaction—the fact that the organism can generate coherence through the coordination of the activities of its parts—attracted the attention of researchers. To understand how this is realized, we need to comprehend how the organism coordinates its parts: the simplest hypothesis is that there is a control center in the brain which can unify and use all the information coming from these parts.

Turing's seminal work offers a crucial element to address this issue, i.e. it shows a possible way to describe a process (a flow of information) in a simple manner and with a unitary code which could also possibly be applied to the way in which the organism executes its operations. This idea was corroborated by Charles Sherrington's discovery in 1879 about how nerve cells, including brain cells, communicate: they exchange signals emitted by distinct entities, i.e. different cells, that are captured after they pass through the intercellular space (Robinson 2001: 32ff). This seemed to be consistent with Turing's idea and corroborated the view that this transmission can be described as a process that occurs in a discrete manner involving a finite number of entities and steps. In this vein, the work of the brain could be compared with that of a computing machine with a finite number of states (Pratt 1987: 232). The exchange of signals itself could be described using a formal and simple code like the one hypothesized by Turing (McCullogh and Pitts 1943) and this gave a decisive boost to the further development of the analogy between an artificial computational machine and an organism viewed as a complex of parts coordinated by communication channels (the nervous system) and a control center (the brain).

These investigations gave rise to others in which Norbert Wiener played one of the leading roles. These developments relied on the idea that the study of communication techniques as techniques for the transmission of the messages would represent a new frontier in the comprehension of the way in which organisms give rise to and control their behaviors (on this see e.g. Rosenblueth et al. 1943). These intuitions form the basis of *cybernetics*—a field of study that concerns communication and communication control in animals (including humans) and in artificial agents (Wiener [1948] 1961; Linguiti 1980)—whose aim is to look for a synthesis between the results for information communication obtained from artificial and from biological systems.

From an epistemological point of view, it is important to observe how the interest in the issue of how the internal parts of the system communicate is focused on a particular view of communication as a technique of transmission of a message between two poles: the one that transmit the information and the one that receives it. The general reference point for developing this perspective was Claude Shannon's theory of communication as the discrete transmission of information—a theory developed in the field of telecommunication engineering which was at the time one of the most advanced approaches in this area. The focus of Shannon's theory is on engineering issues: the only aspects of information transmission it considers are those which are relevant for telecommunication. Thus, the research that used this theory as a basis for understanding how the cognitive system works were led to concentrate only on specific issues about information transmission in biological systems, specifically, issues that were borrowed from Shannon's theory such as whether (and how) the information transmitted inside the cognitive system remains identical to itself or whether it undergoes some change during transmission from one to another area or whether the organism has a system of continuous or discrete reception of signals. Other kinds of questions that were not relevant for a theory of information transmission in telecommunications—mainly issues concerning where the information the organisms have originates from—were by contrast completely overlooked. This had relevant theoretical consequences for the study of the cognitive system, since the lack of interest in these latter questions lead researchers to neglect the limits of Shannon's ideas when applied to psychological and philosophical research on mind.

In line with Turing's ideas, Shannon developed a very simple system for the transmission of a message (content); this message was interpreted as an ordinate sequence of signals in a code which could be interpreted (decoded) by the circuits of an electronic device. In this view, the notion of *message*, meant as *information to be transmitted*, is conceived as something that is codified in a very simple manner and that is transferred through mechanical operations in a way that is neither affected by the semantics (the content) of the message nor affects it in any way. Shannon was trying to solve a theoretical and technical problem in engineering telecommunications at a distance for techniques such as telegraphy and telephony, i.e. how to convey content across a channel without altering this content; thereby allowing the recipient to get the same 'object' transmitted by the sender. The point of this research was to find the best *means* possible to transmit information leaving the signal intact—without affecting or modifying it in any way. To achieve this aim, the content must be decomposed into its minimal constituents when it is transmitted and then it must be recomposed at reception so that its initial structure is perfectly reproduced. The transmission must not alter the message or change/affect its structure. According to Shannon, communication is *effective* when it allows information to be transmitted in the *purest* manner possible, i.e. with the lowest possible degree of distortion (avoiding so-called “ground noise”) that can be caused by the physical characteristics of the transmission channels.

The model at the base of this idea is communication via a telephone. Shannon’s specific aim was to develop a mathematical description of the communicative process suitable for improving engineering techniques for the construction of transmission systems with a degree of distortion close to zero. From this perspective, the ideal code should not be sensitive to either the semantic characteristics of the message or the physical characteristics of the system of information transmission. For this reason, this code must be as simple as possible in order to guarantee successful communication, i.e. to assure that the message remains identical when transmitted from the sender to the recipient. As Shannon effectively summarizes:

> the fundamental problem of communication is that of reproducing at one point either exactly or approximately a message selected at another point. Frequently the messages have *meaning*; that is, they refer to or are correlated according to some system with certain physical or conceptual entities. *These semantic aspects of communication are irrelevant to the engineering problem.* The significant aspect is that the actual message is one selected from a set of possible messages. The system must be designed to operate for each possible selection, not just the one which will actually be chosen since this is unknown at the time of the design (Shannon 1948: 379).

In Shannon’s view, the message is *not* something that is constituted during the process of communication, since this process has a basically *instrumental nature* and is understood as a *passive transmission of information*: the sender transmits a message in a code and this must come to the recipient in such a way as to allow identification as fast and as effective as possible. To have *effective* communication, the message needs to be encapsulated in a simple and universal code. In this perspective, information is just what the code conveys through its abstract and formal sequences of symbols. According to Shannon, good information transmission also requires that the signal be sent in a *discrete*—i.e. *not continuous*—manner, because the transmission is effective only when the information is conveyed precisely piece after piece through a limited number of sending operations. This allows the decoding systems to discriminate the signal that carries the original information from other extraneous signals. In fact, Shannon’s idea of code is built on the notion of the ‘bit’ (the basic unit of information in computing and digital communications) and on the quantity of transferable ‘bits’.

To summarize, Shannon’s view has relevant operational presuppositions: first of all, the message is extraneous to the system that allows for its communication; secondly, this communication consists in the mere transmission of signals (information) between two poles and it does not alter the structure of the information (of the message) conveyed; thirdly, the message itself is indifferent with respect to its communication: the communication as transmission of information must be insensitive with respect to the content that is transmitted. The process consists of a purely formal operation that is not sensitive with respect to its content.

The application of information theory to psychological models played a crucial role with respect to the theories of mental processes based on the human/machine analogy, especially those that considered the mind to be a software with the capacity to transmit and process available information. Applying a notion of information derived from computer data processing to human cognition is not unproblematic and has been criticized since it was first proposed. First of all, in this model information is considered to be something real and to originate from a source external to the subject and to be independent of him/her. Secondly, this idea of information is characterized by a too high degree of abstraction (Tallis 2004: 54–62; Gardner 1985: 384ff). These two features are actually two sides of the same coin; thus, they can be considered jointly. In particular, *they are both the effect of an aseptic and disembodied conception of information which leads to an aseptic and disembodied view of mind* (Wallace et al. 2007).

According to the standard computational model, the information the cognitive system works on is derived uniquely from the environment external to the perceiving subject: information is something available in the external world which can be detected (captured and grasped) by the senses. The senses on the other hand are understood in the classical manner: in line with the standard view that become predominant in the Western tradition, our sensory experience is considered to be characterized by information provided exclusively by the five senses. This information is hypothesized to be transmitted from the senses to the central system without being influenced or modified by the procedures of its transmission and transduction. This view is *aseptic* because it suggests that the information the cognitive system works on is something abstract that can be transmitted by the cognitive system without being ‘contaminated’ in any way by the processing itself. Moreover, this perspective is also *disembodied* because it suggests that the information our cognitive processes rely on is external to the subject and it consists in the external environment. As a consequence, the body is not considered as an informational source that provides us with additional sensory experience that influences (modifies or complements) the information driven from the external world by the senses.

These possibly misleading aspects of the rising trend towards a computational model of mind in cognitive psychology were already pointed out and addressed by one of the grounding fathers of cognitivism: Ulrich Neisser. Neisser was certainly attracted by the theoretical potential of the mind/computer analogy that gained strength during the sixties. However, he was also aware of the limits of this perspective, even though for a long time his critiques were not followed up. An overview of Neisser’s position can already be found at the beginning of one of his fundamental works—*Cognitive Psychology* (Neisser 1967)—in which Neisser offers one of the first and most significant systematizations of the rising cognitivist approach in psychology.

Neisser challenges Shannon's view. He addresses one of the most relevant applications of Shannon theories, i.e. telecommunications and in particular communication by telephone, in order to show that this model is not appropriate for understanding both the (external) communication processes among humans and the (internal) dynamics of information transmission and processing carried out by the cognitive system through perception and thought. Unlike telecommunications in which some informational units are conveyed by neutral and unselective channels without modifying or influencing the object of transmission, in Neisser's view, human communication is not a discrete and passive process through which signals are transmitted from a sender to a recipient. Rather, this is an active process continuously characterized by feedback and by the selection of information that constantly adjusts the interaction between sender and recipient. As Neisser points out at this respect: "a telephone cannot decide which portion of the incoming message is relevant [...] human beings behave very differently, and are by no means neutral or passive toward the incoming information [...] they select some parts for attention at the expense of others, recoding and reformulating them in complex ways" (Neisser 1967: 7). For this reason, the metaphor of formal communication does not apply to human communication; moreover, the elements that mediate human communication cannot be understood in the formal and abstract terms postulated by Shannon's theory.

The same limitations identified for the application of this metaphor to external communication also apply to its usefulness for explaining internal cognitive processes (i.e. the transmission of information from its acquisition by the senses through the various steps of its processing and from one part of the system to another by analogy with software operations). Indeed, in Shannon's model the input and output information is equipollent and the transmission does not influence the content of the message. In this way, the organism would be nothing but a neutral and passive transmitter of information that takes information from the external world, decomposes it for transmission and recomposes it in the original form.

The starting point of Neisser's position is exactly that this is not what happens in humans and more generally in 'living' cognitive systems. Indeed, such systems have no direct access to the external world: what they experience are constructions produced by their own minds, and specifically by their cognitive systems, through the transformation of the information drawn from the external world. In Neisser's words:

> Whatever we know about reality has been mediated not only by the organs of sense but by complex systems which interpret and reinterpret sensory information. The activity of the cognitive systems results in – and is integrated with – the activity of muscles and glands that we call "behavior". It is also partially – very partially – reflected in those private experiences of seeing, hearing, imaging, and thinking to which verbal descriptions never do full justice (Neisser 1967: 3).

The cognitive activities of subjects are described by Neisser using a phrase that rephrases the title of a well-known Freudian work (*Instincts and Their Vicissitudes*): *Stimulus Information and its Vicissitudes* (Neisser 1967: 4). In his work Freud tries to describe the motivational dynamics of human behavior using the notion of instinct; in a similar vein, Neisser's motto sketches the dynamics of our cognitive processes based on ongoing information processing, i.e. on an everlasting transformation of information. In Neisser's view, the very idea of cognition is defined by these dynamics: "the term 'cognition' refers to all processes by which the sensory input is transformed, reduced, elaborated, stored, recovered, and used" (Neisser 1967: 4). Thus, information is continuously transformed and this transformation cannot be conceived as completely independent from the physical—concrete and material (i.e. bodily)—dimension of human beings. The originality of Neisser's work consists in its focus on the bodily component as something that plays an active role in the detection and in the processing of information. From Neisser's point of view, *information cannot be understood as aseptic and disembodied, i.e. as independent and separated from the body, since the body is the means by which information is processed.*

Examined in its entirety, the position expressed by Neisser opens an issue that involves the very concept of information on which cognitive psychology is grounded. Neisser's criticism challenges the idea that the mind/computer (or more generally the human/machine) analogy can be applied to psychology in a strict and rigid manner.[3] Neisser tries to develop a different view of the human cognitive system: even though he relies on the metaphor of cognition as information processing, and considers it an essential element in understanding how the mind works, he also argues that the cognitive system is not merely passive but constantly acts on the information input modifying and influencing it during processing. At the same time, the process of information acquisition cannot be considered a neutral reception of signals as per Shannon's theory.

Neisser's criticism of a purely computational notion of information can be considered the first attempt internal to the cognitive perspective to explain perception and cognition as complex processes carried out by living organisms, i.e. by living bodies that do not merely detect and receive external stimuli, but also actively and dynamically produce information that can then be used by the cognitive system for further processing. Unlike Shannon, Neisser is persuaded that our cognitive activities cannot be properly understood if we neglect the issue of the origin of the information we process and do not investigate where and how this information originates.

Since bodily processes are actively involved in perceptual experience, according to Neisser this bodily information must be taken into consideration as an essential part of the stimuli our cognition works on. Bodily information is a kind of perceptual information that does not originate from the external objects people perceive, but from their own body which is the condition of possibility for perception itself as well as the non-neutral means through which perception is gained. To get

[3]In particular, Neisser expresses strong skepticism about the general idea that the study of formal and computational processes can help us understand specific aspects of human behavior such as emotions: "Unlike men, 'artificially intelligent' programs tend to be single-minded, undistractable, and unemotional" (Neisser 1967: 9).

information on the external world requires, among other things, to know what is 'external' and what is not as well as be able to distinguish an external from an internal environment. This knowledge also allows one, among other things, to be aware of the fact that it is s/he and not someone else who perceives things.

Even though Neisser's view on this in more a suggestion than a fully developed theory, he alerts the rising cognitive research movement of the need to reconsider what information the cognitive system works on, specifically taking into account that information must be acquired by means of the body. His legacy is an essential step towards a reconsideration of the structure of cognitive science that overcomes the limits of the mind/computer analogy and reassigns a proper role to bodily information in any explanation of how information processing in living organisms might actually work.

3 The Body as Information Source: Gibson's Hypothesis of an Integrated Perceptual and Propriosensitive System

Charles Sherrington is the first author who develops a comprehensive and systematic view on the function played by *all* the sensory information we can access, including internal information, with respect to our perception and our perceptual knowledge of the world and of ourselves. Indeed, the studies he carried out between the end of the 19th and the beginning of the 20th century radically modified earlier conceptions of the sensory experience and its composition. Sherrington hypothesizes that the human body as a whole is itself a source of stimuli that can give rise to specific sensations which directly or indirectly contribute to the complex of knowledge we have about the world and ourselves. Bodily information is collected through particular receptors that are—like any other kind of receptors—highly specialized for specific kinds of stimuli.

The attention devoted by Sherrington to internal sensations does not imply that he believed internal sensations to be more important than external perception. Indeed, Sherrington continued to consider the classical five senses—sight, hearing, smell, taste and touch—to be *special senses* (Sherrington 1906). They are described as selectively and mechanically sensitive to a unique, specific kind of information which is available in the external environment: light for sight (more specifically for the retina), soundwaves for hearing (more specifically for the cochlea), etc. However, in his view external stimulation is no longer considered to be the *sole* source of information living organisms have access to. In fact, in addition to the receptors positioned in the organs of sight, hearing, smell, taste and touch, we also have other kinds of receptors with a comparable level of specificity which are positioned deep in the body and along its surface and are directed outwards and inwards. If we consider them in their complexity, these constitute a sensory field that cannot be disregarded when we discuss the origin and nature of our perceptual knowledge.

> The main fields of distribution of the receptor organs fundamentally distinguishable seem, therefore, to be two, namely, a *surface* field constituted by the surface layer of the organism, and a *deep* field constituted by the tissues of the organism beneath the surface sheet [...] But the surface field is further broadly subdivisible. Its subdivisions are two. Of these one lies freely open to the numberless vicissitudes and agencies of the environment; it is co-extensive with the so-called external surface of the animal. It is cutaneous in the widest sense of that term. It possesses as receptive organs not only those of touch, & c., in the skin proper, but also the eye, nose, and organ of hearing. This subdivision of the surface field contrasts with a second subdivision of it, constituted by what is commonly termed the internal surface of the animal, the alimentary or intestinal surface. This latter surface is, it is true, in contact with the environment; but the environment with which it is in contact is a portion of the environment greatly modified from the general environment outside by lying almost completely surrounded by the animal itself. This part of the receptive field of the animal's surface, which is turned inward upon the alimentary contents, may be termed the *intero-ceptive*, in contradistinction to that larger part of the surface which looks outward upon the free environment in general, and the latter may from that circumstance be termed the animal's *extero-ceptive* surface (Sherrington 1907: 469).

With the term *exteroception* Sherrington indicates all the receptors of the special senses (i.e. the classical five senses) as well as other kinds of receptors that are located along the surface of the skin and are directed outwards like e.g. the thermoreceptors (which are sensible for variations in temperature) or the nociceptors (which are sensible for variations in stimuli which are responsible for the feeling of pain). Exteroception is the complex sensory field that results from the conjunction and the integration of all the stimulations captured by these receptors. This represents the widest and most intuitive domain of our sensory experience. Maybe for this reason exteroception (i.e. sensory experience due to stimuli coming from the external word) was considered for a long time to be the most important, if not the only information organisms relied on to form their perceptual knowledge of the world.

> the *extero-ceptive* field, which is most rich in receptive organs, both as regards number and variety. For this to be the case, is in accord with what might be expected. It is this *extero-ceptive* field which, facing outward on the general environment, receives and has received for countless ages the full stream of all the varied agencies for ever pouring on it freely from the outside world (Sherrington 1907: 469).

Nevertheless, in addition to exteroception, through their receptors organisms can also capture other kinds of information. Indeed, according to Sherrington if we consider the internal surface of the organism, we can identify further sources of sensory experience.

First of all, physiological and phenomenal evidence shows that we have another receptive field that is turned inward and captures information on the status of the internal organs and specifically the viscera. Sherrington calls this *interoception*. In his view, interoception captures in particular information on the muscular contractions concerning the voluntary and involuntary muscles related to the internal organs of the body—e.g. the stomach, the intestine etc.

Secondly, there is another kind of internal information organisms can detect which relates to muscular contractions. This is due to receptors which capture the contractions of the muscles linked to the skeletal apparatus which enable the movements of the body. Sherrington calls this *proprioception*.

> The receptors which lie in the deep tissues appear adapted for excitation by changes going forward in the organism itself. These changes work, it appears, largely through the agency of mass with its mechanical consequences of weight and inertia, and also largely through mechanical strains and alterations of pressure resulting from contractions and relaxations of muscles. Therefore, a character of the stimulations occurring in this deep field is that the stimuli are traceable to actions of the organism itself, and are so in much greater measure than are the stimulations of the surface field of the organism. Since in the deep field the stimuli to the receptors are delivered by the organism itself, the deep receptors may be termed *proprio-ceptors*, and the deep field a field of *proprio-ception* (Sherrington 1907: 471).

Interoception and proprioception are both kinds of perception which detect muscular contractions (movements). However, they differ from each other because they are related with different kinds of muscles. *Proprioception* indicates the information derived by the muscles connected with the skeletal apparatus which gives rise to the sensations we experience in relation with the actions and the movements of the organism while *interoception* designate the information derived by the muscles linked to the internal organ (mostly the viscera) which produce the sensation we experience in relation with e.g. digestion, defecation and analogous.

According to Sherrington, the strength of a signal depends on the number of receptors that capture it: in his view the degree to which we are aware of certain sensations depends on whether they were produced by a higher or lower number of receptors. Since the spread and the number of the interoceptive receptors are lower than those of exteroceptors and proprioceptors, the sensations they give rise to are also less intense (Sherrington 1906: 320) and this is the reason why we are often unaware of them. However, this does not mean that they do not exist or that they are unspecific in terms of their quality and localization.

In Sherrington's view, organisms collect perceptual information from all the three kinds of receptors mentioned. The complex of all stimulation captured by these receptors is centrally integrated by the brain. On the basis of Sherrington's description, our experience is therefore the outcome of a systemic process, i.e. it results from the assembly of all kinds of sensations we experience, not only those produced by exteroception, but also those derived from the internal environment of the organism through proprioception and interoception (Sherrington 1907: 475). These sensations are then 'blended' by the brain and made accessible by consciousness (Sherrington 1906, 1941).

This integration produces, among other things, a coherent and unitary representation of what Sherrington calls the '*material me*', i.e. a representation of the body that integrates posture, position in the environment, affective tones connected to bodily states etc. This representation produces a minimal sense of identity consisting primarily of bodily cohesion, coordination of bodily parts and position in the environment and is supported mainly by motoric information about muscular contractions related to any kind of movement we make, including the acts we carry out to acquire external information through the special senses. In fact, if we reflect on what is represented by our consciousness, it becomes clear that we are not only aware of the perceptual objects located in the external environment of the organism (and of their corresponding qualities), but we have also a background awareness of the way in which the organism participates with its micro and macro movements to these perceptual acts and to the acquisition of perceptual knowledge. Indeed, external perception is made possible by all the movements of the body and the final result consists in centralized integration by the brain of both the external information and the internal information related to the processes that made it possible to collect this information.

Moreover, the *material me* also relies on information on our affective tones which are detected through the visceral and muscular contractions we experience by the means of interoception and proprioception.[4] "Sensations derived from the body tissues and organs possess strong affective tone; while sensations of special sense are relatively free from affective tone" (Sherrington 1900: 969). Sherrington's *material me* consists therefore in the self-representation everyone has of him/herself as a living organism: this is always accompanied by an affective tone which—being consciously perceived—make us aware of the general state of our organism.

If we overlook Sherrington's metaphysical speculations on why and how his result can be interpreted in the framework of a strong form of Cartesian substance dualism (Bennett and Hacker 2013: 231–236), his theses can be considered as extremely innovative and original for the research on mind. Indeed, they represent one of the firsts systematic attempts to show that bodily experience has an important function to play with respect to the origin of our perceptual experience in its complexity and therefore also of our knowledge. For this reason, in spite of the author's intention, Sherrington's position also contributes to bridging the gap between mind and body. The focus on internal sensation leads to a change in the way we interpret the role played by the body—by the bodily sensations and by our awareness of them (so-called *proprioceptive awareness*, Gallagher 2005)—in relation to the constitution of our perceptual knowledge. In fact, bodily sensations do not only give us information about the body itself, but they also participate in the organization of the complex of our perceptual experience. In this way, the field of perception—which was traditionally considered to be directed outward toward the knowledge of what is external to us—appears to be more extended and integrated, to include also what happens inside the organism. In this way, bodily information acquires an epistemic dignity that it never had before.

Sherrington's discoveries on the proprioceptive and interoceptive receptors show that our body is organized in a way that enables it to capture signals of what happens outside and inside it and to distinguish between them (see Evarts 1981: 44–45). The description Sherrington offers of this distinction is more detailed and systematic than any other developed in previous philosophical, physiological,

[4]As we anticipate already in Chap 1, Sect. 3 discussing behaviorism and as we will clarify in more specificity below (Chap. 5), emotional states are closely related to bodily modifications such as visceral and muscular contractions.

anatomical and psychological research (see Sherrington 1906). Sherrington's research radically changes the way we can consider the relationships occurring between the complex of the sensations we experience and the constitution of the perceptions we are aware of and use for the development of knowledge. His anatomical and physiological studies lead us to reconsider what happens inside the body—mostly below the threshold of consciousness and of attention, including it in the organism's perceptual field. In fact, *the findings on propriocepton and interoception and on their functions progressively invalidates the (long-lasting) conviction of the philosophical and psychological tradition that the body plays a passive role in the constitution of perceptual objects and knowledge in general.* According to this tradition, perception is essentially something through which we capture information about the world outside our skin and is the result of the organization of the sensory stimuli carried out under the guidance of intellectual functions, while bodily experience plays a marginal or even no role in this information acquisition and processing.

The line of development of Sherrington's research that is most relevant in the context of this study is that conducted by James J. Gibson. Taken as a whole, *Gibson's work can be considered as a particularly original attempt to improve Sherrington's research on internal sensations as forms of perception which leads also to a conceptual reunification of the distinction between proprioception and interoception.* In fact, initially Gibson was skeptical towards Sherrington's distinction between exteroception, proprioception and interoception and especially towards the idea that we have any kind of interoceptive experience at all ("interoceptors […] if they exist in physiological fact, they are not yet understood"—Gibson [1961] 1987). However, a few years later Gibson revisited this characterization and maintained that to account for perception we do need to consider all the kinds of information mentioned by Sherrington. In fact, *he suggests that we should not focus so much on them separately, but we should rather consider their joint contribution* (Gibson [1975] 1987). In his view, Sherrington's distinction should be simplified and traced back to two kind of experiences only, which are respectively external and internal: "Simplifying a distinction made by Sherrington, I propose to use the term perception for an experience of the environment surrounding the body of an animal and the term proprioception for an experience of the body itself" (Gibson [1968] 1987).[5]

The reason why he argues that we should extend the notion of proprioception to also include interoception (i.e. to include all the kinds of perception that keep track

[5]According to Gibson, all bodily sensations are perceptions of the same kind. Gibson welcomes the notion of '*somaethesis*' introduced by Boring (1942), which encompass 'tactual sensibility', 'organic sensibility', 'kinesthesis', feelings of 'pressure', 'muscle tension', 'pain' and 'temperature'. In fact, he considers all these experiences to be part of the same complex perceptual system: "the feelings of the body were so much allied as to be subsumed under one name" (Gibson [1966] 1983: 98). In this sense, in Gibson's view, "Boring's somaesthesis […] includes Sherrington's interoception" (Gibson [1968] 1987). On current philosophical developments concerning *somaesthesis* see Shusterman 2008, 2012.

of what happens in the internal environment of the body where we are directly aware of them or they are only implicitly present) lies in the innovative perspective of perception he puts forward. Gibson maintains that perception does not consist in a passive recording of stimuli, but is rather a complex '*activity*' involving the organism considered as a whole, hence both the 'external' and the 'internal' environment. This activity is an integrated process whose outcome does not depend so much on what sources information originally comes from, but rather on how these various kinds of information interact and are merged together to give rise to our overall experience. In Gibson's view, perception is the result of a synthetic construction which results from an *unbroken flow* of signal exchanges between different organs that are *always in reciprocal relations*. Even though what we perceive appears to us as an organically structured unity, it is actually neither unitary nor simple. At least its formation is not like this. What we define as perception is a mode of our overt attention (Gibson [1966] 1983: 47–58): it is the final result of an active perceptual processing, but it does not reveal anything about the actual process of perceiving: "Perceiving is an achievement of the individual, not an appearance in the theater of his consciousness" (Gibson [1979] 1986: 239).

The studies on the physiology of perception that went back to Sherrington relied on the idea that the sense organs receive stimuli through receptors and these receptors work essentially in a passive way: they are triggered by appropriate stimuli and produce sequences or series of discrete stimulus-information which are then conveyed and processed by the brain to give rise to perceptions. Gibson's view is more complex and it suggests that the content of perception is not determined by one sense or by one organ only, and even less by the receptors which act as the "passive, elementary, anatomical components" of the organs (Gibson [1979] 1986: 53); this is rather the result of the work of a network of organs (and receptors) called the *perceptual system*. A system is a set of highly organized elements that are steadily in reciprocal contact and interaction (Gibson [1966] 1983: 42). Conceived in this way, the act of perceiving does not result—as Sherrington hypothesized—from a central processing of information collected by peripheral receptors and transmitted in a unidirectional and uniform manner to the brain. Perception is rather the product of a systemic organization of the organs: "A system has organs, whereas a sense has receptors. A system can orient, explore, investigate, adjust, optimize, resonate, extract, and come to an equilibrium, whereas a sense cannot" (Gibson [1979] 1986: 245). Gibson's point is that the various forms and modalities of perception do not depend on an individual organ, but result from the work of a system of organs that actively collect and select relevant information to execute and preserve the organism's vital functions and to carry out specific actions in a specified environment. What we perceive is not merely the effect of centralized integration (by the brain) of different stimuli coming from the sense organs; it is rather the outcome of a continuous collection and selection of information.

In Gibson's view, *stimulus and information are not one and the same thing*. Information differs from stimulation in various respects. First of all, the stimuli and the receptors that collect them *do not work in a discrete, but in a continuous manner* (Gibson [1966] 1983: 39ff). The receptors do not receive stimuli in an intermittent way, moment after moment; they are not active only '*in stop and starts*', but they are steadily in a state of solicitation because the environment that produces stimulation is continuously present and changing. Thus, the stimulation is likewise a continuous changing flow which is therefore radically different from information that is conceived as arriving in discrete and stable knowledge units. This leads us to the second remarkable aspect of Gibson's differentiation between stimuli and information. While stimuli are homogeneous and undifferentiated, information is something *relevant* to the subject.

In spite of these differences, stimuli and information are also clearly related to each other. Gibson conceives of information as having a 'differential value': we call 'information' the differences in intensity in the flow of the stimuli. It is this difference in intensity of stimulation that makes it possible for certain stimuli to become salient for perception and thus to be actually perceived (i.e. to reach our awareness as perceptible objects). To give an example taken from visual perception (the field Gibson mostly works on), contrary to Sherrington's hypothesis, the salient information captured by the retina does not consist in the light waves light is composed of. In fact, light and light waves are continuously captured by the ocular photoreceptors. What we actually perceive—what is salient to us and constitutes the information we get from the environment—are the *differences* among various kinds of light waves which are present in the environment and are refracted by objects. If we were in an environment characterized by a constant, intense and homogeneous luminosity, we couldn't distinguish figures, backgrounds, profiles, three-dimensionality, and so on.

Moreover, while the stimuli are something that passively and mechanically hit the receptors, information is something the organism continuously seeks for and selects with the specific intent of actively operating on the external environment. What counts as information is the product of the constant action of the organism in the environment: an organism's life is steadily characterized by micro and macro movements whose outcome is the change of the organism's position and perspective with respect to the environment. A perceptual event that brings about a visual object cannot be the result of an individual sense and of its receptors. It is rather the product of the constant interaction among distinct organs always working together to obtain the best possible conditions to gather information (e.g. when we focus, keep our balance, adjust our posture, etc.). When we see an image, we are not aware of this complex net of information that contributes to constituting it; however, our vision of this image depends on and is made possible by this information complex.

We could never have e.g. visual experiences if vision depended uniquely on the eye and the retina considered in isolation. A much larger part of the organism is involved in the act of vision. Indeed, what allows us to see is the continuous coordination among ocular movements, movements of the head, of the body, contractions in the eye, nervous and cerebral functions. In this sense, what we see is not the product of an individual organ, but of a system of organs which is continuously active and act in reciprocal relation. Even though the main theses developed by Gibson are about vision, he was persuaded that any perception depends on the work of perceptual systems (Gibson [1966] 1983), which put us in actual and active contact with the world.

> Perceiving [...] is a keeping-in-touch with the world, an experiencing of things rather than a having of experiences. It involves awareness-of instead of just awareness. It may be awareness of something in the environment or something in the observer or both at once, but there is no content of awareness independent of that of which one is aware. This is close to the act psychology of the nineteenth century except that perception is not a mental act. Neither is it a bodily act. Perceiving is a psychosomatic act, not of the mind or of the body but of a living observer. The act of picking up information, moreover, is a continuous act, an activity that is ceaseless and unbroken. The sea of energy in which we live flows and changes without sharp breaks. Even the tiny fraction of this energy that affects the receptors in the eyes, ears, nose, mouth, and skin is a flux, not a sequence. [...] Discrete percepts, like discrete ideas, are 'as mythical as the Jack of Spades' (Gibson [1979] 1986: 239–240).

This should clarify in which sense perception in Gibson view's is—as we mentioned—an individual's achievement, i.e. a synthetic act in which the perceptual object is built on the basis of a continuous collection of information and then reaches the awareness of the observer. Thus, even though the perceptual objects we are aware of appear to us as simple units, they are actually very complex: they are formed starting from the stimulation collected by a number of receptors and their qualitative characteristics are derived by the senses through which this stimulation was collected.

Gibson's idea of perception is based on a notion of stimulation which differs from the one defined by the previous psychological theories. First of all, in the traditional theories of perception the senses get and transmit stimuli through the activation of sensory receptors positioned along the organism. Such receptors are activated only when the amount of stimulation (energy) they get exceeds a certain threshold. Otherwise they are at rest. In Gibson's view, it would be a mistake to think that perception results ultimately from exceeding a certain stimulation threshold because perceptual awareness does not seem to depend on a stimulus threshold and above all because it does not depend on individual stimuli.

> An application of stimulus energy exceeding the threshold can be said to cause a response of the sensory mechanism, and the response is an effect. But the presence of stimulus information cannot be said to cause perception. Perception is not a response to a stimulus but an act of information pickup. Perception may or may not occur in the presence of information. Perceptual awareness, unlike sensory awareness, does not have any discoverable stimulus threshold (Gibson [1979] 1986: 56–57).

The awareness of the perceptual object is based on information concerning the object itself, but this information is not the same as the individual sensations produced by the receptors. In fact, the specific quality of a sensation depends on the specific channel which it originates from, while the information provided by the perceptual activity consists of a complex of data concerning the source of the stimulation (i.e. its object) as well as its relationship to the perceiver and his actions and movements during perception. The result of perception is not merely information on the external world, but rather information on the external and the internal environment of the observer blended together: to perceive means to be aware of information concerning both the object and the subject of the perception, so that this information compound reaches awareness as a unity in which the different components cannot be distinguished and taken apart. In this sense, proprioception and exteroception are always complementary and inseparable (see e.g. Gibson [1979] 1986: 116, 151, 183, and 201). As Gibson exemplifies:

> Information exists in a normal ambient array, therefore, to specify the nearness of the parts of the self to the point of observation-first the head, then the body, the limbs, and the extremities. The experience of a central self in the head and a peripheral self in the body is not therefore a mysterious intuition or a philosophical abstraction but has a basis in optical information (Gibson [1979] 1986: 114).

In the traditional psychological and philosophical views, the senses were conceived as unidirectional, simple channels, separated from each other, which serve for the transmission of the stimuli produced by the receptors. On the basis of his notion of perceptual system, Gibson challenges this position and suggests that to understand perception one needs to focus on the whole organism's perceptual activity *and on the blending between exteroceptive and proprioceptive (propriosensitive) information.*

> I maintain that all the perceptual systems are propriosensitive as well as exterosensitive, for they all provide information in their various ways about the observer's activities. The observer's movements usually produce sights and sounds and impressions on the skin along with stimulation of the muscles, the joints, and the inner ear. Accordingly, information that is specific to the self is picked up as such, no matter what sensory nerve is delivering impulses to the brain. The point I wish to make is that information about the self is multiple and that all kinds are picked up concurrently. An individual not only sees himself, he hears his footsteps and his voice, he touches the floor and his tools, and when he touches his own skin he feels both his hand and his skin at the same time. He feels his head turning, his muscles flexing, and his joints bending. He has his own aches, the pressures of his own clothing, the look of his own eyeglasses – in fact, he lives within his own skin (Gibson [1979] 1986: 115).

Thus, in Gibson's view perception is the product of a continuous and not-discrete collection of information on the external and on the internal environment of the perceiver. The perceptual system never stops working. Its task does not merely consist of passively detecting packets of stimuli or plain sensations coming from the various, specific receptors positioned along the body when they are properly activated. "The perceptual capacities of the organism do not lie in discrete anatomical parts of the body but lie in systems with nested functions." (Gibson [1979] 1986: 205) And further: "The established theory that exteroception and proprioception arise when exteroceptors and proprioceptors are stimulated will not do. The doctrine of special channels of sensation corresponding to specific nerve bundles has been abandoned." (Gibson [1979] 1986: 238)

This is one of the most important aspects of Gibson's revision of the notion of proprioception which was originally put forward by Sherrington. As we considered above in this section, with the term *proprioception* Sherrington indicates the specific capacity some receptors have to capture and to transmit a specific kind of information collected within the organism. To coin the neologism 'proprioception', he puts together the Latin words *proprius* (whose literal meaning is 'own') and the Latin verb *capere* (which can be translated as 'capture') to indicate that the perceptual capacities of the organisms are not only sensible to the stimuli taken from the external world, but they are also able to capture stimuli generated by what happens in their internal environment. These stimuli give rise to specific sensations related to the condition of the body and thus to a form of bodily perception of the organism's acts and muscular movements. As Evarts underlines: "The Latin word *proprius*, meaning *own*, provided a prefix which called attention to the fact that the organism's *own* acts created the adequate stimuli for these receptors" (Evarts 1981: 44). However, in Sherrington's view, not all internal stimuli are considered as proprioceptive since he distinguishes between proprioception and interoception. Proprioception captures motoric information generated by muscle contraction, while interoception captures states related to the internal organs, in particular, to visceral contractions. The sensations produced by these two kinds of receptors are then integrated by the brain in a way that better resembles a juxtaposition then a synthesis.

Gibson's view on this is quite different. He considers perception as the result of the action of systems of organs which are in continuous and incessant activity. In his view this applies not only to external but also to internal perception. Thus, the perception of the internal states of the body is also "an overall function, common to all systems, not a special sense" (Gibson [1966] 1983: 320). For this reason, Gibson does not distinguish between interoception and proprioception, but uses the word 'proprioception' to indicate both kinds of information. To be precise, 'proprioception' means something even more comprehensive: it indicates the continuous contribution of the corporeity of the perceiving subject to perceptual processes; it concerns the way in which the individual explores the world, looks for/selects information and detect/feels the changes that occur inside him/her. Proprioception and proprioceptive (propriosensitive) awareness are more than the background of our perception, they are the condition of possibility for it. "The continuous act of perceiving involves the coperceiving of the self. At least, that is one way to put it. The very term perception must be redefined to allow for this fact, and the word proprioception must be given a different meaning than it was given by Sherrington" (Gibson [1979] 1986: 240).

According to Gibson, proprioception is a constitutive part of perception in the sense that "exteroception is accompanied by proprioception" (Gibson [1979] 1986: 141). Thus, proprioception is also a part or complement of the awareness we have of what we perceive in the external word (Gibson speaks of "egocentric awareness"); its object is a kind of self-perception or better—in Gibson's terminology—of ego-reception or sensitivity to the self: "In my view, proprioception can be understood as egoreception, as sensitivity to the self, not as one special channel of sensations or as several of them" (Gibson [1979] 1986: 115). As 'sensitivity to the self', proprioception is the result of a perceptual system that provides in various ways information on the global state of the subject as a living system, as an organism who lives in his/her own skin and who keeps track of the information about his/her internal environment.[6]

4 'Proprioception' as Propriosensitive Information

Gibson's view on what proprioception might be and on the function it could play with respect to perception constitutes the first step in the direction of a possible, even more radical revision of the notion of proprioception put forward by past theories and it provides some new suggestions regarding the relationship between proprioception and exteroception. *Specifically, 'proprioception' and 'proprioceptive information' could be redefined as wide-embracing notions including all forms of experience people have of their bodily states*. According to this view, *'proprioception' would denote any kind of propriosensitive information and would incorporate* kinesthetic, *sensorimotor and postural information which we are consciously aware of as well as sensations generated from and related to the general state of the body and its parts and organs*. Since this broad definition of the notion differs from the way in which it has commonly been understood in the field of neuropsychology (where Sherrington's differentiation between interoception and proprioception still apply), when we use it in this work we will put it in inverted commas. In other cases, we will use the term (also of Gibsonian origin) propriosensitive information.

Even though the literature still lacks a specific characterization of bodily information and the way in which it is represented (Alsmith and de Vignemont 2012), it seems undeniable that we have some 'proprioceptive' awareness, even though—especially when we are focusing on the external world—it is not usually the most salient ingredient of our conscious experience and rather remains in the background. If the direction indicated in Gibson's work is correct, *'proprioceptive' awareness accomplishes a fundamental function in carrying out any kind of cognitive operation, including those related to perception*. In fact, this is the hypothesis we will try to support in this chapter along with the idea that 'proprioception' is an essential component of how we represent ourselves to ourselves as well of how we represent any other objects in our experience (Gallagher 2005); or even more radically, the idea that proprioception (propriosensitive information) constitutes the condition of possibility for 'humanlike' cognition.

[6]In this sense, proprioception is a constitutive component of all the systems of our organism, to include the postural system, the system that coordinates orientation and locomotion, the appetitive, performative expressive and semantic systems. Any of these systems involve proprioceptive information (Gibson [1966] 1983: 57). Since in Gibson's terms proproception indicates mainly bodily awareness (see Gibson [1974] 1987, [1975] 1987), each of these systems contribute to forming the internal perspective of the first person that humans experience as a result of the embodiment of their cognitive system.

To explain the implications of this view, take for example the visual representation of a distal object like an apple. This includes elements that are not explicitly known by the subject, but accomplish a fundamental function in the constitution of the percept. First of all, all the perceptions we have are '*relativized*' in the sense that they are experienced as *relative to the perceiving subject*, i.e. they are perceptual *experiences* belonging to the subject who is experiencing them. When I see an apple, I know that this is the apple I am seeing, that the perspective I am seeing it from is *my* perspective and that my perception is influenced by the acts I perform in order to see it (such as e.g. orienting my body in a certain position, focusing on the apple, feeling the tension related to the movement I would need to perform to pick it up, etc.). In this sense we can say that—in a wide Gibsonian sense—we continuously and tacitly use kinetic and kinesthetic information in performing perception (Sheets-Johnstone 2009). Moreover, our perceptions are also '*perspectival*': when I see e.g. an apple, I always see it from a certain perspective and in a certain condition: I co-perceive how far it is from me, what direction I would need to move into reach it and what kind of movement I would need to make to grasp it. And the sensations coming from my bodily position and states are relevant for the way I seek information to determine (my sense of) the perspective from which I see and consider the object (Sheets-Johnstone 2010). Furthermore, this propriosensitive information allows me to locate the origin of visual object (the apple) in the external world, i.e. outside the limits of my body, and to 'orient' and to 'position' it in space according to coordinates that are relative to myself as the observer. In this sense, propriosensitive information constitutes—or better co-constitutes—the *frame of reference* for what is perceived and represented (Gallagher 2009). Moreover, perception also depends on *the state* of the perceiving subject: the way in which I experience my bodily states during the perceptual act contribute to determine what I see. If I am e.g. hungry when I experience the apple, it will capture my interest and attention in a more immediate way and thus I will perceive some of its characteristics (related e.g. to its spatial location like proximity vs. distance) in a more immediate and definite way (Sheets-Johnstone 2010). This applies to both actual perceptions and imaginary perceptual acts.

This integration of propriosensitive and exteroceptive information does not occur only for vision. Also in the case of something we perceive through another sense, e.g. through the acoustic, the olfactory, the gustatory or the haptic system, we are aware that *we* are the subjects of that perception (that it is us who are perceiving the stimuli), that the origin of the stimulation is located and oriented in a specific position outside our body; and we perceive the stimulus in a way that it is shaped by our bodily states. Thus, 'proprioception' *enriches* the representations we form of the external world with additional information: this does not only make them much more specific and detailed, but it also connects them univocally to the perceiving subject (determining e.g. their ownership and perspective). Interpreted in this way, 'proprioception' is always involved in perception and the reception of information from the external world can no longer be considered a purely passive or even a neutral process. Since propriosensitive information is continuously produced by the body and it permeates any other kind of information processing, the representations that people form of the external world are never neutral: because they include information about the perceiving subject, they are always 'perspectival'—the perceiving subject is in a way always co-represented with them.

These considerations allow us to go back to the notion of 'proprioception' (and that of propriosensitive information, which we use as a synonym) and to further specify it in a way that is relevant for the following part of this work. We use 'proprioception' to describe the information produced by the perceptual system (meant in Gibson's sense) when this is primarily directed towards the internal environment of the observer. This 'internal environment' should not be understood in some substantialistic sense, but rather *in terms of the changes occurring within the body and their effects on what the body can catch and perceive*. Following Gibson, like every kind of information produced by the perceptual system, these changes will also be *consciously accessible by the perceiver, at least to a certain extent and under certain conditions*. In fact, 'proprioceptive' (propriosensitive) information on the processes occurring in the body have a recessive nature with respect to the other kind of representations people have and use for thinking, including exteroceptive representations, in the sense that it remains mainly in the background of our experience (O'Shaugnessy 1995) and it constitutes something like a psychical *basso continuo* which is always present and available to the cognitive system even though it is rarely in the focus of our attention.

The reason why propriosensitive information remains in the background of our awareness is that our attention is usually directed outwards. However, 'proprioception' cannot be considered as a source of information which is extraneous with respect to perception. If 'proprioception' were separated from perception, we would have two sources of information that were *contrasting* and *competing* with each other in the constitution of what is perceptually salient for the subject. And this is not the case. As is emphasized e.g. by Brian O'Shaughnessy, bodily awareness—i.e. the availability of propriosensitive information—is essential for the constitution of *perceptual scenes*. Indeed, bodily information is a precondition for any kind of finalized action since this allows us to perform actions and sequences of movements that are congruent with perceptual flow. If the body moved in a dysfunctional manner with respect to perception, we wouldn't be able to gain the information we need to act properly. Moreover, if propriosensitive information were not immediately integrated with perception, we would need to readjust our position and orientation through voluntary and conscious actions and the flow of perception would be continuously interrupted (O'Shaugnessy 1995).

Perception is a form of knowledge related to attention and attention is intrinsically selective. Since usually the perception of what is happening outside our skin is the most relevant to us (at least if everything in the internal environment is working well), our attention most often is directed primarily at exteroceptive information—on what we are perceiving in the external world—than to propriosensitive information. However, even though the 'proprioceptive' component of our experience remains mostly in the background of our awareness, it can become the specific object of attention when we focus on our performance of certain perceptual acts. Also in this instance the availability of propriosensitive information does not generate any conflict with respect to perception; it makes us rather more aware of our actions and movements and of our involvement in perception. Indeed, if we *reflect* on the representation of the apple we have, we can find traces of our co-representation, e.g. in the perspective of the visual representation we have and in the awareness that it is we who are perceiving the apple. Thus, 'proprioception' accompanies and is complementary to exteroceptive perception in the sense that it constitutes the background of the perceptual act as well as a necessary condition for the success of perception; while the traits (i.e. the perceptual features) of this background information can be identified when we reflect on our perceptual act and go beyond what is more salient. These observations are consistent with Gibson's general view on the function of perception. Living organisms look for information in order to ascertain what objects are in the external environment and to act in it. This is the reason why our attention is mainly oriented outwards to the things that are directly relevant for our actions.

Still, 'proprioception' is not always only a background support for external perception. Sometimes it becomes salient and is not accompanied by any corresponding exteroceptive act as for example when we suddenly experience a sensation of pain or we become aware that we are hungry. In these cases, 'proprioception' shifts to the foreground of our attention and we directly perceive our bodily states. Of course even when 'proprioception' is the specific element that draws our attention, it always remains less determinate than perception. Indeed, unlike exteroception, propriosensitive information does not concern something like distal, (more or less) structured 'objects' but rather some kind of internal stimulus. Since internal stimuli are intrinsically vague, the perception we have of them is necessarily vague too. Still, this vagueness does not imply that this perception we have is false or that it is function is irrelevant or marginal with respect to the identification of the object. In fact, this experience is the necessary condition for a person to identify his/her own bodily states and to classify them in some way: it is necessary to identify e.g. that we are in pain or hungry. The subject who has a bodily experience knows that s/he is perceiving something about his/her body and s/he can extract many relevant data from it.

As a matter of fact, as we will clarify in more detail later (in the next section as well as in Chap. 2, Sects. 3–6), in spite of this vagueness, if we reflect on propriosensitive experience, we can individuate other background elements that at first did not stand out and that support the identification and classification of the state. The example of pain is paradigmatic in this respect: pain is indeed a *multidimensional* experience (Melzack and Wall 1996) that feels to us like a simple and unitary phenomenon, even though it incorporates information coming from different sources which can somehow be differentiated in the phenomenal experience we have. We feel pain has a certain intensity (this signal comes from the so-called nociceptive system); however, at first this pain might feel as vague: we know it is something we experience in the body, but we cannot immediately localize it exactly or identify it as having certain characteristic properties. However, if we pay attention to it, we might be able to be more precise: to locate more precisely the source of the pain (e.g. not only where exactly the pain comes from, but also whether it is more superficial or more deep) and to relate it to specific sensations like e.g. some impairment in movements, muscle tension or analogous aspects. Furthermore, we can pinpoint some more qualitative features of the pain (for example, whether it is e.g. drilling, burning, pulsating, throbbing, or penetrating). The same applies for other bodily states such as e.g. hunger: hunger cannot only be identified as a specific sensation different from others that are located in the same area—the stomach—but it can also be further specified by its precise features, e.g. it can be identified as more or less urgent or directed towards something specific, etc. The fact that we can distinguish among various characteristics of our internal states supports the idea that 'proprioception' is not fully indeterminate and that it is instead a form of perception, in the sense that propriosensitive experience as a whole provides us with actual knowledge about the internal 'environment' of the body. The 'object' of 'proprioception' (its reference) is, as with perception, the stimulus information that caused it; i.e. its original source (in the 'internal' environment): more specifically it is the bodily condition, state or process detected by the perceptual system through its internal monitoring.

The hypothesis that living beings are equipped with a *propriosensitive perceptual system* that makes information available to the cognitive system about the general state of the body and that this system appears to be an essential and determining factor in the successful outcome of any cognitive processing is confirmed by many recent research lines developed in the fields of cognitive psychology and the neurosciences.[7] Neurophysiological research on multisensoriality is one of the most important developments based on Gibson's views and especially his idea of the perceptual *system*. This work suggests that—even though percepts appear to our awareness as simple, unitary and immediate and thus we have the impression that they are produced by a unique channel—they are actually the result of complex synthetic processes at a low level which operate on stimuli coming from different sensory channels. If we consider the case of the apple again, since in the case of this perception visual information is dominant, we have the impression that the representation of the apple consists of one kind of information only derived from the visual receptors. Research on multisensoriality shows however that this view is wrong and that the perceptual activity cannot be traced back to the processing of one kind of sensory signal only (Ghazanfar and Schroeder 2006; Calvert et al. 2004).

Even though studies on multisensoriality have mainly focused on the analysis of the functioning of what we called exteroceptive perception and are mostly not interested in describing conscious experience, and they frame their achievements merely in terms of what information contributes to generating a perception, they are of great interest for the perspective we are trying to develop here at least for two

[7]The hypothesis that the information provided by the various systems that collect information on the internal states of the body (mainly *proprioception* meant in the narrow sense as information on body position and body movements; *interoception* and *nociception*) is de facto integrated and that it is therefore appropriate to consider it unitarily has been put forward e.g. by Feinberg (2009) and by Craig (2015). On this see below Sect. 5.

main reasons. On the one hand, they confirm the relevance of the notion of the perceptual system also for the field of neuroscience. On the other hand, they also open further lines of investigation on the forms of aware experience that humans have of their body. This primarily concerns the explicit and the tacit/implicit *images* we have of our body—which are called the *body image* and *body scheme* respectively (Gallagher 1986, 1995; de Vignemont 2009); the control that we have over our movements and the capacity to *orient ourselves in space* (Berthoz [1997] 2000; Berthoz and Petit [2006] 2008); the sense of *ownership* we have towards ourselves, i.e. towards the living body we are and towards events occurring within the body itself and its parts (Tsakiris et al. 2007; de Vignemont 2013).

Theories on multisensoriality make an important contribution to understanding what role these kinds of representations play in terms of our capacity to conceptualize our identity (our sense of self). And they also offer at least indirect support for the idea that living beings do have a propriosensitive perceptual system understood as a global collector of information on the states of the body and its parts. They also offer indirect confirmation that this information continuously penetrates cognitive operations leaving traces of its presence at different levels. Still, this presence is not always immediately available, even though it guides cognitive processing since it is intimately related to the development of a sense of self. In Gibson's view, the sense of self also plays a fundamental function in both our perceptual and cognitive processing. If this is the case, then we should conclude that the information provided by 'proprioception' is fundamental for all our mental acts: for the constitution of the perspective from which we perceive the objects in the world; for the constitution of the space we act in, for the way in which we represent ourselves in the world, for the way in which we perceive ourselves as entities in the world; for the way in which we perceive what happens outside us; for the way in which we perceive what happens inside us, etc. In this work we will follow this line of research and we will use the term 'proprioception' or, equivalently, 'propriosensitive information' in the sense described here to include all kinds of bodily information that is available to us and that our cognitive system may use to carry out its operations and to process information.

5 Bodily Mapping and Propriosensitive Monitoring

The idea that living beings continuously map their internal states and use their bodily experience as an informational source to build representations of their internal states, of themselves and of the external world is also supported by work in the field of neuropsychology. At least two particularly prominently positions deserve mention here, those of Antonio Damasio and A.D. (Bud) Craig.

Among Antonio Damasio's starting points we find the idea that the mind is a "spectacular consequence of the brain's incessant and dynamic mapping [...]. The mapped patterns constitute what we, conscious creatures, have come to know as sights, sounds, touches, smells, tastes, pains, pleasures, and the like" (Damasio

2010: 307). External as well as internal information is acquired by the same means, or more specifically through the same vehicle—the body—and it is integrated to form the unitary perception of the world we experience.

> Because of this curious arrangement the representation of the world external to the body can come into the brain only via the body itself, namely via its surface. The body and the surrounding environment interact with each other, and the changes caused in the body by that interaction are mapped in the brain. It is certainly true that the mind learns of the outside world via the brain, but it is equally true that the brain can be informed only via the body. The second special consequence of the brain's body aboutness is no less notable: by mapping its body in an integrated manner, the brain manages to create the critical component of what will become the self (Damasio 2010: 97–98).

In Damasio's view, exteroception is nothing but an internal perception of our body being affected by external stimuli and in this sense exteroception is in itself a form of bodily information. Of course, usually when we perceive something we are not so much aware of our body being affected by this something, but rather aware of the stimuli in the external environment that caused this effect. This is the result of how perceptual processes work: they tend to hide the body itself as much as possible, so that we can focus on the external objects of our perception. Still, this effect is only brought about by a shift of attentional focus and it is due to the fact that mostly we need to direct our attention to the outside would. Even so, the body remains always in the background of our awareness: not only does bodily information continue to impinge on us, even though in a silent manner, but a small change of the attentional shift from the outside to the inside suffices to bring the body back to the foreground and to make us aware of our bodily feelings.

> In the beginning, there was no touching, or seeing, or hearing, or moving along by itself. There was, rather, a feeling of the body as it touched, or saw, or heard, or moved. To a considerable extent, this arrangement would have been maintained. It is appropriate to describe our visual perception as a "feeling of the body as we see," and we certainly "feel" we are seeing with our eyes rather than with our forehead. [...]. It is true that the attention allocated to the visual processing itself does tend to make us partly unaware of the body. However, if pain, discomfort, or emotion set in, attention can be focused instantly on body representations, and the body feeling moves out of the background and into center stage. We are actually far more aware of the overall state of the body than we usually admit, but it is apparent that as vision, hearing, and touch evolved, the attention usually allocated to their component of overall perception increased accordingly; thus the perception of the body proper more often than not was left precisely where it did, and does, the best job: in the background (Damasio 1994: 232–233).

In Damasio's view, all the changes occurring within our body are continuously mapped by the brain as they evolve. The way in which the brain 'informs us' of the outcomes of this mapping is through awareness, viz. through the *feelings* we experience. "That process of continuous monitoring, that experience of what your body is doing while thoughts about specific contents roll by, is the essence of what I call a feeling" (Damasio 1994: 145). These feelings are at the basis of any perception: they are the precondition for and the vehicle of any kind of perception, external as well as internal.

As Damasio puts it: "The human brain is a born cartographer, and the cartography began with the mapping of the body inside which the brain sits" (Damasio 2010: 68). These maps are not only registrations of changes actually occurring in the body due to external and external stimuli, they can also be constructed through the recall of past memories and they recreate past feelings even in the absence of the actual stimuli. These maps keep track of all that happens inside the body as well as in the outside word. They are not mere copies of what we experience, that passively record some data, but the active, integrated product of all the kinds of experiences we have. We become aware of these maps in the form of perceptual images and once they are formed can also be re-evoked to the mind (reconstructed) as recalled images. This reconstruction can also be used to form images of possible future states and to make plans about them (see e.g. Damasio 1994: 94ff). These perceptual images allow us to be aware of what happens outside and inside us; they are, among other things, the precondition for using verbal symbols.

> The nonverbal kinds of images are those that help you display mentally the concepts that correspond to words. The feelings that make up the background of each mental instant and that largely signify aspects of the body state are images as well. Perception, in whatever sensory modality, is the result of the brain's cartographic skill (Damasio 2010: p. 75).

Damasio's suggestion that the brain is a cartographer which maps all the bodily changes occurring inside the body exhibits relevant analogies with the Gibsonian idea of a global perceptual system which records, processes and integrates all the kinds of sensations that we experience related to both our external and our internal environment. Moreover, being also directed inwards, this mapping also explains why we are aware of our bodily states. In this sense, it works as a 'proprioceptive' (i.e. propriosensitive) system that makes us aware of what is happening inside our body. Even though internal ('proprioceptive') information remains mostly in the background of our experience, on certain occasions—when the bodily changes are remarkable and relevant for the organism's survival—it becomes salient; when this happens, we perceive bodily changes directly and through them gain information about our somatic states (e.g. we are in pain, we are hungry etc.) and our emotions.

According to Damasio, any kind of feeling related to something that happens inside our body is due to changes in body states. These cause automatic physiological reactions and corresponding mental experiences which in turn trigger corrective physiological responses. Their role is to assure "the maintenance of the body's physiology within an optimal homeostatic range" (Damasio and Carvalho 2013: 143). They have a regulatory function: they provide information on the state of the body in order to support appropriate responses that assist with life management. These feelings result from a combination of sensations that depend on the changes that happen inside the body. As Damasio exemplifies:

> To grasp what I have in mind, I ask the reader to imagine a state of pleasure (or anguish) and try to itemize its components by making a brief inventory of the varied parts of the body that are changed in the process: endocrine, cardiac, circulatory, respiratory, intestinal, epidermic, muscular. Now consider that the feeling you will experience is the integrated perception of all such changes as they occur in the landscape of the body. As an exercise,

> you can actually try to compose the feeling and assign values of intensity to each component. For each instance that you imagine, you will obtain a different quality (Damasio 2010: 106).

According to Damasio's description, internal states are not a-specific, they are not just undetermined sensations or states of arousal, but quite specific feelings that can be characterized and distinguished by virtue of their characteristic features. Every bodily change on the "landscape of the body" and the intensity of this change gives rise to a specific quality which we experience and identify as such. It is on the basis of this qualitative experience that we get acquainted with what is happening exactly inside us.

Bodily feelings are the vehicle through which we perceive anything that occurs in our body, not only feelings of somatic states (like pain, hunger, thirst, etc.) but also emotions (e.g. happiness, sadness, anger, fear, disgust etc.). In Damasio's view emotions (meant in the sense of salient emotional episodes) are particular kinds of bodily feelings that we can perceive and that depend on changes of bodily states. They are triggered by something (a situation or a thought) that evokes a bodily response that is salient enough to catch our attention and shift it from the outside world to ourselves (Damasio 2010: 119ff).

> All emotions use the body as their theater (internal milieu, visceral, vestibular and musculoskeletal systems), but emotions also affect the mode of operation of numerous brain circuits: the variety of the emotional responses is responsible for profound changes in both the body landscape and the brain landscape. The collection of these changes constitutes the substrate for the neural patterns which eventually become feelings of emotion" (Damasio 2000: 51).

In Damasio's perspective, those particular feelings which we call 'emotions' accomplish the function of informing us in an immediate and pre-conceptual manner about the relationship between ourselves as living organisms that need to stay alive and to preserve vital functions and the objects and events that constitute our external environment. Our knowledge of the emotion we are experiencing—and thus also of whether something we are experiencing is positive or potentially dangerous for us and of how we should react—goes through the body: it is the bodily feelings and the conscious experience we have of them that tells us (makes us aware) what emotion we are experiencing (we will go back to the issue of whether bodily feelings can uniquely identify emotions in Chap. 5).

Like bodily feelings, emotional states also affect external perception: depending on how we feel (e.g. happy, sad, angry, fearful, disgusted etc.), the outside world is perceived differently, i.e. the perception of things and the thought processes related to them are modified. No matter how hard we try, we cannot ignore these feelings and separate our perceptions and thinking processes from them. Moreover, external perception is influenced by other bodily feelings which are analogous to emotions even though they are not salient, short emotional episodes that drive the attention to the inner states of the individual. Indeed, according to Damasio there are bodily feelings that tend to remain in the background of our experience—and for this reason should be called background feelings—which accompany any kind of perception and 'paint it' (provide it) with a specific hedonic tone. They determine how we feel during a certain period of time. As Damasio specifies:

> [...] background feelings are neither too positive nor too negative, although they can be perceived as mostly pleasant or unpleasant. In all probability it is these feelings, rather than emotional ones, that we experience most frequently in a lifetime. We are only subtly aware of a background feeling, but aware enough to be able to report instantly on its quality. [...] The background feeling is our image of the body landscape when it is not shaken by emotion. The concept of "mood," though related to that of background feeling, does not exactly capture it. When background feelings are persistently of the same type over hours and days, and do not change quietly as thought contents ebb and flow, the collection of background feelings probably contributes to a mood, good, bad, or indifferent (Damasio 1994: 150–151).

Damasio's background feelings are thus feelings like well-being or malaise, calm or tension (Damasio 2000: 50) which are the precondition for having moods.[8]

The common denominator of all bodily feelings—be they feelings corresponding to somatic states, emotions or background feelings—is exactly that they are *feelings*, in the sense that *they are something we consciously experience, which is characterized by a particular quality and which we detect through a form of perception directed inwards*. Different bodily states correspond to different feelings (they feel differently) and this is the reason why we can differentiate among them and identify them. For Damasio, we know that we are in a particular state because we have conscious access to the bodily information corresponding to that state. In his view, our 'introspective capacities', meant in the sense of the capacities we have to report what we feel inside us, are related to our ability to verbally convey the conscious sensations produced by our bodily states.

Further support for Damasio's neuropsychological hypothesis on the existence of an internal (brain) mapping that accomplishes the function of what we have called propriosensitive system is also provided by the recent work of A.D. Craig. Craig provides neuroanatomical evidence that we have a system for the representation of the self which consists in a continuous mapping of our bodily states in which all the signals coming from the different areas of the body and traditionally characterized as proprioception (sense of position and movement), interoception ("the sense of the physiological condition of the body" (Craig 2003)—e.g. hunger, thirst, visceral functions, cardiorespiratory control, temperature, itch, sensual touch, muscular and visceral sensations, vasomotor activity etc.) and nociception (perception of pain) *are merged*. The integration of all these signals constitutes the

[8]Damasio also distinguished between background feelings and background emotions (Damasio 2000, Chap. 2). Strictly speaking, the real analogy should be between existential feelings and background emotions. Background emotions are a subset of background feelings: while background feelings describe generically all kinds of bodily feelings (including also those directed towards the body for example being satiated or experiencing an almost imperceptible pain in the foot while sitting etc.), background emotions only denote those background feelings that determine our relationship with the external world and others. However, the really relevant element of this analogy is that both authors consider bodily feelings as the common background for all kinds of internal states: emotional as well as physical.

deeper and original core on which the conscious awareness of all our internal states, be they emotional or physical (e.g. hunger, thirst, itch, burning sensation, but also anxiety, relaxation and the like, etc.), is grounded.

> The key stage of integration at the core of this model is a coherent representation of all feelings at one moment, which I referred to as a 'global emotional moment'. This construct can also be called 'the sentient self' (Craig 2010: 563).

Craig is persuaded that this global awareness of our bodily condition is realized by a specific brain area—i.e. the insula—which provides sensory representations of the affective feelings from the body, integrates these representations with other data coming from various other sources (Craig 2002, 2008) and instantiates all feelings the subjects perceives which are the basis of the *sentient self* or *material me* (Craig 2009, 2010, 2015). In particular, consistently with Damasio's view—which he mentions explicitly—Craig suggests that "the anterior insular cortex [...] contains interoceptive representations that substantialize (that is, provide the basis for) all subjective feelings from the body and perhaps emotional awareness [...]." (Craig 2009: 59) By virtue of the work carried out by this part of the brain, we gain conscious access to our 'interior life' meant as the integrated signals coming from all parts of our body. This gives rise to a kind of 'interoceptive awareness' which describes the global representation of the internal perception we have of our body. Since Craig considers the insula to be the locus where all of this information is integrated, in his view the traditional separation between different mechanisms such as interoception, nociception or proprioception no longer makes sense. Thus, we can speak in general of internal perception.

In fact, even though Craig lists interception first, in Gibsonian fashion it seems more appropriate to speak generally of '*proprioception*' and of a '*proprioceptive*' *or propriosensitive system*. The way we use this term in this work diverges from the usual, technical definition of proprioception which only denotes information provided by the muscular and skeletal systems. The way we understand 'proprioception' is more similar to the notion of interoception as described by Craig. However, by using the old Gibsonian term 'proprioception' we intend to recall Gibson's idea of a general and integrated system of perception that is both propriosensitive and exterosensitive. Furthermore, the word 'proprioception' suggests that we can consider internal perception as something unitary and undifferentiated: if we are interested in understanding how bodily signals are consciously perceived, we do not need to distinguish between the various internal channels or sources that might carry different bodily signals, because these gain access to consciousness as an integrated whole.

Even though Damasio and Craig's views exhibit relevant differences with respect to Gibson (and with respect to each other), globally they can be interpreted as an updated extension of Gibson's original intuition about the nature, function and organization of perception as the outcome of integrated perceptual systems that are continuously active and that are directed toward both the external and the internal world of the subject. Furthermore, they suggest that humans can feel their body from the inside and that these feelings are the permanent background of our experience as living creatures: they permeate and influence all kinds of perceptions that we can have. Disregarding the specificities of their views, the position that emerges from their analyses suggests that living organisms are equipped with an internal perceptual system that is propriosensitive and detects, integrates and makes available to consciousness the information produced by bodily changes. This underpins the idea that humans (as opposed e.g. to computers) are equipped with a propriosensitive ('proprioceptive') system that produces sensory information about their bodily states which they are consciously aware of, even though it is recessive and vaguer than exteroception. Indeed, as we discussed above, bodily sensations remain mostly in the background of our awareness. And yet, if we draw our attention to them, many characteristics can be brought more sharply into focus: the complex of sensations produced by these modifications become more salient and this allows us to conceptualize ourselves as beings equipped with an 'internal world'.

The complex of the sensations generated by bodily changes is represented in our mind at a low level of awareness and it remains in the background of our experience. This propriosensitive system produces representations of (information on) the internal environment of the subject by synthesizing the signals coming from the various bodily parts in a unitary and stable manner and providing a constant, dynamical mapping of internal states and their changes. These representations are always present to our cognitive system, even though they are usually in the background and thus beyond the usual focus of selective attention. For this reason, the representations we have of our internal states are *vague*. For a long time, this vagueness has been interpreted in terms of an intrinsic ambiguity because of which it was impossible to explain in a conceptually stable manner how we get the knowledge we have of our internal states. Well known examples of this view are the criticisms of introspection made by both the psychological and the philosophical traditions (see, a.o., Boring 1953; Nisbett and Wilson 1977; Lyons 1986; Schwizgebel 2012). However, even though we experience the changes occurring in our body in a rather indefinite manner, this perception is neither a-specific nor cognitively impenetrable. As a matter of fact, when these states become particularly salient and/or when we pay specific attention to them we become explicitly aware of their content and we can reflect on them. This reflection allows us to trace back the origin (the localization) as well as the specific quality of these internal representations.

In fact, a similar phenomenon happens in the structure of visual perception with regard to the opposition between figure and ground. When we see, we see primarily (we focus primarily on) the figure—i.e. (on) what is in the foreground. Still, this does not mean that the information in the background is not present. On the contrary, the background information represents the condition for constituting the distal percepts in a sharp and clear manner. On the basis of this analogy, 'proprioception' can be seen as a system that continuously provides information on the states of activation of the body at a low level of awareness, giving us the possibility of having at any given time a background knowledge of our situation as living organisms. The propriosensitive system offers a continuous mapping of the changes occurring in the body and the sensations detected through this are perceived as *somatic events*. Even though these are perceived only in a vague manner, they still have a *content* and more specifically a *reference* which consists in the bodily states or bodily changes that have been mapped by the system. This system produces representations of the internal environment of the body that keep track of their origin. The conscious information we have on our bodily signals inform us which parts of the body have changed their levels of activity and how. Thus, we can use this information to identify the bodily changes we are experiencing. Moreover, since we have conscious access to this kind of information, it can be used for higher cognitive processing that also involves e.g. conceptualization and language: we can convey tacit information using explicit, verbal forms of knowledge about our internal states. Finally, this kind of information influences and structures every other kind of experience we have and its content can be reconstructed starting from the contribution it makes to other kinds of perceptions.

The relevance of these last aspects becomes clear when we consider the perspective of the anti-mentalistic approach stemming from logical empiricism, behaviorism and functionalism (see Chap. 1), which suggests that internal states cannot be grasped directly, and we need to develop a third-person approach to determine what they are. The actual nature of internal states can be understood only if we develop a way to avoid subjectivism and describe them on the basis of intersubjectively observable—i.e. external—elements. Because qualitative experience and more generally internal sensations have a univocally subjective nature, the anti-mentalistic tradition has for a long time imposed the view that internal and qualitative states should not be identified directly on the basis of their intrinsic properties through something like internal ostension. They should be rather identified in a relational manner using the set of observable events (i.e. situations and behaviors) occurring outside the body in correspondence with the internal states. An emotion like e.g. 'love' should not be identified or described on the basis of how it feels to be in love, but rather in terms of the observable condition in the external world that 'goes with' love.

The idea of a 'proprioceptive' system (i.e. of a propriosensitive perceptual system) we tried to support in this section challenges this approach and suggests that human perception and cognition cannot be addressed starting exclusively from a third person perspective. On the contrary, the general perspective that emerges from the views we considered above suggests that the internal, qualitatively characterized information produced by the body needs to be taken into account in order to provide an adequate description of how the mind works because it performs a critical function both in terms of how we experience the world and how we think.

If the hypothesis about the existence of a propriosensitive system of the kind we outlined is correct then we can conclude that the amount of information (at least tacitly) available to our cognitive system to individuate, recognize and classify internal states is much larger than has been traditionally admitted. Further, that information we gain about our bodily states is central for developing our knowledge of both our internal as well as external world. The idea of introspection thus gains new meaning as a capacity to focus on the internal information which is consciously available to us and form some conceptualization of it in spite of its vagueness. Finally, this hypothesis constitutes a first step in overcoming the intrinsic abstractness and formalness that characterize classical cognitive theories. It emphasizes the fact that knowledge is produced by living organisms and that the functions of biological organisms are of primary relevance in understanding how information and knowledge are acquired and processed by the cognitive system.

6 Bodily Feelings and Emotional Experience

In the previous sections of this chapter we pointed out that 'proprioception' contributes to perception and provides us with information on the internal states of the body (i.e. it reveals what is happening in the body at a certain moment and where this is happening). On the basis of Damasio and Craig's views, we considered whether (and why) emotions may be nothing but bodily feelings and that the awareness we have of our emotions derives from the bodily feelings we experience when we have an emotion. From this perspective, emotions reveal something about our relationship with what is happening outside us: they tell us whether what is happening in the external environment at a certain moment is positively or negatively related to us and they also 'color' our perception which is influenced by the way we 'feel'. In fact, emotional experience is of crucial relevance for a discussion of our hypothesis of an integrated propriosensitive perceptual system for at least two reasons that are also related to each other. On the one hand, it offers us some more elements for clarifying more precisely what propriosensitive information is: i.e. what kind of information the propriosensitive system provides, how specific it is and what it is used for. On the other hand, a discussion of emotions as a kind of internal experience can help us address the issue of whether and how propriosensitive information plays a role with respect to cognition and what this role exactly consists in.

The authors that offer a particularly relevant contribution as we start to discuss these points and to introduce the specific analysis of emotional experience which will be carried out in Chap. 5 are Peter Goldie and Matthew Ratcliffe. In spite of the differences between their positions, they share the idea that the knowledge we have of our emotions and more generally of our internal states relies, at least in part, on internal information produced by the body which is available to our awareness. Their views offer relevant clues in support of the idea that humans are equipped with a propriosensitive system which provides us with essential information to carry out any cognitive process as well as suitable insights on how we can identify these internal states on the basis of the information provided by the propriosensitive system.

Goldie's view belongs to a line of research called "perceptual account of bodily feelings" (Goldie 2002: 236) which opposes the traditional, philosophical and psychological view that bodily feelings (sensations) are radically different from perceptions in the sense that they cannot give rise to any knowledge. According to this traditional view, apart from perception, sensations do not produce stable and coherent knowledge, e.g. representations referring to something specific, that can be re-identified on different occasions. Sensations are epistemically unreliable because they are not precise enough to support the constitution of representations with stable properties. In opposition to this perspective, the perceptual accounts of bodily feelings argue that bodily sensations are specific enough to allow us to identify events that occur in us (Armstrong 1968; Crane 1998). These events can be identified quite precisely through temporal and spatial coordinates: we know when and approximately where in the body they occur (Martin 1995). Goldie applies this view to emotions as a particular kind of bodily sensations or feeling we experience and suggests that it is the feelings involved in an emotion which allow us to know what emotion we are experiencing.

Consider the following two examples given by Goldie: the experience of an agonizing pain in the elbow on the one hand and that of the fear of a lion on the other. In the case of the pain, we know that we are experiencing it because we have specific bodily feelings that inform us about the pain, its intensity and its localization at the elbow. Thus, these feelings are everything we need to know we are in pain. The case of the fear of a lion is similar, but not identical. Here our knowledge that we fear the lion derives also from the sensation of fear we experience; however, the case for fear differs from that of pain, in that our knowledge of fear is not exhausted by the sensation we experience. In Goldie's view, there is more to know about fear. When we experience an emotion like fear we feel in a certain way but we also know the reason for this feeling, i.e. we know *toward* what this feeling is directed: we know that we fear *the lion*.

Somatic states like pain (or like hunger, itch, spasm etc.) are not directed toward anything other than themselves: their only object is the body condition in a certain moment, e.g. the agonizing pain in the elbow. Unlike somatic states, emotions are more complex bodily states which involve both a bodily feeling and an external object towards which the emotion is directed. For this reason, Goldie maintains that "emotions involve two kinds of feelings" which he calls "bodily feelings" and "feelings towards" (Goldie 2002: 235). The notion of *bodily feeling* describes the specifically perceptual—i.e. the bodily—component of the emotions. In the case for example of fear, this could consist, among other things, of hair rising on the back of our neck (Goldie 2002: 235–236). However, in Goldie's view bodily feelings alone are too vague to determine what emotion we are experiencing. They become recognizable as one specific emotion only when they are accompanied by—or more specifically blended with—an external state, object, person or event *toward* which they are directed. "Bodily feelings alone cannot reveal to you what your emotion is about; as we have seen, the most they can reveal is that you are feeling an emotion about something or other, which has a certain determinable property. Feelings towards, on the other hand, are directed towards the object of one's emotion as such —for example, feeling fear towards the lion" (Goldie 2002: 241). Thus, the notion of *feeling toward* describes our awareness of the object toward which an emotion is directed. In this view, emotions are somatic states with an additional ingredient consisting of feelings which inform us that the somatic state is directed toward something specific in the external world.

In his examination of emotional experience, Goldie maintains that both kinds of feelings that constitute emotions are *intentional* (in Brentano's sense), i.e. they have a specific content. This is hardly surprising if we consider the aspect of emotions characterized by our feelings *towards* an external object: we fear a *lion*; we love *someone*; we are angry about *something,* etc. These kinds of feelings are clearly intentional because they are about the object they are directed toward. The most remarkable part of his view lies in the hypothesis that the aspect of emotions consisting in bodily feelings also has a content.

> A bodily feeling or sensation, the feeling from the inside of the condition of one's body, is intentional in the sense that the feeling is directed towards an object, one's body, as being a certain way or as undergoing certain changes. I will call this, as others have done, the perceptual account of bodily feelings. When you feel an agonising pain in your elbow, the object of the sensation is your elbow which feels a certain way: agonisingly painful. Similarly, when you feel the prickly sensation of the hairs going up on the back of your neck, the object of the feeling is the hairs on the back of your neck which feel a certain way: prickly, as if they were rising. When intentionality is thus understood, in terms of directedness towards an object rather than in terms of "aboutness", bodily feelings are unproblematically intentional, being directed towards a part of one's body in a certain location (Goldie 2002: 236).

Thus, bodily feelings have a content in the sense that they are *directed towards a part of one's body*, i.e. that *they indicate the location in the body certain sensations come from*. This means that they offer a mapping of the internal condition of the body (we know that the pain is in our elbow, that the hairs on the back of our neck are going up etc.). And in Goldie's view these sensations are one of the two kinds of information we use to identify the state we are experiencing. Even though Goldie remains in the line of the anti-mentalistic tradition in considering the external element (the feeling towards) as the most important for determining emotions, he suggests that living beings are able to perceive our "bodily changes from the inside" and that this is the "epistemic route" to the knowledge that we are experiencing an emotional episode (Goldie 2002: 237).

This internal perception captures our internal sensations and provides us with conscious information about our bodily conditions at a given time: "Bodily feeling involves consciousness – from the inside, so to speak – of the condition of your body, or of changes to it, such as muscular reactions (including changes in facial expression), hormonal changes, and changes to the autonomic nervous system; I will call these bodily changes" (Goldie 2000: 51). These internal perceptions are not as precise as the perceptions of external things; for this reason, the knowledge we can have of them is also less clear and precise than in external perception: 'proprioceptive' representations do not have the 'sharp profile' and the specific form of 'self-subsistence' that characterizes representations of objects in the external world. Emotional episodes are perceived as a-specific, but also as inseparable from ourselves. The first characteristics indicate that—even though emotional feelings are not wholly ambiguous—we cannot represent what is happening inside us in a clear and distinct manner. The second characterization indicates that emotional states cannot be represented as something other than ourselves; this way of experiencing emotions is due to their bodily origin, in the sense that we perceive emotions first of all as related to the body, i.e. as being bodily states, and the body is itself experienced as our own body, i.e. as inseparable from us.

Goldie's idea that bodily feelings are characterized by a specific *directness towards a part of one's body* suggests that Goldie sees a relationship between the aware perception of the bodily feeling and attention. As the word itself suggests, *directedness* is related to the attentional focus, because it indicates the fact that we can concentrate on specific bodily signals and this make us aware of the sensations we are experiencing in those specific bodily parts. Directedness can therefore be considered as a particular property of *any kind of 'proprioceptive' state since 'proprioceptive' feelings are always recessive and they remain mostly in the background of our conscious awareness: these become central in our perception or even an object of explicit knowledge only when they capture our attention and when they become the object of reflective consciousness* (Goldie 2000: 62ff). This idea that living beings permanently experience sensations related to the internal condition of their body, but that these sensations come to the foreground of their awareness and become the object of some explicit representation (knowledge) only when their attention is directed toward them could be considered as the general way in which the propriosensitive perceptual system works.

This nature and the effects of this permanent background awareness of our bodily condition can be further investigated and specified using e.g. Matthew Ratcliffe's description and explanation of the so called *existential feelings* which also offer an additional argument in support of the idea that human beings are equipped with a propriosensitive perceptual system that detects any kind of internal information and that makes it available for cognitive processing. Ratcliffe points out that the classical discussion on emotions neglected a number of emotional phenomena which are not included in the "standard list of 'emotions'" but make nevertheless "a considerable contribution to the structure of experience" (Ratcliffe 2005: 46–47).

The phenomena Ratcliffe is interested in resemble, at least vaguely, Damasio's background feelings and are feelings related to our experience with the world: "[…] the feeling body […] is a framework through which world-experience is structured. Even when one is not explicitly aware of the body, it still functions as a structure – giving background to all experience. For example, one can have a sense of 'up', 'down', 'left' and 'right' without being explicitly aware of one's bodily position" (Ratcliffe 2005: 52). Among the feelings that Ratcliffe focusses on there is the "sense of belonging to the world" (Ratcliffe 2012), "sense of reality" (Ratcliffe 2008), "sense of familiarity" (Ratcliffe 2009a) and other more complex feelings concerning our relationship with the world and with other people:

> People sometimes talk of feeling alive, dead, distant, detached, dislodged, estranged, isolated, otherworldly, indifferent to everything, overwhelmed, suffocated, cut off, lost, disconnected, out of sorts, not oneself, out of touch with things, out of it, not quite with it, separate, in harmony with things, at peace with things or part of things. There are references

> to feelings of unreality, heightened existence, surreality, familiarity, unfamiliarity, strangeness, isolation, emptiness, belonging, being at home in the world, being at one with things, significance, insignificance, and the list goes on. People also sometimes report that 'things just don't feel right', 'I'm not with it today', 'I just feel a bit removed from it all at the moment', 'I feel out of it' or 'it feels strange' (Ratcliffe 2008: 68).

The existence of these kinds of feelings is not only supported by everyone's personal experience, important evidence also comes from specific psychiatric illnesses where existential feelings are *altered* and lead to a change in our relationship with the world and/or with other people (see e.g. Ratcliffe 2008, 2009b; Colombetti and Ratcliffe 2012). When one is e.g. depressed, s/he feels detached from the world, disconnected, isolated and unable to interact with the world and with others, as if s/he was observing the world like an external spectator (see e.g. Ratcliffe 2015: 31–32; Ratcliffe and Stephan 2014). Delusions such as e.g. schizophrenia are accompanied by a "profound shift in the sense of reality" and by "a pervasive disconnection from the world and other people". Patients become unable to anticipate what they are about to experience, so they feel a sense of uncertainty, doubt, tension and anomaly and a disruption or fragmentation of self (see e.g. Ratcliffe 2015: 259–260).

Illnesses like these show in a particularly salient manner how our everyday experience is characterized in the background by specific (bodily) feelings which determine very important aspects our experienced relationship with the world and with others. As long as things go well and we are healthy, we tend to overlook these feelings. Their role and importance becomes clear, however, in cases in which—even though our perception remains per se unaltered—the experiences we have radically change in their phenomenology and are qualitatively colored in an anomalous way that can influence and modify the entire course of our thoughts and behaviors.

This mention of psychiatric illness can also help us differentiate between moods and existential feelings. In fact, existential feelings are deeper than moods: this does not mean that they must necessarily be more enduring, since both moods and existential feelings can also shift from moment to moment (Ratcliffe 2010: 367); the point is rather that existential feelings do not simply concern how we feel in a certain moment, but they "are a background which comprises the very sense of 'being' or 'reality' that attaches to our world experiences." (Ratcliffe 2005: 48; see also Ratcliffe 2008, 2010) While we can separate ourselves from our moods—consider e.g. when we apologize, for example, for being in a bad mood—our existential feelings are inseparable from us because they determine the way we perceive and relate to the world and to others (Ratcliffe 2005: 57). Psychopathological illness in which patients undergo fundamental changes in the way they experience themselves, other people or the world show that our usual sense of reality relies on specific '(existential) feelings', which—if altered—lead to huge disturbances in the normal condition.

Existential feelings are background feelings that shape our experience of the world, our actions and our thoughts. They are *felt*, and this means that "they are bodily states of which we have at least some awareness" (Ratcliffe 2008: 2).

However, the fact that they are a kind of—or maybe they are brought about by—bodily states does not mean that they are just "feelings of the body" or that their object consists in bodily feelings (Ratcliffe 2012: 28). They are rather "background orientations through which experience as a whole is structured" (Ratcliffe 2008: 2). In spite of remaining in the *background* of our experience, they play a primary function with respect to the structuring of any other kind of experience.

> [Existential feelings] are ordinarily part of the background structure of experience, constituting ways of finding oneself in a world that shapes more specific experiences. Nevertheless, they are phenomenologically available, as is evident from the various, usually metaphorical descriptions employed to communicate them. So they are part of the structure of experience, rather than an experientially inaccessible contributor to that structure. However, there may be a thin line between noetic and prenoetic aspects of existential feeling. For example, the role of feeling in constituting our sense of reality, which I will discuss now, is perhaps something that is hidden beneath everyday experience and can only be made phenomenologically explicit through reflection upon highly unusual states of oneself or others (Ratcliffe 2005: 53).

Ratcliffe takes the notions of 'prenoetic' and 'noetic' from Gallagher (2005). Prenoetic means that "it shapes experience without itself being an object of awareness or part of the structure of awareness", while 'noetic' indicates "an awareness of one's body, which might not be at the centre of one's attention but is still accessible through phenomenological reflection" (Ratcliffe 2005: 52). Existential feelings have both noetic and prenoetic aspects and thus they are only partially accessible to consciousness in a direct way; other features of them are accessible only indirectly through an analysis of their contribution to other cognitive phenomena. Thus, existential feelings are *recessive* feelings (in the sense of this notion that we discussed before) that can be accessed at least in part through reflection and that can, in part, be reconstructed starting from the role they play as structuring elements of our experience.

Ratcliffe distinguishes existential feelings from other forms of direct awareness of our bodily states as explicit objects of our attention. If I pay attention e.g. to my heart rate, then I become aware of it, because I bring this state to the foreground of my experience. This is a feeling in the body that I am aware of as a feeling of something happening in my body. The situation is not the same when we consider existential feelings: if I feel hopeless, for example, this feeling is certainly in my body (is certainly brought about by my body), however, it is not a feeling concerning my body, but it is much more a feeling concerning something outside myself, i.e. it is something concerning my relationship with the world and with other people. "There is a distinction between the location of a feeling and what that feeling is *of*. A feeling can be *in* the body but *of* something outside the body. One is not always aware *of* the body, even though that is where the feeling occurs." (Ratcliffe 2005: 46) In this sense, we do not become aware of our sense of being hopeless as a bodily feeling—we do not feel it as something like reduced heart rate, reduced pressure, weakness or anything analogous—but become aware of it as an attitude towards the world.

In this respect, Ratcliffe (2005: 49) goes back to Goldie's notion of directness and modifies it in an important direction. Goldie maintains that there are two kinds of directedness. (i) Some feelings like e.g. an agonizing pain in your elbow are just bodily feelings; they are directed toward the body (the body is their object). (ii) Other feelings like e.g. your fear of a lion are 'feelings toward' because they are directed toward an external object. Even though Goldie often makes clear that these feelings might be two inextricably blended components of the emotional experience, he insists that they need to be kept separated. Regarding this aspect, Ratcliff embraces a more radical position and suggests that bodily feelings and feelings towards are one and the same thing: "[…] bodily feelings just are feelings towards. Some are feelings towards the body or parts of it and others are feelings towards things outside the body" (Ratcliffe 2005: 49). In Ratcliffe's view, this distinction is a mistake because we do not experience emotions (usual emotional episodes as well as existential feelings) as two different things, i.e. as a self-directed feeling in a specific part of our body on the one hand and as a feeling directed to something else in the world on the other. We experience emotions as one unitary phenomenon: they are bodily in the sense that they rely on bodily feelings, however they are not perceived as a feeling in our body, but they become part of the way in which we perceive the world: they "are inextricable from the structure of world experience" (Ratcliffe 2005: 49).

Ratcliffe distinguishes emotional experience from the perception of our own body. He maintains that *the bodily feeling that characterizes any kind of emotion is not the perception of a body state*. Giving special consideration to existential feelings, he tries to show that they are not feelings of our body, but rather bodily feelings that contribute to structuring all our experience, including the experience of the external world. They constitute the condition of possibility for having further experiences in the world (Ratcliffe 2015: 33ff). To explain this aspect, Ratcliffe uses the example of touch: touch differs from vision since we cannot differentiate the experience we have of ourselves and the one we have of the external object, but the first is the condition for the second. The experience of ourselves in touch structures the tactile experience itself, while the location of the tactile feeling on the body "does not determine what it is a feeling of" (Ratcliffe 2005: 50). In the same way, existential feelings are bodily feeling but we do not experience them primarily as sensations in our body, but as ways in which the world appears to us. They are part of our intentionality in the sense that they structure it. These considerations might now be enough to explain all three main theses concerning the nature of bodily feeling (including emotions as well as existential feelings) Ratcliffe argues for:

(1) Bodily feelings are part of the structure of intentionality. They contribute to how one's body and/or aspects of the world are experienced;
(2) There is a distinction between the location of a feeling and what that feeling is of. A feeling can be in the body but of something outside the body. One is not always aware of the body, even though that is where the feeling occurs;
(3) A bodily feeling need not be an object of consciousness. Feelings are often that through which one is conscious of something else (Ratcliffe 2005: 46).

Since bodily feelings are a means through which we gain information, they have content. Even though they have no explicit 'object' of their own, they are not 'invisible' or 'cognitively irrelevant' because they influence, or even determine the way in which we perceive any other thing. The perception we have of our body—our bodily feeling—is a non-neutral vehicle for any other kind of perception, which contributes to structuring the objects we perceive; and we cannot approach perception and perceptual objects without considering the contribution and the 'filtering' performed by bodily feelings. Indeed, they are the condition of possibility for perceiving anything and they characterize the way in which and through which we perceive. Indeed, they influence what we perceive since internal and external information is blended together. Existential feelings are a clear example of how feelings towards things outside the body influence the way we perceive these things: how you feel affects not only how you see things, but also what you pay attention to—i.e. how you select information in the external world. And this 'how you feel' is the global effect of the bodily information which we become aware of in the form of bodily feelings.

In spite of the differences between Goldie and Ratcliffe, taken together their position give us relevant clues as for why and how emotional experience might be considered a form of 'proprioception'. They suggest that emotional experience in all its forms relies on bodily feelings and that it is not radically different from the experience we have of somatic states like pain or hunger but is rather a different point along the same continuum (we will go back to this in detail in Chap. 5). To know what emotions we are experiencing is not radically different than to know that we are hungry: in both cases, we could not acquire this knowledge if we didn't have the bodily feelings corresponding to it. Even though internal perceptions are not as precise as the perceptions of external things, bodily sensations are specific enough to allow us to identify events that occur in us. Like other kinds of bodily feelings, emotional experience can also remain in the background of our attention. Nevertheless, even in this case it plays a primary function with respect to the structuring of any other kind of experience.

References

Alsmith, A. T., & de Vignemont, F. (2012). Emboding the mind and representing the body. *Review of Philosophy and Psychology, 3*(1), 1–13.

Anderson, M. L. (2003). Embodied cognition: A field guide. *Artificial Intelligence, 149*(1), 91–141.

Armstrong, D. M. (1968). *A materialist theory of mind*. London: Routledge.

Barsalou, L. W., Niedenthal, P. M., Barbey, A. K., & Ruppert, J. A. (2003). Social embodiment. In B. H. Ross (Ed.), *The psychology of learning and motivation* (pp. 43–92). San Diego, CA: Academic Press.

Bennett, M. R., & Hacker, P. M. S. (2013). *History of cognitive neuroscience*. London, New York: Wiley-Blackwell.

Berthoz, A. ([1997] 2000). *The brain sense of movement* (G. Weiss, Trans.). Cambridge, MA: Harvard University Press.

Berthoz, A. & Petit, J.-L. ([2006] 2008). *The physiology and phenomenology of action* (C. Macann, Trans.). Oxford: Oxford University Press.

Boring, E. G. (1942). *Sensation and perception in the history of experimental psychology*. New York: Appleton Century Croft.

Boring, E. G. (1953). A history of introspection. *Psychological Bulletin, 50*(3), 169–189.

Calvert, G. A., Spence, C., & Stein, B. E. (Eds.). (2004). *The handbook of multisensory processes*. Cambridge, MA: MIT Press.

Carruthers, P. (2006). *The architecture of the mind*. Oxford: Oxford University Press.

Colombetti, G., & Ratcliffe, M. (2012). Bodily feeling in depersonalization: A phenomenological account. *Emotion Review, 4*(2), 145–150.

Craig, A. D. (2002). How do you feel? Interoception: The sense of the physiological condition of the body. *Nature Reviews Neuroscience, 3*(8), 655–666.

Craig, A. D. (2003). Interoception: The sense of the physiological condition of the body. *Current Opinion in Neurobiology, 13*(4), 500–505.

Craig, A. D. (2008). Interoception and emotion: A neuroanatomical perspective. In M. Lewis, J. M. Haviland-Jones, & L. F. Barrett (Eds.), *Handbook of emotions* (3rd ed., pp. 272–288). New York: Guilford Publications.

Craig, A. D. (2009). How do you feel—Now? The anterior insula and human awareness. *Nature Reviews Neuroscience, 10*(1), 59–70.

Craig, A. D. (2010). The sentient self. *Brain Structure and Function, 214*(5–6), 563–577.

Craig, A. D. (2015). *How do you feel? An interoceptive moment with your neurobiological self*. Princeton: Princeton University Press.

Crane, T. (1998). Intentionality as the mark of the mental. In A. O'Hear (Ed.), *Current issues in philosophy of mind* (pp. 229–251). Oxford: Oxford University Press.

Damasio, A. R. (1994). *Descartes' error. Emotion, reason and the human brain*. New York: Avon Books.

Damasio, A. R. (2000). *The feeling of what happens. Body and emotions in the making of consciousness*. Boston. MA: Mariner Books.

Damasio, A. R. (2010). *Self comes to mind. Constructing the conscious brain*. New York: Pantheon Books.

Damasio, A. R., & Carvalho, G. B. (2013). The nature of feelings: Evolutionary and neurobiological origins. *Nature Reviews Neuroscience, 14*(2), 143–152.

de Vignemont, F. (2009). Body schema and body image—Pros and cons. *Neuropsychologia, 48* (3), 669–680.

de Vignemont, F. (2013). The mark of bodily ownership. *Analysis, 73*(4), 643–651.

Evarts, E. V. (1981). Sherrington's concept of proprioception. *Trends in Neuroscience, 4*(1), 44–46.

Feinberg, T. E. (2009). *From axons to identity: Neurological exploration of the nature of the self*. New York: Norton.

Fodor, J. A. (1983). *The modularity of mind*. Cambridge, MA: MIT Press.

Fodor, J. A. (2000). *The mind doesn't work that way*. Cambridge, MA: MIT Press.

Fodor, J. A., & Pylyshyn, Z. W. (1988). Connectionism and cognitive architecture: A critical analysis. *Cognition, 28*(1–2), 3–71.

Gallagher, S. (1986). Body image and body schema: A conceptual clarification. *Journal of Mind and Behaviour, 7*(4), 541–554.

Gallagher, S. (1995). Body schema and intentionality. In J. Bermudez, B. Marcel, & N. Eilan (Eds.), *The body and the self* (pp. 225–244). Cambridge, MA: MIT Press.

Gallagher, S. (2005). *How the body shapes the mind*. Oxford: Oxford University Press.

Gallagher, S. (2009). The intrinsic spatial frame of reference. In H. L. Dreyfus & M. A. Wrathall (Eds.), *A companion to phenomenology and existentialism* (pp. 346–355). Chicherester: Wiley.

Gardner, H. E. (1985). *The mind's new science*. New York: Basic Books.

Ghazanfar, A. A., & Schroeder, C. E. (2006). Is neocortex essentially multisensory? *Trends in Cognitive Sciences, 10*(6), 278–285.

Gibbs, R. W. (2005). *Embodiment and cognitive science*. Cambridge: Cambridge University Press.

Gibson, J. J. ([1961] 1987). Outline of a new attempt to classify the senses and the sensory input. In J. Pittenger, E. S. Reed, M. Kim & L. Best (Eds.), *James Gibson's purple perils. A selection of James J. Gibson's unpublished essays on the psychology of perception*. http://www.trincoll.edu/depts/ecopsyc/perils/.

Gibson, J. J. ([1966] 1983). *The senses considered as perceptual systems*. Westport, CT: Greenwood Press.

Gibson, J. J. ([1968] 1987). On the difference between perception and proprioception. In J. Pittenger, E. S. Reed, M. Kim & L. Best (Eds.), *James Gibson's purple perils. A selection of James J. Gibson's unpublished essays on the psychology of perception*. http://www.trincoll.edu/depts/ecopsyc/perils/.

Gibson, J. J. ([1974] 1987). An insoluble puzzle of epistemology. In J. Pittenger, E. S. Reed, M. Kim & L. Best (Eds.), *James Gibson's purple perils. A selection of James J. Gibson's unpublished essays on the psychology of perception*. http://www.trincoll.edu/depts/ecopsyc/perils/.

Gibson, J. J. ([1975] 1987). Note on proprioception in relation to somaesthesis, self-awareness, and introspection. In J. Pittenger, E. S. Reed, M. Kim & L. Best (Eds.), *James Gibson's purple perils. A selection of James J. Gibson's unpublished essays on the psychology of perception*. http://www.trincoll.edu/depts/ecopsyc/perils/.

Gibson, J. J. ([1979] 1986). *The ecological approach to visual perception*. Hove: Psychology Press.

Glenberg, A., & Robertson, D. (2000). Symbol grounding and meaning: A comparison of high-dimensional and embodied theories of meaning. *Journal of Memory and Language, 43* (3), 379–401.

Goldie, P. (2000). *The emotions: A philosophical exploration*. Oxford: Clarendon Press.

Goldie, P. (2002). Emotions, feelings and intentionality. *Phenomenology and the Cognitive Sciences, 1*(3), 235–254.

Harnad, S. (1987). Category induction and representation. In S. Harnad (Ed.), *Categorical perception: The groundwork of cognition* (pp. 535–565). Cambridge: Cambridge University Press.

Harnad, S. (1990). The symbol grounding problem. *Physica D: Nonlinear Phenomena, 42,* 335–346.

Hodges, A. (1983). *Alan Turing. The enigma of intelligence*. London: Unwin.

Linguiti, G. L. (1980). *Macchine e pensiero. Da Wiener alla cibernetica*. Milan: Feltrinelli.

Lyons, W. E. (1986). *The desappearence of introspection*. Cambridge, MA: MIT Press.

Mandler, J. M. (1998). Representation. In D. Kuhn & R. S. Siegler (Eds.), *Handbook of child psychology. Volume 2: Cognition, perception, and language* (5th ed., pp. 255–308). New York: Wiley.

Martin, M. (1995). Bodily awareness: A sense of ownership. In J. L. Bermúdez, A. Marcel, & N. Eilan (Eds.), *The body and the self* (pp. 267–289). Cambridge, MA: MIT Press.

McCullogh, W., & Pitts, W. (1943). A logical calculus of the ideas immanent in nervous activity. *Bulletin of Mathematical Biophysics, 5*(4), 115–133.

Melzack, R., & Wall, P. D. (1996). *The challenge of pain*. London: Penguin.

Neisser, U. (1967). *Cognitive psychology*. New York, London: Psychology Press.

Newell, A., & Simon, H. A. (1972). *Human problem solving*. Englewood-Cliffs, NJ: Prentice-Hall.

Nisbett, R. E., & Wilson, T. D. (1977). Telling more than we can know: Verbal reports on mental processes. *Psychological Review, 84*(3), 231–259.

O'Shaugnessy, B. (1995). Proprioception and the body image. In J. Bermudez, B. Marcel, & N. Eilan (Eds.), *The body and the self* (pp. 175–203). Cambridge, MA: MIT Press.

Paivio, A. (1990). *Mental representations. A dual coding approach*. New York: Oxford University Press.

Pratt, V. (1987). *Thinking machines. The evolution of artificial intelligence*. Oxford, London: Basil Blackwell.

Pylyshyn, Z. W. (1973). What the mind's eye tells the mind's brain: A critique of mental imagery. *Psychological Bulletin, 80*(1), 1–25.

Pylyshyn, Z. W. (1980). Computation and cognition: Issues in the foundations of cognitive science. *Behavioral and Brain Sciences, 3*(1), 111–169.

Ratcliffe, M. (2005). The feeling of being. *Journal of Consciousness Studies, 12*(8–10), 45–63.

Ratcliffe, M. (2008). *Feelings of being: Phenomenology, psychiatry and the sense of reality*. Oxford: Oxford University Press.

Ratcliffe, M. (2009a). Belonging to the world through the feeling body. *Philosophy, Psychiatry and Psychology, 16*(2), 205–211.

Ratcliffe, M. (2009b). Existential feeling and psychopathology. *Philosophy, Psychiatry and Psychology, 16*(2), 179–194.

Ratcliffe, M. (2010). The phenomenology of mood and the meaning of life. In P. Goldie (Ed.), *Oxford handbook of philosophy of emotion* (pp. 349–371). Oxford: Oxford University Press.

Ratcliffe, M. (2012). The phenomenology of existential feeling. In J. Fingerhut & S. Marienberg (Eds.), *Feelings of being alive* (pp. 23–54). Berlin: De Gruyter.

Ratcliffe, M. (2015). *Experiences of depression. A study in phenomenology*. Oxford: Oxford University Press.

Ratcliffe, M., & Stephan, A. (Eds.). (2014). *Depression, emotion and the self: Philosophical and interdisciplinary perspectives*. Exeter: Imprint Academic.

Robinson, J. D. (2001). *Mechanisms of synaptic transmission: Bridging the gap (1890–1990)*. Oxford: Oxford University Press.

Rodríguez, D., Hermosillo, J., & Bruno, L. (2012). Meaning in artificial agents: The symbol grounding problem revisited. *Minds and Machines, 22*(1), 25–34.

Rosenblueth, A., Wiener, N., & Bigelow, J. (1943). Behavior, teleology and purpose. *Philosophy of Science, 10*(1), 18–24.

Schwizgebel, E. (2012). Introspection, what? In D. Smithies & D. Stoljar (Eds.), *Introspection and consciousness* (pp. 29–47). Oxford: Oxford University Press.

Searle, J. (1980). Minds, brains and programs. *Behavioral and Brain Sciences, 3*(3), 417–457.

Shannon, C. (1948). A mathematical theory of communication. *Bell System Technical Journal, 27* (379–423), 623–656.

Shapiro, L. (2011). *Embodied cognition*. New York: Routledge.

Sheets-Johnstone, M. (2009). Animation: The fundamental, essential, and properly descriptive concept. *Continental Philosophy Review, 42*(3), 375–400.

Sheets-Johnstone, M. (2010). Body and movement: Basic dynamic principles. In S. Gallagher & D. Schmicking (Eds.), *Handbook of phenomenology and cognitive science* (pp. 217–234). Berlin: Springer.

Shepard, R. N., & Cooper, L. A. (1982). *Mental images and their transformations*. Cambridge, MA: MIT Press.

Sherrington, C. S. (1900). Cutaneous sensation. In E. A. Schäfer (Ed.), *Textbook of physiology* (Vol. 2, pp. 920–1001). London, Edinburgh: Young J. Pentland.

Sherrington, C. S. (1906). *The integrative action of nervous system*. Yale: Yale University Press.

Sherrington, C. S. (1907). On the proprio-ceptive system, especially in its reflex aspects. *Brain, 29* (4), 467–485.

Sherrington, C. S. (1941). *Man on his nature*. New York: MacMillan.

Shusterman, R. (2008). *Body consciousness. A philosophy of mindfulness and somaesthetics*. New York: Cambridge University Press.

Shusterman, R. (2012). *Thinking through the body: Essays in somaesthetics*. Cambridge: Cambridge University Press.

Tallis, R. (2004). *Why the mind is not a computer*. Exeter: Imprint Academic.

Thompson, E. (1997). Symbol grounding: A bridge from artificial life to artificial intelligence. *Brain and Cognition, 34*(1), 48–71.

Tsakiris, M., Schütz-Bosbach, S., & Gallagher, S. (2007). On agency and body-ownership: Phenomenological and neurcognitive reflections. *Consciousness and Cognition, 16*(3), 645–660.

Turing, A. M. (1936). On computable numbers, with an application to the Entscheidungsproblem. *Proceedings of the London Mathematical Society, 42*, 230–265.

Turing, A. M. (1950). Computing machinery and intelligence. *Mind, 59*(236), 433–460.
van Steenberghen, F. (1946). *Ontologie*. Louvain: Publication Universitaires de Louvain.
Varela, F. J., Thomson, E., & Rosch, E. (1991). *The embodied mind: Cognitive science and human experience*. Cambridge, MA: MIT Press.
von Borman, C., Franzen, W., Krapiec, A., & Oeing-Hanhoff, L. (1972). Form und Materie (Stoff). In J. Ritter (Ed.), *Historisches Wörterbuch der Philosophie* (Vol. 2, pp. 978–1030). Basel: Schwabe.
Wallace, B., Ross, A., Davies, J., & Anderson, T. (2007). *The mind, the body, and the world. Psychology after cognitivism?* Exeter: Imprint Academic.
Wiener, N. ([1948] 1961). *Cybernetics, or control and communication in the animal and the machine*. Cambridge, MA: MIT Press.
Wilson, R. A. & Foglia, L. (2011). Embodied cognition. In E. N. Zalta (Ed.) *The Stanford Encyclopedia of Philosophy*. http://plato.stanford.edu/archives/fall2011/entries/embodied-cognition/.

Chapter 3
Semantic Competence from the Inside: Conceptual Architecture and Composition

Abstract Concepts are essential constituents of thought: they are the instruments we use to categorize our experience, i.e. to classify things and group them together in homogeneous sets. Here we define concepts as the internal mental information (representations) that allows us, among other things, to master words in natural language. By analyzing the way in which individuals master word meanings we explore a number of hypotheses regarding the nature of concepts. Following Diego Marconi's research, we differentiate between two kind of abilities that underpin lexical competence—so-called 'referential' and 'inferential competence'—and we suggest that, in order to support these abilities, concepts must also include two corresponding kinds of information, i.e. inferential and referential information. We point out that the most widely used and acknowledged theories of concepts do not make this distinction, instead broadly characterizing the information used for categorization in terms of propositionally described feature lists. However, we show that while feature lists can explain inferential competence, they do not account for referential competence. To address the issue of referential competence we examine Ray Jackendoff's hypothesis that to account for the possibility of linking perceptual and conceptual information we need to assume the existence of a (visual) representation that encodes the geometric and topological properties of objects and bridges the gap between the percept and the concept. Furthermore, we analyze the extension of this work by Jesse Prinz who introduced the notion of a *proxytype*, a perceptual representation of a class of objects that incorporates structural and parametric information related to their appearance. However, as we point out, proxytypes can only explain the relationship between perception and concepts with respect to instances that can be perceived through the senses and that belong to the same class by virtue of their physical similarity. We suggest that this notion be extended to include larger conceptual classes. To accomplish this, we further develop Mark Johnson, George Lakoff and Jean Mandler's idea of a *schematic image* and argue that conceptual representations include a *perceptual schema*. Perceptual schemata are non-linguistic, structured *experiential gestalts* (patterns or maps) that make use of information taken from all sensory modalities, including body perception. They accomplish a quasi-conceptual function: they allow us to recognize and to classify different instances. In this work, we hypothesize that

S. Dellantonio and L. Pastore, *Internal Perception*, Studies in Applied Philosophy, Epistemology and Rational Ethics 40, DOI 10.1007/978-3-662-55763-1_3

perceptual schemata are an essential component of concepts, but not identical to them. Instead, we suggest that concepts include both perceptual and propositional information with perceptual schemata providing the 'perceptual core concept' that grounds related propositional information.

1 Concepts and Meanings: Why They Are Not One and the Same Thing

Concepts are essential constituents of thought: they are the instruments we use to categorize our experience, i.e. to classify things and group them together in homogeneous sets. Categorization is necessary for memorization, since it would be impossible to remember all the single experiences we have had and all the single objects we have experienced. Furthermore, categorization is indispensable for the acquisition and processing of knowledge, since it allows us to attribute common properties to objects belonging to the same class and to assume they obey the same laws. If one knows for example that an object belongs to a certain class, one can predict that this object will behave more or less like the other objects that belong to the same class and will have more or less the same properties they have (if we know e.g. that a certain object is an apple, even though have never seen this particular apple before, we will know—among other things—that it is edible, sweet, that it has no hidden thorns, etc.). If one observes a noteworthy diversity in the behavior of the new object or its properties in comparison with the behavior or with the properties of other objects belonging to the same class, this will affect either how s/he categorizes the object or his/her knowledge of the class itself, and will lead to a revision or to a further specification of this class.

While the importance of concepts with respect to cognitive functions is widely and unanimously recognized, there is still strong disagreement regarding how concepts are structured and what information they consist of (see e.g. Dellantonio and Pastore 2006, 2009). Not only is there no shared theory on the origin of concepts and the kind of information they are composed of, but different research traditions and different studies also rely on different definitions of the very notion of *concept*. This section aims to clarify the notion of concept used in this work and how this notion itself relies on the notion of *meaning*, which is just as controversial and often confused or even identified with that of concept.

The most widely shared position on concepts in the recent cognitive tradition can be summarized in Susan Carey's words as follows: "Concepts are units of thought, the constituents of beliefs and theories, and those concepts that interest me here are roughly the grain of single lexical items. Indeed, the representations of word meanings are paradigm examples of concepts. I take concepts to be mental representations—indeed, just a subset of the entire stock of a person's mental representations" (Carey 2009: 5). In the field of cognitive science concepts are mainly understood as subjective mental representations, and more specifically as a particular subclass of mental representations which is "subpropositional" (Laurence and Margolis 1999) and "potentially accessible to conscious thought" (Mandler 2004: 4) and which match (somehow) linguistic meanings. A large number of studies investigate concepts using single words in the language: they start from the assumption that there is a direct correspondence between the way a speaker applies a word (e.g. 'dog') and the way in which his/her concept DOG leads him/her to classify dogs.[1] In such cases, we speak of 'lexicalized concepts', leaving open the possibility that the class of concepts is wider than the class of words and also includes non-lexicalized concepts. Complex natural language expressions like 'black cat' are mainly considered to correspond to composite concepts like BLACK + CAT, even though it is also possible that there are conceptual units larger than words that can be described using complex linguistic expressions. The fact that the literature usually distinguishes between composed and component concepts does not imply that component concepts (i.e. the conceptual units) must be simple. On the contrary, with the exception of the research line called 'conceptual atomism', led by Fodor (1987, 1994, 1998), conceptual units are also usually considered to be complex elements, which are composed of different kinds of information (Laurence and Margolis 1999).

The link between the words of a natural language and concepts is not only due to the fact that concepts are mainly studied using words. In fact, studying concepts on the basis of words is usually considered possible and appropriate because concepts are thought to be those mental representations that make word mastery possible. If a person didn't have the concept of CAT that allows him/her to group together in the same class all objects that are cats and therefore to distinguish cats from those objects that are not cats, and to assign to them the specific properties that s/he has learned belong to cats, s/he wouldn't be able to master the word 'cat', as a linguistic label which (in English) indicates precisely the class of cats.[2]

However, the idea—widely accepted by cognitive scientists—that concepts support the semantic mastery of words, tends to induce the problematic and partially misleading belief that word meanings depend on and originate from concepts and that therefore, concepts and word meanings are one and the same thing. As, for example, Laurence and Margolis state: "[…] it's common to think that words in natural languages inherit their meanings from the concepts they are used to express" (Laurence and Margolis 1999: 4). To come to a definition and to a differentiation of the notions of 'concept' and of 'meaning' it is useful, first of all, to recall the position of Hilary Putnam who—in his famous article *The meaning of 'meaning'*—distinguishes between two different approaches to the issue of meaning:

(A_1) a sociolinguistic approach that addresses the question of how meanings are built up over time by a social and linguistic group (Putnam 1975: 227ff);

---

[1]On this point see e.g. the classical studies that systematize and summarize the recent debate on concepts: Smith and Medin (1981) and Margolis and Laurence (1999).

[2]In this work, concepts (and categories, see Chap. 4) are indicated with small capitals; for word meanings we use inverted commas.

(A_2) an 'individualistic' and 'psychologistic' approach that focuses on the semantic competence of individuals and tries to explain how this competence develops (Putnam 1975: 245ff).

(A_1) A sociolinguistic approach to language that investigates the criteria used by a linguistic community to determine e.g. that a certain instance doesn't belong to the conceptual class of artifacts, but is classified as a plant and more specifically as an elm; or that a certain animal is called a 'tiger' instead of a 'lion'; or that 'water' denotes water; or that the word 'arthritis' denotes only inflammation of the joints, etc. An approach of this kind does *not* take into account the semantic competence of individual speakers who often use words in an inappropriate manner, without having a complete mastery of them.[3] It only considers the procedures used by a linguistic community to develop—thanks to its experts—application criteria for the word that fixes the correct use of them according to the community.[4]

(A_2) An individualistic and psychologistic approach to meaning is not interested in the processes by which the rules governing the application (the use) of words are set and evolve over time, but rather addresses the way in which individuals come to master the words they use. While a sociolinguistic explanation of the rules governing the application e.g. of the word 'fruit' can resort to botany and clarify that the word applies to the products of the modifications of the ovary following pollination, an individualistic and psychologistic explanation of meaning focuses on the kind of information that an individual learns and uses to apply the words s/he masters and e.g. to distinguish the instances which are fruits from those which are not. The information a subject uses to classify something as a 'fruit' can diverge also radically from the conditions governing the application of the word as they are identified from a sociolinguistic point of view. The potential discrepancy between the conditions for application of a word abstractly considered and the information individuals use to classify things results, among other things, in the possibility that a person may master a word only partially or mistakenly. In this sense, meaning as it is understood in the sociolinguistic approach accomplishes a normative function with respect to the individual use of words.

Putnam's differentiation between the sociolinguistic and the individualistic-psychologistic approaches to meaning was used as a starting point for clarifying the notion of 'concept' by Ray Jackendoff, who posited a distinction

[3]Among the most discussed examples of partial semantic competence there is the case of the speaker who confuses aluminum with molybdenum (Putnam 1975: 205ff and 230ff) and elm trees with beech trees (Putnam 1975: 205ff and 230ff, discussed again in Fodor 1994: 33ff) or the case of the person who thought that arthritis might afflict also the thigh, in spite of the fact that arthritis is defined as a disease which afflicts joints only (Burge 1979: 79ff). For an overview on this debate see Dellantonio (2007).

[4]This way of presenting Putnam's thesis takes into account only the social component of his positions (which is centered on the idea of the division of linguistic labor), while it entirely disregards the indexical (essentialist) element that goes with it in Putnam's view.

between, (B_1) *externalized concepts* (*E-concepts*) and (B_2) *internalized concepts*, (*I-concepts*) (Jackendoff 1992: 22ff).[5] As Jackendoff maintains:

> There is a fundamental tension in the ordinary language term *concept*. On one hand, it is something out there in the world: 'the Newtonian concept of mass' is something that is spoken of as though it exists independently of who actually knows or grasps it. Likewise, 'grasping a concept' evokes comparison to grasping a physical object, except that one somehow does it with one's mind instead of one's hand. On the other hand, a concept is spoken of as an entity within one's head, a private entity, a product of the imagination that can be conveyed to others only by means of language, gesture, drawing, or some other imperfect means of communication (Jackendoff 1992: 22).

(B_1) In expressions such as 'the Newtonian concept of mass', the word 'concept' is used in an externalized sense; i.e. 'concept' is meant in the sense of E-concept, since it refers to concepts as objective, socially defined entities, whose application criteria has been fixed by experts, and which therefore exist independently from being thought about by any particular individual or from the way in which they are understood by individuals. (B_2) Nevertheless, there is also another use of the notion of concept, which is described through the notion of I-concept. This qualifies concepts as "mental resources that make possible human knowledge and experience of the world" (Jackendoff 1992: 22) or, in other words, as internal information individuals need to master a language.

If one overlooks the details of the specific explanations that Putnam and Jackendoff develop respectively for meaning in a *sociolinguistic sense* and meaning in an individualistic-psychologistic sense (Putnam) and for *E-concepts* and *I-concepts* (Jackendoff), one can acknowledge that the distinction introduced by these two authors expresses basically the same idea, even though the first focuses on the notion of meaning, while the second focuses on the notion of concept. Both concepts and meanings might be understood, on the one hand, as superindividual (i.e. abstract and intersubjective) phenomena and, on the other hand, as individual, mental entities that explain how humans can come to learn a language from a semantic point of view. Exactly because of this basic ambiguity, it is useful, as a first step towards terminological clarification, to disentangle the double pair of synonyms (A_1–B_1; A_2–B_2): we will use the word 'meaning' solely to indicate the abstract (sociolinguistic) dimension of semantics, i.e. Jackendoff's *E-concepts*; on the other hand, we will use the word 'concept' to refer to what has been described as an internal mental entity thanks to which individuals can develop semantic competence.[6]

This differentiation between 'meanings' and 'concepts' leads to the definition of two kinds of entities which are the basis for semantic competence and the semantics of natural languages, respectively: one has an internal and representational character; the other has an external and social character. This conclusion is in agreement with a thesis in the philosophical literature sometimes called "dualism of content" (Kripke 1979; Loar 1985) or "two-factor theories of semantics" (e.g., Block 1986),

[5]Jackendoff's distinction between *I-concepts* and *E-concepts* is built on Chomsky's differentiation between *E-language* and *I-language*: see e.g. Chomsky (1986: 19–24).

[6]Putnam would probably not agree with an explanation of semantic competence in terms of internal mental entities. For a discussion see e.g. Dellantonio (2007: 62–79).

according to which semantics must be approached starting from two different perspectives: the one internal and the other external to the subject. Indeed, on the one hand, we need to account for meaning in its social and shared dimension; on the other, we need to address the issue of concepts as internal mental representations, which our individual competence with meaning is based on. Given this clarification, it is also possible to provide a first definition for the specific scope of this investigation. We are interested in concepts in a specific, cognitive sense that approaches Jackendoff's *I-concepts*. This work investigates concepts as internal mental resources (as internal mental information) used, among other things, for learning and mastering a natural language.

2 Concepts and Meanings: Mutual Relationships

Even though concepts and linguistic meanings are not the same thing, they are connected by a deep affinity as far as their function is concerned. As e.g. Barsalou points out: "The semantics of natural language are closely related to the human conceptual system. Although lexical meanings are not identical to concepts, the two have much in common and influence each other extensively" (Barsalou 2012: 239–240). Both concepts (as mental entities) and meanings (as social entities) individuate classes of things that constitute their referent. The word 'cat' identifies the class of objects that, according to English language and according to the zoological classification underlying it, are cats; analogously, the concept of CAT individuates the class of objects that an individual recognizes as corresponding to the word cat (in his/her language). However, this link between concepts and meanings does not imply any strong correspondence between them.

First of all, it is possible or even plausible that conceptualization starts before and independently of language learning. In fact, even in the hypothetical case in which an individual never had the possibility to acquire a socially codified language, it would be plausible to assume that s/he would still be able to carry out categorizations of the concrete objects around him which are useful for him and that s/he would be able e.g. to distinguish plants from animals and from inorganic material or edible from inedible things or dangerous from harmless animals, etc.

Secondly, even when an individual develops a concept in relation to a natural language in the attempt to come to master the words of that language, one cannot take for granted that there will be a one-to-one correspondence between concepts and meanings. In many cases, indeed, individuals learn the meanings of their languages only partially or mistakenly. It is possible, for example, that a person might not make a distinction between elms and beeches, yet nevertheless have a vaguer concept which allows him/her to distinguish these two kinds of tree taken together from other kinds of trees which have a different overall aspect (for example, pines). Furthermore, we might suppose that this person associates this broader concept with the English word 'elm'. In this case, however, his/her concept will be deviant with respect to the word meaning as it is defined by the English language, since his/her concept includes both elms and beeches. We could say that this person links the word 'elm'—or, more precisely the letter sequence 'e-l-m'—to a concept which could be more precisely described as 'broad-leaved, tree with a tall trunk', which plausibly picks out some macroscopic common exterior features of elms and beeches (i.e. the overall shape, the trunk, the leaves etc.). Similarly, it is possible for a person to identify a certain pain in the thigh as 'arthritis', even though the socio-linguistic definition of the word states that arthritis is an inflammation of the joints only (which therefore cannot affect the thigh).[7] In this case, too, the person will have mastered the word in a deviant manner since s/he will link the sequence of letters 'a-r-t-h-r-i-t-i-s' to a concept that does not pick out the same reference (the same extension) as the word. His erroneous use of the word reveals however something important for the investigation of the concept s/he associates with the word 'arthritis', i.e. that the conceptual identification of the reference is carried out primarily by relying on a certain kind of pain, rather than on the basis of the information 'inflammation of the joints only'.

In a perspective like the one we pursue in this work which aims to account for semantic competence on the basis of two different notions—the one individual and internal to the subject: the concept; the other social and external to the subject: the word meaning—cases like these can be explained by linking individual semantic competence to the concepts, considered as Jackendoff's *I-concepts*. In this case, individual competence with respect to the meaning of a word depends on the *I-concept* the subject associates with that word and depends on the degree of matching between the individual concept and the social meaning of the word.

This suggests that—even though the concepts 'chase' the meanings of the words, in the sense that speakers tend to develop concepts which pick out the same referents identified by words, in order to correctly use the word meanings as they have been defined by their linguistic community—concepts cannot be studied as if they were identical to meanings. Concepts can indeed identify the instances that belong to them on the basis of vague and 'deviant' representations which have nothing in common with the precise and cogent definitions carried out by linguistic experts and codified in dictionaries which provide rules for the social use of words in a natural language. For this reason, the study of concepts cannot coincide with the study of meanings and the criteria governing the use of words. Concepts must rather be investigated in their specific nature: in particular, we need to account for the fact that concepts often produce classifications of instances that are partially independent from language, even though they still assure (a certain level of) linguistic mastery. At the same time, since concepts are the means through which we acquire and master our language, they must, at least to some extent, match the linguistic meanings of the language which is mastered: i.e. the concepts that people develop must produce classifications that are largely analogous to those underlying

---

[7]The example of the confusion between elm and beech trees was originally introduced by Putnam (1975: 205ff and 230ff), and then discussed again by Fodor (1994: 33ff). The case of the person who thought that arthritis might afflict also the thigh was introduced by Burge (1979: 79ff). For an overview of this debate see Dellantonio (2007).

the linguistic meanings they are exposed. Therefore, even though concepts are relatively independent from linguistic meanings, because they need to match these meanings to support language mastery, they are also influenced by meanings and by the classifications language produces.

The idea that language (specifically meanings) influences the structure of thought (i.e. the concepts people develop) was widely discussed in psychological and philosophical research during the first half of the 20th century. This was one of the central theses of linguistic determinism in philosophy (Sapir [1921] 2004; Whorf [1956] 2012), but also in other approaches such as behaviorism in psychology which maintained that thought is not structured 'from the inside'—on the basis of some mental structure, but inherits its structure from the outside, and specifically from language.[8] When these views fell into crisis and cognitivism became established as the new major line of research in psychology, the problem of the influence of language on thought was relegated to the background by researchers.[9]

The cognitivist tradition contributed in a decisive manner to the slump of (both philosophical and psychological) behaviorism, reaffirming the principle that individual minds actively structure thoughts (in developing mental contents). This general position also has many consequences for the role of language with respect to thought: language is no longer conceived as the means to structure mental contents. On the contrary, since language learning is viewed as a process of "hypothesis formation and confirmation" (Fodor 1975: 35) regarding the classifications underlying lexical meanings, to learn a language, human beings must have pre-linguistic capacities to classify the instances they perceive (Fig. 1).

> What the organism has to do in order to perform successfully is to extrapolate a generalization […] on the basis of some instances that conform to the generalization […]. The game is, in short, inductive extrapolation, and inductive extrapolation presupposes (a) a source of inductive hypotheses […] and (b) a confirmation metric […]. There are, of course, many, many ways of fleshing out the details of this kind of model. […] But, however the details go, what seems entirely clear is that the behavior of the organism will depend upon the confirmation relation between the data and the hypothesis, so that accounts of its behavior will require information about how, in the course of learning, the data and the hypotheses are represented (Fodor 1975: 37).[10]

This capacity to classify instances in homogeneous classes is hypothesized to rely on a representational structure which allows us to group things together according to some criterion. This representation allows us to identify in the external world all and only those objects which fulfill the criteria underlying these classifications. To understand the meaning of a word one must have understood the classification

[8]For an overview of linguistic determinism and its relation to behaviorism see. e.g. Pinker (2007: 44–73).

[9]On this see also Carruthers and Boucher (1998: 2).

[10]The problem of how the learner narrows down the number of possible hypotheses—which is potentially limitless—is not considered in this work. For a brief overview see, e.g., Tomasello (2001).

Fig. 1 Language learning through hypothesis formation and confirmation

criteria underlying that word: i.e. s/he must have developed a representation that allows him/her to pick out all and only those objects the word refers to. In this sense, it cannot be language which bestows structure on thoughts, but thoughts must be already structured before and independently from language in order to allow language acquisition. However, the fact that thought must have some structure before language acquisition does not imply—at least in principle—that language learning cannot influence, at least in part, the structures of thought previously developed. In fact, language might have a retroactive power on thought and lead to the formation of specific concepts.

After being overlooked for a long while, in recent times the problem of the influence of language on thought is again beginning to be discussed in cognitive and philosophical research and many authors belonging to the cognitive tradition have pointed out different ways in which linguistic meanings contribute to shaping our concepts.[11] Firstly and more trivially, language allows us to communicate our thoughts and therefore to transmit them across history giving rise to a collective and cumulative knowledge that could never be reached by a single creature during his/her lifetime.[12] This means that language allows us to acquire new beliefs about

[11]A wide range hypotheses have been put forward in the literature as to whether language influences thought, with some researchers positing no influence whatsoever (see e.g. Fodor 1994, 1998) and others suggesting radical influence. As is well-known, the latter perspective is represented by the so-called Sapir-Whorf Hypothesis (SWH) according to which the semantic or syntactic structure of language determines the speakers' way of categorizing the world (Sapir [1921] 2004; Whorf [1956] 2012). Views on how exactly this influence or determination might occur are also far from being unanimous. For a short overview, see Baghramian and Carter (2016). We are not interested in taking a position in this debate. Here we just assume a minimum and hardly deniable level of influence of language on categorization due to the fact that language acquisition leads us to classify things on the basis of the principles that rules the use of words in our language.

[12]See e.g. Jackendoff (1996) and Tomasello (1999), especially pp. 55–56.

how things are. Secondly and less trivially, the language one learns influences the kinds of categorization one makes, at least in the sense that words "serve as a catalyst in object categorization from infancy" and function as "invitations" to categorize things in specific ways (Waxman 1999: 269; Anggoro et al. 2008).[13] If a language does not have a certain word, the child is not spurred to categorize the objects according to the application rules of the word, i.e. both to form a concept that maps to the word and to fix (remember) this same categorization for future uses. Thirdly, words drive people to carry out a very wide and differentiated range of categorizations, which are far more precise than the ones that a creature without a language could perform (Pinker and Jackendoff 2005: 212).[14] Finally, when humans acquire words, they don't only learn to categorize objects in a way that corresponds to the application rules of the word (i.e. to the word meaning), they also learn certain relationships among words (which words are e.g. subordinate, superordinate, antonyms, meronyms, etc. with respect to other words), and therefore they learn certain relationships among the objects denoted by the words, implicit in the semantic structure of the language they acquire (Clark 1993; Deacon 1997; Miller 1991; Miller and Fellbaum 1991). Most of the word meanings we acquire have been developed across time by the various linguistic communities as a result of their understanding of the relationships that persist between things. Therefore, by learning a language, we also learn the specific relationships between the things implicit in the meaning structure of that language. This further explains how language helps us make precise and differentiated categorizations.

These observations should illustrate and clarify in what sense concepts 'chase' meanings: linguistic learning leads the structure of thought to approximate that of language. A person will e.g. develop the concept of PINE because his language contains the word 'pine' and classifies trees accordingly. If learning is successful, the concept of pine this person develops will further allow him/her to identify as 'pines' all and only those instances that his language classifies as such. For this reason, even though concepts do not correspond to meanings, their study cannot disregard that of meanings. An analysis of concepts must rather investigate them in relation to meanings, and come to distinguish which components of concepts are taken from language and which, on the contrary, depend on the way in which the cognitive system structures information in order to carry out classifications prior to and independently from language.

[13]On the same topic, also see e.g. Waxman and Lidz (2006), Waxman and Markow (1995), Goss (1961) and Weiskopf (2008).

[14]For a more radical thesis see also Bermúdez (2003: 150).

3 Lexical Semantic Competence Fully Loaded

If concepts are defined as that class of mental representations that support our comprehension of words (or, more generally, linguistic expressions) than language can be defined as something like a window on the conceptual system, through which we can observe the composition of concepts. Since conceptual structure and conceptual composition must be able to support lexical semantic competence, an analysis of the way in which individuals master word meanings can provide indications which help to specify a number of hypotheses regarding the nature of concepts.

When can we say that an individual has full semantic competence with respect to a word? What is the kind of information an individual must have available and must be able to acquire and/or reach in order to develop full lexical semantic competence?

Starting already in the 1980s, philosophical and psychological research has tried to answer this question by distinguishing between two kinds of competence: the first primarily concerns the capacity to identify the things that words denote; the other is mainly related to the ability to link words to other words (i.e. define them through their interactions).[15] This double dimension of semantic competence has been systematized in a particularly effective way by Diego Marconi in his book *Lexical Competence* (1997).

Marconi's discussion starts with an analysis of the semantic competence exhibited by artificial systems. In particular, his intent is largely analogous to that pursued by Searle through his famous example of the Chinese Room: he is interested in defining the peculiarity of *human* semantic competence as opposed to the performance of artificial systems (Marconi 1987, 1991; Searle 1980). Marconi notes how—through the manipulation of symbols—artificial systems can easily simulate the human capacity to link words with others and to infer specific informational content of propositional kinds from others (e.g. if something is a dog, then it is a mammal; if something is a dog, then it is a domestic animal; if something is a mammal, then it nurses its pups and doesn't lay eggs; if something is a domestic animal, then it is not ferocious, etc.). However, artificial systems cannot simulate the capacity humans have to recognize the instances in the external world that correspond to words (for a more detailed discussion see Chap. 2, Sect. 1) (Fig. 2).

To realize a capacity of this kind the artificial system would have to be connected with the external world at least via a pattern recognition device that allows it to identify objects, situations, events, etc. Because artificial systems are reasonably successful at simulating the one way in which people master words, but not the other, Marconi argues that these two ways should be dissociated and considered as the expression of two different and distinct human capacities which he calls

[15]Among the first systematic works on these differences are, e.g. Moravcsik (1981), Johnson-Laird (1983) and Harman (1987).

Fig. 2 Recognition of instances denoted by a word

(1) *inferential competence* and (2) *referential competence*. A subject will have a full mastery a word, when s/he exhibits both inferential and referential competence.

(1) First of all, to grant that a person masters a certain word, s/he needs to show that s/he is able to connect this word with others in his/her language. A person will be considered competent with respect e.g. to the word 'cat', if s/he is able to say that cats are animals, four-footed, that they are pets, carnivorous, that they have a tail, hair, and so on. This competence is called 'inferential' because it supports our capacity to work out inferential links such as: if x is a cat, then x is an animal; if x is a cat, then x is a carnivorous; if x is an animal, then x is a living thing, etc. This capacity is based mainly on linguistic and conceptual information and also relies on knowledge:

- of properties that cannot be extracted from the object through the senses (from the perception of a cat we can extrapolate the information that cats have four feet, hair and a tail, but not e.g. that they are feline);
- of hierarchical relationships among classes of objects (e.g. with respect to the class of cats, Siamese is a subordinate, while animal is a superordinate class);
- of the relationships existing between the objects of this class and those belonging to other classes (e.g. cats fear dogs);
- of the function that the objects of a certain class can accomplish (cats can be used to hunt mice), etc. About this capacity Marconi specifies: "it is the ability to manage a network of connections among words, underlying such performances as semantic inference, paraphrase, definition, retrieval of a word from its definition, finding a synonym, and so forth" (Marconi 1997: 59).

The idea that inferential capabilities are essential for the mastery of language is widely recognized not only by philosophers, but also by cognitive scientists. As pointed out e.g. by Barsalou in his recent presentation of the main views proposed by contemporary cognitive science on the human conceptual system:

> Once an individual has been assigned correctly to a category, a multitude of useful inferences follow from associated conceptual knowledge that go beyond what has been perceived thus far for the individual. Imagine perceiving and categorizing an unfamiliar individual as a *cat*. Useful inferences about the individual's structure, behavior, and internal states include that the cat has teeth and claws, that it can purr and scratch, and that it could be hungry and grateful. Useful inferences about relevant actions that the perceiver could perform follow as well, such as being gentle to the cat, petting it, and touching it. Many other potentially useful inferences also follow, including that the cat had a mother and father (potentially relevant for breeding purposes) and so on. Once integrated conceptual knowledge about *cat* becomes active during categorization, a variety of associated inferences follow (Barsalou 2012: 241–242).

(2) Even though most of the studies that address linguistic mastery from a cognitive point of view consider at least implicitly almost exclusively this kind of competence,[16] as Marconi shows semantic competence includes also another and at least as important competence which consists in the capacity to "select the right object or circumstance in response to a given word" (Marconi 1997: 2). This is the capacity that allows people to identify which objects in the world correspond to a certain word. It is on the basis of this competence that we can e.g. identify 'that thing' sleeping on the mat right now as a cat and answer questions like: 'Is there something on the mat right now?' or 'Is there a cat in this room?' Referential competence "underlies such performances as naming, answering questions concerning the obtaining situation, obeying orders such as 'Close the door!', following directions, etc. These performances are partially based on the ability to *recognize* objects and actions" (Marconi 1997: 64).

Inferential and referential competence are partially interconnected. In fact, inferential competence can support at least in part our referential competence, while our referential competence can support at least in part our inferential competence. If a person has well-developed inferential competence with respect to 'cat', s/he will probably be able to recognize a cat even though s/he has never seen one before. Indeed, s/he could identify a cat using the inferential properties s/he knows belong to cats. On the other hand, if a person has referential competence with respect to 'cat', s/he will also have at least a minimal inferential competence regarding this term: indeed, (assuming s/he does master a language) s/he will be able to report at least some properties of cats that are related to the way they look, e.g. that they have four feet, they have hair, they have a tail, etc.[17]

---

[16]This is true of all the different computational models of word meanings that have been developed in contemporary cognitive research: see e.g. Lappin (2005).

[17]The link between referential and inferential information presupposes that the person has a language. We will discuss this more precisely below.

Even though these two competencies are interconnected, they are nevertheless also functionally distinct. As a matter of fact, while inferential competence is intrinsically linguistic, because it describes the capacity to connect words with other words, referential competence goes beyond the dimension of language and, as Marconi also points out, "is not a purely linguistic ability; under a certain description it can even be regarded as *non*linguistic" (Marconi 1997: 64). This competence can also be conceived of in terms of the capacity (which can also be attributed to creatures that do not have a language) to classify things together as members of the same class on the basis of certain similarities and to recognize that a newly perceived instance belongs to one class rather than another. In fact, it is highly plausible that animals or a hypothetical Robinson Crusoe, who has grown up alone on an island, would be able to group things together according to general classes that are relevant for their survival and that they would be able to distinguish e.g. cats from snakes and from trees (even though they might categorize these things according to very rough classes).

Moreover, this nonlinguistic capacity can be considered the starting point for semantic competence, at least in its referential aspect, which consists in the capacity to identify specific instances as members of particular classes and to connect each class with the right name (word): i.e. for example to identify an instance as a member of a specific class (e.g. the class of cats) and to connect this class with the 'right name' ('right' in terms of the taxonomies developed by the various natural languages): e.g. with the English word 'cat', with the Italian word 'gatto', or with the German name 'Katze', depending on which language we are considering. This does not mean that *linguistic* referential competence is not qualitatively different from *nonlinguistic* referential competence. In fact, as we already pointed out, language learning modifies the way things are grouped together making it far more specific, accurate and differentiated. The point is more that referential competence shifts the attention from language itself to different cognitive capacities that are necessary to develop a language and that concern, among other things, perception, object recognition and categorization.

Previously we described concepts as those mental representations that support our semantic competence. On the basis of this analysis we described what semantic competence is like. We can now state that, to perform this function, concepts must account for both referential and inferential competence and must therefore be made of or be linked to information of the referential and the inferential kind. To simplify the issue, we can speak of 'referential' and of 'inferential information'.

Inferential information is of a linguistic kind and depends on the specific, explicit and propositional knowledge everyone has about objects; thus, it consists of linguistically conceived properties that a person associates with the things s/he knows. In this sense, to be able to support semantic competence the inferential dimension of a concept (i.e. a conceptual representation) must (also) include inferential information of a propositional kind. If e.g. a person is able to say that cats are mammals, carnivorous and feline, then his/her concept of CAT must include the information MAMMAL, CARNIVOROUS and FELINE. This leaves the door open both for the idea that this information is part of the concept and that the conceptual unities are connected to each other through inferential links.

On the other hand, *referential information* is of a perceptual origin and must allow one to group things together also in a nonlinguistic fashion: it must allow us e.g. to identify all and only the cats, differentiating them from other kinds of instances. Referential information can also be described linguistically through properties like 'is small', 'has a long tail', 'has hair', but to explicitly know what the concrete features of cats are, it is probably neither necessary nor sufficient to recognize cats (about this, see the critiques of feature lists in the next section). The capacity to recognize cats and plausibly also to describe their specific features must rely on perceptual information about cats of a rather complex kind. In the next sections in this chapter and—in some respects—throughout this work we will focus on this kind of information and discuss its nature and function with respect to different kinds of concepts.

4 "A Pile of Bird Features Does not Make a Bird"

Subjective concepts (subjective conceptual representations) support both the referential and inferential components of semantic competence. As we observed previously, these two competences interact with each other. Sometimes inferential competence can even function as a means for the formation of referential competence: think e.g. of a person who sees the mushrooms called *amanita muscaria* for the very first time in his/her life; even though s/he never experienced these mushrooms directly before and therefore never developed referential competence allowing him/her to recognize them, s/he might be able to identify them correctly anyway, if s/he has inferential competence with respect to this word that also includes information about the characteristics of *amanita muscaria*—like [*amantita muscaria*: mushroom; red cup with white spots; white stem]. However, despite this, referential competence is more basic than inferential competence since it constitutes the contact point between language (concepts) and the world (referents). Referential competence describes the cognitive ability of the human mind to recognize and discriminate objects (events, actions, etc.) in the world and it is therefore intrinsically tied to perception; and mainly—as Marconi (1991: 19, 1995: 147, 1997: 74) suggests—to vision, since for average subjects vision constitutes the most important perceptual channel for the recognition of referents in the external world.

Even though there is not a univocal and unanimous explanation of the relationship between semantics and perception (vision), many studies carried out in the field of cognitive psychology and neuropsychology show that the processes of object recognition, categorization and naming are structured in extremely articulated and complex ways (for an overview, see e.g. Ullman 2000). Without entering into detail on the various theories, a summary description of referential competence understood as the capacity to recognize the objects denoted by words suggests that for it to be realized at least the following conditions are required:

(i) the cognitive system must be able to compare incoming perceptual information with conceptual representations, which have been previously developed and stored in the semantic system (in semantic memory), in order to establish whether what we perceive presently falls under a certain concept stored in memory; and
(ii) conceptual representations stored in memory must be linked to sequences of letters, i.e. to words: more specifically to corresponding 'name representations'. If this link is missing, the verbal dimension of referential competence would get lost. In fact, in the absence of words, referential competence would take the form (mentioned in the previous section) of the non-verbal capacity to group perceived instances together according to some relevant general criteria.

To compare incoming perceptual information and conceptual representations as described in (i), the cognitive system must first of all be able to carry out preliminary processing of unstructured, proximal perceptual stimuli in order to grant them a stable structure as individual distal objects. In other words, starting from unstructured experiential data, a representation of the *individual* objects that are presently perceived must be formed. This representation is defined as *token-representation* since it is about a specific object token such as the neighbors' dog Fido who is in front of my door right now, or the flower-pot on the desk of my colleagues in the office (about the definition of token-representations or token concepts see e.g. Jackendoff 1983: 41–56 and 109–127). A token representation is used, among other things, to recognize the same object at later occasions and/or in different situations. As many authors point out (see e.g. Kluender 1994; Kluender and Kiefte 2006: 171ff), the process that leads to the constitution of a token-representation is extremely similar to, if not identical with, the one that leads to the categorization of classes of objects (like dogs or pots in general).

Among the major processes that need to be carried out by the cognitive system to realize a token conceptual representation there are: (a) the individuation of single objects by separating them from other adjacent objects; (b) the achievement of object constancy, which requires that the hidden parts of the object be restored; the shape and the size of the object must be reconstructed independently from its viewing conditions, since it might be perceived from an unusual angle or from different distances which make the object appear bigger or smaller; (c) the mereological processing of the object information on the basis of which the structure of the object and of its parts is identified (for an overview see e.g. Ullman 2000: 1–35).

Furthermore, the categorization of a distal object as an instance of a general concept (a so called *type-concept*: e.g. the categorization of Fido as a DOG; or the categorization of the flower holder on the desk of my colleagues as a POT) requires that the token representation of the individual object is comparable with the type representation of the class previously stored in semantic memory: i.e. it requires that token and type representations have a similar structure and consist in the same kind of information (Jackendoff 1983: 77–94). A hypothesis that is often considered in order to explain this similarity is that type concepts (*also*) consist in perceptual information analogous to that which allows the recognition of the token concepts.

This information concerns the general physical and perceptual features of the objects with regard to shape, edges, or structural configuration codified according to dynamic spatial and geometric principles such as those proposed by Biederman (1987, 1990) on the basis of Marr's theory of vision (Marr 1982) (we will come back to this point in the next section).

However, the information which token and type representations are made up of cannot consist uniquely of perceptual features (the literature on this point focusses mainly on vision, therefore these perceptual features are actually mainly visual features). In fact, if so, humans wouldn't exhibit any inferential competence with relation to the words connected to the concepts they develop and they could not rely on inferential competence to retrace the objects denoted by a concept starting from an inferential description, as with the example of *amanita muscaria*. In addition, they would not be able to develop semantic competence in relation to words (for example, 'mammal' or 'bachelor'), whose acquisition cannot rely uniquely on perceptual information, but depends—at least in part—on inferential information. Inferential information compensates, so to speak, for all the aspects of categorization that cannot be traced back to the mere concrete and perceptual features of objects.

To support full semantic competence, the type representation must also include information about the non-accidental features of the objects that are not related to their form and their perceptual features (see also Jackendoff [1987] 1994: 163–212). First of all, it must minimally include information on the function of the objects: the concept of POT must e.g. include the information CONTAINER, since an object will not be categorized as a pot if it cannot fulfill the function of holding something inside. This information can be of an inferential kind and concern the explicit knowledge people have of the function of the objects. However, at least sometimes, this information can also be connected to referential competence and be perceptively induced from the form of the objects. In fact, as Gibson's classical affordance theory suggests, humans do not perceive only the physical properties of objects, but also their 'affordances'—i.e. what the objects can be used for (Gibson 1977). Affordance can support categorization since it can lead one to group together under one class all objects that have a certain form which appears appropriate for fulfilling a specific function.

Secondly, the type representation must include inferential information on general epistemological, non-observable properties of the object belonging to a certain class, which is usually called 'dictionary-information' or 'encyclopedia-information'. Think e.g. of the case of 'mammal' or 'bachelor': to master the word 'mammal' or 'bachelor' a person must know that mammals are animals which nurse their offspring and that bachelors are adult, unmarried men. Having this knowledge is indispensable for classifying e.g. dogs as mammals or a specific man as a bachelor. In cases like these, the possibility of referentially recognizing that an object falls under a certain concept is *inferentially mediated* since the identification of an instance as a member of a certain class depends on the knowledge of some general, non-observable properties of the object. In this sense, inferential, propositional knowledge of the objects might also be essential with respect to categorization. Even though referential information (concerning the perceptual properties of the objects) can indeed be considered as more basic than inferential information, inferential information supports referential competence in cases (such as 'bachelor' or 'mammal') in which the perceptual (e.g. visual) properties are not sufficient to support categorization.

Since both referential and inferential information contribute to the categorization of objects and since it is difficult to precisely discriminate between them and to identify the specific function each of them plays with respect to categorization, theories of concepts often do not explicitly distinguish between these two kinds of information, but characterize all information used for categorization in terms of feature lists propositionally described. The general idea on which theories of concepts rely on is that for categorization to be accomplished the properties of the token must be compared with the properties of the type in order to establish whether the token is an instance of the type or not. To give a very general example, we can assume that a concept like BIRD can be described as composed of information like *animate*, *feathers*, *wings*, *head*, *legs*, eyes, *lays eggs*, *flies*, *nests*, *eats worms* etc.; and that analogously, a concept like APPLE can be described as composed of information like *seeds*, *sweet*, *you eat it*, *round*, *skin*, *juicy*, *grows on trees,* etc.[18] To establish whether a particular instance falls under the concept of BIRD or of APPLE, the properties of this instance must be analyzed: if it indeed exhibits the properties *animate*, *feathers*, *wings*, *head*, *legs*, eyes, *lays eggs*, *flies*, *nests*, *eats worms* and *flies*… or the properties *seeds*, *sweet*, *you eat it*, *round*, *skin*, *juicy*, *grows on trees*, ... it will be classified as a bird or as an apple respectively.

With the exception of Fodor's atomistic view,[19] all other main families of theories of concepts proposed in recent times in the field of cognitive science share the idea that concepts are composite entities, made up of different kinds of information, which serve to identify classes of objects belonging to their domain. They also share the tendency to describe the information concepts are composed of in terms of lists of discrete and propositionally described features, which include without differentiation both inferential and referential information (Fig. 3).

Despite these commonalities, the various theories of concepts also exhibit relevant differences. One of the most important differences concerns how they conceptualize and describe the relationship between a token and type and the process

[18]These examples of feature lists are taken from Eleanor Rosch and colleagues (Rosch et al. 1976). Some of the features have been left out, while *animate* has been added to the list describing the features of birds. In fact, it is the knowledge that birds are animals that allows us to distinguish real birds from e.g. toys which resemble birds. This piece of information will be relevant in the next chapter.

[19]According to atomistic theories, concepts cannot be thought of as complex entities, made of different kinds of properties or parts that are useful for the identification of tokens. Atomism assumes that there is a direct relationship between type and token, and that the type concept is immediately activated by the perception of the token. This view is based, however, on a strong nativist position which did not meet with the approval of other cognitive approaches to concepts. For an explanation of atomism see e.g. Fodor (1987, 1998, 2003). For an overview of critiques to Fodor's atomism see e.g. Laurence and Margolis (1999: 59ff) and Dellantonio (2007: 247–282).

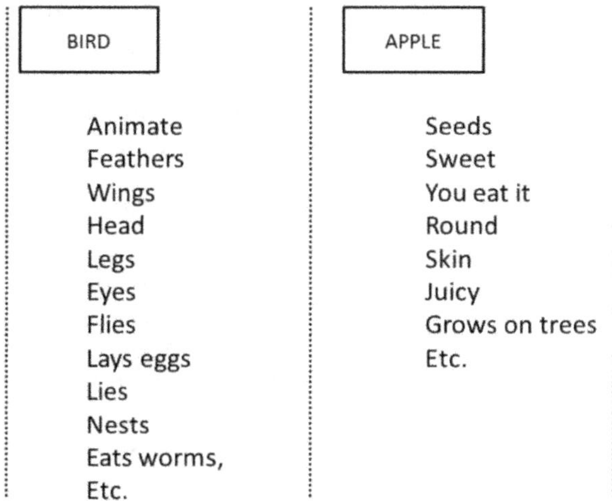

Fig. 3 Examples of concepts described as feature lists

by which a token is compared to a type: more specifically, they assume different kinds of information are included in a type, and used to identify tokens.

The so-called *classical theory* (which is grounded in an approach to meaning based on definitions) maintains that concepts have a strict definitional structure and that the information concepts are made of is structured in the form of conditions that are individually necessary and jointly sufficient to determine whether a certain instance falls under a concept or not (or whether a certain concept falls under another concept from a higher level). According to classical theory, a concept like e.g. BACHELOR is formed by the information that defines what a bachelor is and can be described in terms of the features *human*, *male*, *adult*, *not married*. In order for a certain instance to fall under a concept, this instance must necessarily exhibit all the features that characterize the concept; having these properties is sufficient to establish membership. This means that an instance will be categorized as a bachelor if and only if it is *human, male, adult,* and *not-married* (for an overview on classical theories—also called definition views—see e.g. Laurence and Margolis 1999: 8ff).

Other families of theories consider some consequences of the classical view to be highly implausible and try to develop alternative solutions to them. In particular, they point out that the evidence we have on the way categorization works seems to contradict the idea—implicit to definitionalism—that concepts necessarily produce a discrete partition among classes. This would imply that it is always possible to determine whether an instance falls under a certain concept or not and that no instance is a 'privileged' member of a class, but that all the members belong equally to it. In particular, prototype and neoclassical theory challenge this conclusion, showing that categorization is both not discrete (concepts gradually fade into others —some objects cannot be univocally categorized as instances of a specific concept: e.g. a small, round container can be categorized either as a 'cup' or as a 'bowl') and not 'equal' (some instances are more typical for a class and more univocally categorized as members of that class: a robin is e.g. a more typical instance of BIRD than a penguin); therefore, a different explanation of categorization is needed (for an overview of this criticism see e.g. Murphy 2004: 11–40).

In contrast to the classical view, *prototype theories* dismiss the thesis that conceptual structure must include necessary features for the identification of instances. According to prototype theory, an instance is classified as a member of a concept if a sufficient number of its properties match the features included in the structure of the concept. This theory maintains that it is impossible to establish a priori *what* and *how many* features the instance must share with the concept in order to be classified as a member because the different features have different 'weights' depending on how prototypical they are with respect to the class and they assume therefore a greater or smaller importance in different cases. If e.g. one needs to evaluate whether a certain instance—let's suppose a robin—is a bird, the features FLIES and FEATHERS are surely those with the higher weight and they might also be sufficient for its categorization, since they are highly typical (i.e. almost all kinds of birds have them) and at the same time highly distinctive of birds (i.e. they don't typically belong to any category but the bird category). If, on the contrary, one needs to evaluate whether a different instance like e.g. a penguin is a bird, the number of properties that will turn out to be sufficient for the classification will be higher: since penguins lack all the highly typical and distinctive properties of birds (FLIES and FEATHERS), a higher number of less typical and less distinctive properties will be needed to carry out the classification. One of the macroscopic consequences of prototype theory is that it promotes a graduated theory of concepts that radically diverge from the classical view, according to which, in some cases (i.e. in the case of *a* typical instances), it will not be possible to univocally establish whether an instance belongs to one concept or to another since the number of features the instances shares with the concepts will not be sufficient to establish membership univocally (on this point and its consequences see e.g. Goldman 1993: 128ff).

Prototype theories maintain that concepts have a prototypical structure: the best example of the category is called the 'prototype'. In spite of this commonality, these theories come in several varieties. In particular, there is little agreement on how exactly prototypes and the features prototypical structures are composed of should be conceptualized: whether they should be thought of as something perceptual like an image or as a verbally expressed set of features (for a discussion of this, see e.g. Murphy 2004: 41ff). However, since the main problem prototype theory classically aims to address is not the relationship between perception and conceptualization—i.e. not how the comparison between an actually perceived object and a concept stored in memory is concretely carried out—but rather which aspects of an object categorization relies on and which weight each of these aspects assumes under which circumstances, generally prototypes and prototypical structures are described and discussed in terms of verbally expressed feature lists.

Analogous considerations apply to the so-called *neoclassical theories* which can be considered an attempt to reconcile the classical and prototype views. Like the prototype view, the neoclassical theory assumes that categorization criteria are graded and that different features have different weights with respect to categorization depending on how typical and distinctive they are. However, unlike prototype theories neoclassical theories agree with classical theories in specifying that categorization relies on individually necessary conditions, even though these are not considered to be also jointly sufficient to carry out the categorization. A necessary condition for a certain quality to be categorized as RED is that it fulfills the criteria of being a color; however, being a color is not also a sufficient condition for the quality to be an instance of RED (about this see e.g. Jackendoff 1983: 121–122).

A partially different explanation of categorization is given by the so-called *exemplar view*. In general, the exemplar view maintains that the concepts stored in semantic memory do not (only) consist of an abstract summary of features which are shared by a class, but semantic memory (also) contains specific salient exemplars of the concepts, that the subject has experienced.[20] When a person categorizes an instance e.g. as a bird, s/he does not (only or primarily) rely on a general conceptual representation of BIRD, but will also compare the perceived instance with one or more exemplars of birds stored in memory.[21] More specifically, the instance will be compared with various specific exemplars of different classes stored in memory (dogs, trees, birds, etc.) and it will be classified as a bird only if it turns out to be more similar to one of the exemplars of bird stored in memory than to the exemplars of other classes. Like the prototype view, various versions of the exemplar view have been proposed, differing mainly with respect to how exemplars are understood: if as something perceptual-like mental images or in terms of feature lists. The versions that conceive of exemplars as something analogous to images have the problem of explaining how these images can be used to identify instances that are different from each other as members of the same categories (see e.g. Barsalou 1992). On the other hand, feature based approaches do not address the question of how features can be extracted from perceptual experience (more on this will be specified later in this section).

A last and, in some respects, complementary main family of theories of concepts is the *knowledge view* (also called *theory–theory* or *theory view*). The *knowledge view* quite closely resembles some of the philosophical theories of meaning related to holism and to the inferential (or conceptual) role of semantics,[22] according to which the meaning of an expression depends on the inferential or conceptual links that this expression enjoys with all other expressions of the same linguistic system. Similarly, the knowledge view maintains that to identify the content of a concept and therefore also the information it consists in, the expressions must be considered as a component of a complex theory about the world endorsed by the individual who has that concept. Its content will therefore depend on the relationships that the

[20]This theory was first proposed by Medin and Schaffer (1978).

[21]Most radical versions of this view maintain that there isn't any summary representation codifying a whole class of objects (e.g. a summary representation for all birds), while milder versions of this view maintain that categorization relies on both summary representations and specific exemplars: see e.g., Smith and Medin (1981).

[22]For an introductory presentation see e.g. Pagin (2006) and Greenberg and Harman (2006).

concept entertains with the other parts of the theory (see e.g. Carey 1985, 1991; Murphy and Medin 1985). In this sense, the structure of the concept consists of all inferential relationships that this expression enjoys with other concepts in the system to which it belongs. The content of a concept like e.g. DOG, in other words, the knowledge that a person has about what a dog is—will be determined by all the inferential relationships that this concept maintains with other pieces of knowledge in the same system: it will depend e.g. on what this person knows about animals in general, about domestic animals in particular, about mammals, about the canine race, about dogs' faithfulness and about faithfulness in general, etc.

Even though it is only a summary, this description of the general traits of the main families of theories of concepts shows that their primary focus is directed toward the issue of establishing what role/weight the various features (i.e. the various pieces of information) have in relation to the identification of instances as members of specific classes: whether complete feature match constitutes the necessary and sufficient condition for categorization to occur; whether some feature matches are necessary; whether only a sufficient number of feature matches is needed; whether some matches have more weight than others; whether the features used for categorization are general or extracted from specific exemplars or whether they depend on the theories we have about the world. Because of this specific focus, these theories are not specifically concerned with the question of how concrete the feature matches may be and simply describe the confrontation between instances and concepts using *propositionally expressed, discrete feature lists* which are supposed to characterize both the salient information concepts consist in and the properties of the instances. If the concept of BIRD can be described through the features [animate, feathers, wings, head, legs, eyes, lays eggs, flies, nests, eats worms etc.] and the instance in front of me now exhibits the properties of being animate, having feathers and flying, then this instance is a bird.

Issues like:

- *How are these features concretely identified—i.e. how are these feature lists 'drawn up'?*
- *How are they extracted from perceptual experience?*
- *Are they all of the same kind or are they acquired from different sources (e.g. some though perception and others though explicit learning)?*
- *How does the match concretely occur—is it a match among perceptual or among linguistic elements?*

are often left unanswered or not even discussed.

These issues raise numerous problems. First of all, as e.g. Gregory Murphy points out: "A pile of bird features does not make a bird—the parts need to be tied together in just the right way" (Murphy 2004: 48). There are unlimited ways in which feathers, wings, legs, and a head … could be tied together that do not form a bird (e.g. the legs could come out of the head; the feathers could be attached to each other forming something like an Indian headdress, etc.). Therefore, the availability of feature lists does not per se explain how the features can be used to recognize/classify a bird when we perceive one.

Secondly, it is not clear how and when these features can be learned. As e.g. pointed out by Sharon Armstrong and colleagues in a section entitled *The sum of the features is not the whole concept*:

> In addition to the fact that separable bird features don't seem to do the job in describing the bird concept, there is the question of whether proposed bird-features are, as required by a feature theory, somehow more primitive components of the concepts they describe—little meaning atoms that combine in differing ways to form the multitude of concepts in our mental world. But if so, why hasn't anyone found them? Shouldn't one expect the many words in the language to be describable by a (smaller) set of more primitive words, corresponding, however crudely, to these meaning atoms? Perhaps we should, but dictionaries seem to tell us otherwise. In a dictionary, most words are defined in terms of each another, with most words—unfamiliar ones excepted—taking the role of being defined on some occasions and being definers on others. It is as if all the words made their living, by taking in each other's washing (Armstrong et al. 1983: 303).

The point of this argument is mainly that a person needs to already know what e.g. a head, a leg, a wing, a feather, etc. is in order to identify an instance as a bird on the basis of the fact that it has a head, two legs, two wings, many feathers, etc. But how can we explain the formation of concepts such as HEAD, LEG, WING, FEATHER? And why should they be more basic than BIRD? Analogous considerations aimed at clarifying more precisely the question raised by Armstrong and colleagues are also proposed by Jesse Prinz:

> The features alleged to compose prototypes often seem no more primitive than the prototypes they compose. For example, the prototypical snake is dangerous, the prototypical chocolate cake has many calories, and the prototypical yacht is expensive. Explaining how we come to have the concept DANGEROUS seems more difficult than explaining how we come to have the concept SNAKE. Prototype theorists typically give no reason for thinking the features they name in describing the contents of prototypes are primitive or reducible to primitives. They offer no theory of what the primitives are or how they are acquired. Thus, their theory of acquisition is inadequate (Prinz 2002: 59).

Even though Prinz's observations are specifically concerned with prototypical views, they apply to all theories of concepts which aim to explain categorization on the basis of propositionally described feature lists. Further, they constitute one way of addressing the so called grounding problem, which is common to the entire classical cognitive approach (see Chap. 2, Sect. 1). Feature lists leave unanswered the question as to how features are formed and how they might be used to produce categorization. From this point of view, the question about the origin and the nature of features—or, more precisely, of the kind of information we use to produce categorization—is of concern for all families of theories of concepts, and is basic with respect to the choice of the one or the other as the best explanation for categorization. This is the reason why the analysis we carry out in this work focusses mainly on this question instead of investigating the various families of theories of concepts. The questions that will be raised are: *What information do concepts consist of? How can they be formed? How can they support our semantic competence?*

In order to support individual lexical semantic competence, the information concepts consist of must be appropriate to guarantee full semantic mastery: in the light of the description previously given, this information must assure both referential and inferential capacity. This allows us to draw some conclusions on the structure of concepts that also intersect with some aspects of the theories of concepts previously sketched.

The lexical semantic competence a person has—we suggested—is characterized by an inferential component which consists of the capacity to connect a word with others in his/her language. In order to support this competence, concepts must include inferential information of a linguistic kind. The inferential information included in a concept depends on the specific, explicit and propositional knowledge that a specific person has about the objects belonging to that concept and will therefore vary from one subject to another. This idea that conceptual structure must allow us to account for inferential competence (in the broadest sense) opens the door to a view of concepts—specifically with respect to the inferential dimension of concepts—consistent with the knowledge view.

Semantic competence is also characterized however by referential competence. To account for this aspect of semantic competence we need to assume that concepts contain, in their structure, a different kind of information in addition to the inferential one: a perceptual and non-linguistic kind of information. The hypothesis put forward by this work distances itself from the main families of theories of concepts, at least with respect to feature lists. We start from the premise that concepts also consist of referential information which cannot rely on propositional, or discrete features like the ones described in the previous examples (BIRD = animate, feathers, wings, head, legs, eyes, lays eggs, flies, nests, eats worms and flies etc.; APPLE = seeds, sweet, you eat it, round, skin, juicy, grows on trees etc.). As the critics we discussed above point out, referential competence and the capacity to perceptually identify instances in the external world is difficult to explain in terms of complex and non-structured properties of this kind. To understand how identification of instances occurs and what kind of information is used for this identification, we must examine how the identification and the recognition of these properties is carried out through perception at a prelinguistic level.

For this reason, the view put forward in this work suggests that conceptual representations consist of two kinds of information which are functionally distinct —inferential and referential information—which are of the linguistic and pre-linguistic kinds respectively. These kinds of information interact in supporting lexical semantic competence. For the purposes of this work it is not of primary interest to discuss whether they are joined into a unique representation or whether they form different types of representations which are linked to each other. The investigation will start from the hypothesis that these different kinds of information are jointly used by the conceptual system to produce full lexical semantic mastery. The main focus of the investigation will concern the problem of defining in what cases referential (perceptual and prelinguistic) information must necessarily be part of a concept in order to assure full semantic mastery of the corresponding word and what is the role and the nature of this kind of information in these cases.

5 Perceptual and Conceptual Representations: Grounding Concrete Concepts

Contemporary cognitive research tends to make a sharp distinction between two kinds of representations that are considered to be radically different from each other, i.e. perceptual and conceptual representations. The reasons for strict separation can be traced back to a number of theoretical assumptions that characterize classical cognitive theory.

(i) The first assumption concerns the way in which the architecture of the cognitive system is conceived. According to a widely shared view which developed from Fodor's well-known modularity theory (Fodor 1983), conceptual representations are codified in the central system (in which thought is processed and which semantic memory is part of), while perceptual representations are the product of modular input systems, each of which processes a particular type of information and works separately and in isolation from other subsystems and from the central system.

(ii) The second assumption is tightly bound to the architectural hypothesis described in (i) and has to do with the format of representations. While representations in semantic memory are usually considered to be amodal (i.e. independent from any specific sensory modality), representations produced by other systems are thought to be modality-specific, i.e. they depend on the system they have been produced by and have modality specific qualities related to those systems (see e.g. Barsalou 2012: 247; on this issue see also Chap. 2, Sect. 1).

(iii) The third assumption is also concerned with the format of representations, but more specifically, with the fact that conceptual representations codify information in a propositional (i.e. digital) way, while perceptual representations codify information in a continuous (i.e. analog) manner (Dretske 1999).

It is in particular this last distinction that describes the main theoretical motivation for the differentiation between perceptual and conceptual representations (i.e. the reason that is also philosophically the most salient, since it doesn't depend on any particular hypothesis regarding the architecture of the cognitive system). The primary function of perception is to record information. This recording is not just passive; on the contrary—as we saw in the previous section—the sensory input is processed in many complex ways. Despite this, perception alone cannot give rise to structures supporting the categorization of instances. In other words, perception does not in itself allow for the recognition and identification of instances, as well as their classification in homologous groups and their association with the specific knowledge available about them.

Perception processes data with their differences and specificities: perceptual acts are unique; even the very same object is always perceived in different ways. It is conceptualization that has the function of classifying and identifying the perceived objects according to specific groupings which might be associated with the particular knowledge available about them. As e.g. Barsalou also points out:

> The distinction between a recording system and an interpretative system is central to characterizing conceptual systems […]. A recording system captures information about a situation by creating attenuated (not exact) copies of it. Conversely, a conceptual system interprets the entities perceived in an experience or in a recording of one. […] To achieve interpretation, the conceptual system binds specific individuals in perception to knowledge about components of experience in memory. This is essentially the process of categorization. A system that only records perceptual experience does not categorize individuals in this manner. Instead, it simply records them in the holistic context of an undifferentiated scene (Barsalou 2012: 239–240).

This functional separation of perception and conceptualization has been effectively described by Fred Dretske in terms of a difference in the way they encode information: while perception produces an analog encoding of information, concepts produce a digital encoding of information. Analog encoding of information is continuous and a-specific, while digital encoding of information is discrete and highly specific. To go back to Dretske's example:

> To illustrate the way this distinction applies, consider the difference between a picture and a statement. Suppose a cup has coffee in it, and we want to communicate this piece of information. If I simply tell you, 'the cup has coffee in it,' this (acoustic) signal carries the information that the cup has coffee in it in digital form. No more specific information is supplied about the cup (or the coffee) than that there is some coffee in the cup. You are not told how much coffee there is in the cup, how large the cup is, how dark the coffee is, what the shape and orientation of the cup are, and so on. If, on the other hand, I photograph the scene and show you the picture, the information that the cup has coffee in it is conveyed in analog form. The picture tells you that there is some coffee in the cup. The shape, size, and color of the cup, and so on (Dretske 1999: 137).

The point of this example is mainly that analog encoding carries more information, actually it carries a lot of information, but—since this information is per se a-specific and indeterminate—it cannot be used as such for further cognitive operations. For these operations we need to extract specific elements of information from the analog encoding. All irrelevant features of perceptual information should be disregarded and only the relevant aspects should be picked out: the loss of information allows us to systematically ignore the differences between various analogous stimuli, and to focus only on relevant similarities, thereby generalizing the stimulus and producing a classification of the incoming information. As Dretske specifies: "Until information has been lost, or discarded, an information-processing system has failed to treat different things as essentially the same. It has failed to classify or categorize, failed to generalize, failed to 'recognize' the input as being an instance (token) of a more general type" (Dretske 1999: 141). From this point of view, perception—as an analog encoding of information—has the function of delivering large amounts of information, while conceptualization transforms this information, ignoring the differences and passing from the particular to the general. "Until information has been extracted from this sensory structure (digitalization), nothing corresponding to recognition, classification, identification or judgment has occurred—nothing, that is, of any conceptual or cognitive significance" (Dretske 1999: 153).

Even though there is certainly a functional difference between perception and conceptual thought, the distinction need not be understood as absolute or as devoid of points of contact. In this respect, many mediating positions are possible. However, mediating position are not widespread in cognitive research which is, on the contrary, mainly inclined to the view that these processes are entirely distinct and are by nature completely different from each other. As e.g. Barsalou points out:

> In contrast to pre-twentieth century thinking, modern cognitive scientists began working with representational schemes that were inherently nonperceptual. [...] These new representational schemes differed from earlier ones in their relation to perception. Whereas earlier schemes assumed that cognitive representations utilize perceptual representations [...], the newer schemes assumed that cognitive and perceptual representations constitute separate systems that work according to different principles (Barsalou 1999: 578).

A rigorously 'separatist' position is maintained in particular by the so-called symbolic views of the mind that, in a way or another, hark back to Fodor's hypothesis of a language of thought (Fodor 1975, 2008) and maintain the absolute non-involvement of the central system (which operates by manipulating language-like amodal symbols) with the processing of modal and domain specific information carried out by peripheral sensory modules. As Barsalou points out: "Rather than extracting a subset of a perceptual state and storing it for later use as a symbol, an amodal symbol system transduces a subset of a perceptual state into a completely new representation language that is inherently nonperceptual" (Barsalou 1999: 578).

As Jean Mandler in her article *Representation* suggests, symbolic views on mind are particularly widespread and are so widely accepted in cognitive research because the idea that a conceptual system is language-like is both intuitive from the point of view of an adult, who has already acquired language and the ability to use it for thinking and also because it is convenient for explanatory purposes.

> As adults we are so imbued with language that it just seems natural to us that the mind should consist of symbolic representations and their manipulation. There are symbols (roughly words) for each of the objects we can think about. These may or may not be decomposable into parts (e.g. semantic features), but vis-á-vis thinking or reasoning, they are the basic units that are put together to form complex propositions. In this view, thinking consists of manipulating symbols in sentence-like expressions (Mandler 1998: 258).

This description is not only 'natural', but also allows us to explain various aspects of thought.

> Like language in general, this form of representation is productive; one can combine symbols in ever-longer strings and in ever-new ways. [...] In this kind of system, both programs and representations can be written in the same language. In addition, the rules to combine symbols are typically recursive so it is easy to embed statements within each other. [...] Reasoning is straightforward using symbolic languages; unambiguous rules of inference are easily stated as premises and conclusions. These are some of the reasons that the higher level cognitive processes such as logical reasoning, mathematical reasoning,

> planning and so forth tended to be modeled by symbol systems of one type or another. Typically, in such systems it is easy to learn more information, as long as the information is couched in the appropriate language (Mandler 1998: 258).

However, in spite of these explanatory advantages, there are even stronger reasons to think that a too sharp separation between perception and the conceptual system cannot be maintained. First of all, a problem arises concerning the way in which conceptual representations conceived as entirely different from perception could be formed and about how a natural language can be learned. Secondly, as e.g. Ray Jackendoff points out, one of the fundamental problems with a symbolic theory of conceptual and linguistic thought is to explain how we can speak about what we perceive: i.e. how sensory information is linked to the conceptual system (Jackendoff [1987] 1994: 193ff, 1990). In fact, to speak about what we perceive, perceptual information must be compared to (i.e. linked to) conceptual information and there must be some kind of representation that bridges the gap. The problem with a too sharp separation between perception and concepts is that we have no means to explain the capacity to recognize and identify the objects belonging to a certain concept. Going back to the terminology we introduced previously: (amodal) symbol systems have many powerful resources to explain (among other things) inferential competence since they can perfectly account for the ability to produce categorical inferences, but they offer no means to explain referential competence, which is not only an essential component in the adult capacity to master language, but is also plausibly the starting point for conceptual formation and linguistic learning.

Jackendoff investigates this problem in connection with vision (which he considers to be a privileged source of information for categorization) and the issue of the link between perceptual information of a visual kind and the conceptual system. The main focus of his analysis is on concepts which refer to concrete physical objects (that can be perceived through vision). Jackendoff's analysis starts from the description of visual processing given by David Marr (1982), according to which the most central level of visual representation is the 3D model representation, where the stimulus is represented as a distal object (i.e. as a viewer-independent, well-formed object). "In order for an organism to accomplish visual identification and categorization, independent of language, there must be a form of visual representation that encodes geometrical and topological properties of physical objects" (Jackendoff 1996: 44).[23]

This representation is considered to be volumetric (three-dimensional), organized around the coordinate axes of the object that specify the positions of its parts, and hierarchically structured—i.e. composed of parts and parts of parts according to a 'tree structure' that keeps track not only of the reciprocal relations among the parts, but also of their reciprocal spatial proportions, points of attachment and all possible angles of attachment of the parts.

[23]For a more detailed presentation of this hypothesis, see Jackendoff [1987] 1994: 193–212.

> This level is conceived as *volumetric*: objects are represented as occupying volume in space, by contrast with the surface representations of the lower levels. In addition, it is object-centered rather than viewer-centered: it makes explicit the shape and size constancy of objects, regardless of the viewer's position. It is therefore the level most appropriate for encoding an object's shape in long term memory, so that it might be recognized on subsequent occasions. [...] What is particularly interesting about the 3D structure of an object is that it is hierarchical: it represents the three-dimensional structure of objects not just in terms of holistic shape (it is not a 'statue in the head') but rather in terms of a hierarchical decomposition of the object into parts and parts of parts (Jackendoff [1987] 1994: 174).

And further: "In each case of elaboration the configuration of parts is specified in terms of the coordinate system defined by the axis of the main part in the configuration" (Jackendoff [1987] 1994: 175) (Fig. 4).

As Jackendoff emphasizes, this representation is something different from a mere "statue in the head" that serves as a geometrical "image of a prototypical instance." Indeed, while a hypothetical prototypical statue of an instance would encode only the spatial properties of a *particular* instance, the spatial representation which Jackendoff appeals to encodes parametric information, i.e. information that describes space not only in a continuous and topologically structured way, but in a way that is also able to account for the overall spatial and motor configuration of the objects. Since it can keep track of a whole range of spatial properties regarding, e.g., the possible proportions among the parts, angles of attachment, etc., this spatial representation can be used to categorize a whole range of instances sharing the same geometrical, topological and movement properties. "[...] the parameters specifying angle of attachment of the various parts [...] can give a range of possible values, corresponding to the possible configuration of the joints. On the other hand, certain aspects will be kept rigid, for example the point of attachment" (Jackendoff [1987] 1994: 176). Although this is not mentioned by Jackendoff, it is important to note that a representation of this kind is not only useful for describing real objects,

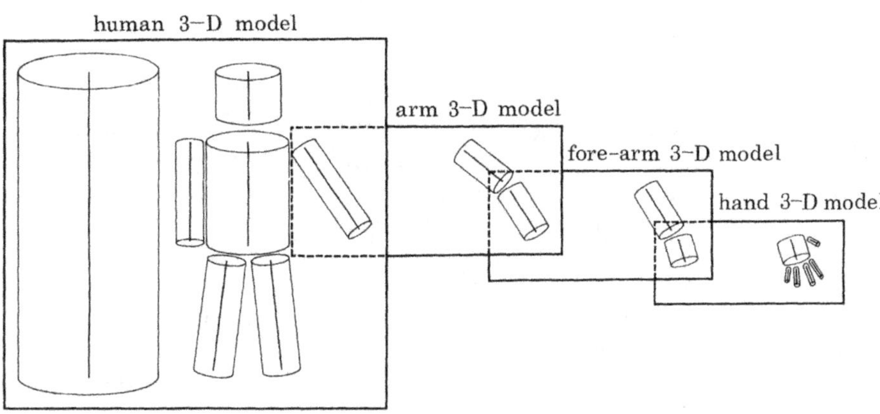

Fig. 4 Diagram illustrating the organization of shape information in a 3D model description. In: Marr and Nishihara 1978: 278, Fig. 3

but also possible objects, i.e. objects which are composed out of parts of real objects, like e.g. a unicorn which can be built imaginatively through the organized combination of spatial parts derived from other perceived instances.

Jackendoff thinks of a perceptual representation as mainly composed of visual information (since in his view vision dominates other sensory modalities when it comes to categorization): they are the contact point between the perceptual and conceptual system. "The overall hypothesis that emerges from this consideration is that the 3D model […] is one of the central interface languages of the computational mind. It is a general purpose representation for all tasks involving spatial cognition, and language, vision, touch, action, and the body sense can all influence it and make use of it" (Jackendoff [1987] 1994: 196). Even though the 3D model is essential for categorization, in Jackendoff's view it is something different from a concept. In fact, it is not a conceptual representation yet, but still a perceptual representation, even though of the highest possible kind. Even theories which do not adhere to the symbolic view of mind proposed by classical cognitive science, nevertheless make use of this notion of a highly elaborate perceptual representation similar to the 3D model. Therefore, they maintain that there is a continuum between perception and conceptualization, and that a conceptual representation is nothing more than a perceptual representation of this kind.

A similar hypothesis was put forward e.g. by Prinz (2002) whose position relies at least in part on Barsalou (1987). Like Jackendoff, Prinz pursues the idea that categorization must be based on representations of a perceptual kind which are developed starting from the 3D representations of Marr (and Irving Biederman) which he calls an "*object model*".

> Hereafter, I use the generic term 'object model' to subsume these viewpoint-specific representations as well as the viewpoint-invariant representations that Biederman defends. According to Marr (1982), object models can be highly structured. Imagine approaching a gorilla from a distance. Your visual system first represents the basic arrangements of its torso and extremities. As you draw near, you discern the parts making up its limbs and head. A closer look reveals its surface detail, such as facial features, wrinkles, and fur. Marr speculates that such perceptual states can be grouped together into a hierarchical representation (Prinz 2002: 141–142).

Object models can be considered as token representations of objects that provide a structural description of what they represent. While Jackendoff thinks they essentially consist in exclusively visual information, Prinz maintains that they are formed starting from information coming from different sensory modalities. "Each modality furnishes the mind with a multitude of representations tuned to various aspects of the environment" (Prinz 2002: 143). However, except for this difference, Prinz's object model essentially corresponds to Jackendoff's 3D representation.

The difference between Prinz and Jackendoff (which is actually substantial from the point of view of the general theory of concepts that it gives rise to) is that Prinz rejects the symbolic view of representation embraced by most classic cognitive scientists, including Jackendoff, and pursues a perceptual conception instead,[24] according to which the conceptual (amodal) representations hypothesized by symbolists are in fact perceptual (multimodal) representations. This position is largely analogous to those of Lawrence Barsalou, according to which "a common representational system underlies perception and cognition, not independent systems" (Barsalou 1999: 578). To support this view, Prinz has to show that something like the 3D representation/the object model can indeed operate as a conceptual representation. As stated e.g. also by Barsalou "As long as perceptually based theories of knowledge are viewed as recording systems, they will never be plausible, much less competitive. To become plausible and competitive, a perceptually based theory of knowledge must exhibit the properties of a conceptual system" (Barsalou 1999: 582).[25]

However, there is a very relevant reason why Jackendoff maintains that the 3D representation—the object model, to use Prinz's terminology—*is not in itself a conceptual representation*. In fact, we can have a structured perception of an object —and represent it e.g. as composed of parts, and of parts attached to each other in a specific way—without having categorized it as an instance of a certain concept. Suppose we come across a gorilla and we form a perceptual representation which catches the geometrical and topological aspects of its body (the parts it is composed of, the attachment angles, the proportions among the parts, the possible movements of the parts, etc.). A representation of this kind cannot support a conceptual understanding of what a gorilla is or of what an ape is (and even less a conceptual understanding of 'mammal' or of other more theoretical terms—we will come back to this later). The possession of the concept of gorilla presupposes that one has *stable parameters which define how a gorilla in general can be* (how big, how small, which features can vary and how, etc.). These parameters allow one to establish whether the token object model s/he has perceptually formed in a certain instant of time while looking at an object matches the type concept of a gorilla, i.e. the general features s/he has learned belong to gorillas. These parameters must also lead to set thresholds between potentially analogous/continuous groupings and allow us e.g. to establish that a certain object representation matches the features of

[24]Prinz is so anxious to demonstrate his distance from the symbolic tradition that he refuses to call the perceptual representation by the name Barsalou uses—i.e. 'perceptual symbol system'. Since this term includes the word 'symbol', he thinks it might mistakenly suggest a connection with the classical model. He uses the names 'object models' or, as we will see 'proxytypes' instead.

[25]Barsalou's solution for this problem in primarily meant to answer specific psychological issues related to: the format of the symbols that support conceptualization (they are hypothesized to be perceptual and modal); the nature of these symbols (they are conceived as unconscious neural representations); the way in which perceptual symbol system can carry out functions that are classically assigned uniquely to a-modal and language-like symbols (e.g. categorical inferences, productive thought, type-token mappings, implementation of propositions and analogies). Our approach is rather concerned with more philosophical issues such as the nature and the structure of concepts. Since our aims are more consistent with Prinz's view, in the following section we will focus primarily on his work, even though in the literature perceptual theories of knowledge are more commonly associated with Barsalou.

a gorilla, while a certain other, structurally analogous representation, matches the features of an orangutan or of a chimpanzee.

Prinz agrees that the function of the concept is not to record information, but to form classes and lead to the identification and to the recognition of the perceived instances as members of a specific class. In fact, he defines concepts as a "detection mechanism" (Prinz 2002: 149) which compares the token representations of presently perceived objects with sets of features saved in memory that provide the criteria to determine which object is an instance of which concept.

> When an object is perceived (or named or described), we represent it using a set of features. Those features are matched with features in our memory networks. The similarity between a perceived object and a memory network is measured by summing the diagnostic weights of all features that match a representation in that network. An object is recognized as falling under the category whose memory network contains a representation with which it has the highest similarity (provided that the similarity rating exceeds a critical threshold). For example, if a perceived object is more similar to a representation in a dog network than to any other representation, it gets identified as a dog (Prinz 2002: 161–162).

Since the object representation contains structural information about an object, but does not produce conceptual classes and since—unlike Jackendoff—Prinz is interested in showing that conceptual representations are in fact perceptual representations (i.e. consist solely of perceptual information), he hypothesizes the existence of another, more complex kind of perceptual representation—built on the basis of the object representation—called a *proxytype* which serves the purposes of categorization. "On the proposal I am recommending, concepts are proxytypes, where proxytypes are perceptually derived representations that can be recruited by working memory to represent a category" (Prinz 2002: 149). Prinz hypothesizes that our long term memory stores many overlapping proxytypes: which one is recalled in a specific thinking process depends on what are we thinking about; it depends on the context in which we are thinking:

> Context determines what proxytype is used in working memory on any given occasion. If one is looking for dogs in the arctic tundra, one can call up a representation of a typical sled dog. If one is looking for a guard dog, one can use a representation of a more ferocious breed. Sometimes these proxytypes already exist in long-term memory, and sometimes they must be constructed. If one reads a news report about a dog that is 5 feet tall, one probably constructs a new representation on the fly (see Barsalou 1987). This flexibility makes proxytypes ideal for tracking things (Prinz 2002: 149–150).

The idea that our concepts consist of many overlapping proxytypes of the objects they refer to leads to a problem analogous to that of object representation. In fact, the hypothesis that e.g. our concept of DOG is composed of many proxytypes related to specific types of dogs—sled dog, Doberman Pinscher, wolf dog etc.—which are recalled depending on what one is thinking about, leaves at least the following key issues unresolved:

(a) To identify that all these proxytypes belong to (are exemplars of) the same concept of DOG, one must have a general and non-contextual concept of dog.

(b) To identify that a specific dog is a sled dog or a Doberman Pinscher, or a wolf dog, one must have the general concept of SLED DOG, DOBERMAN PINSCHER and WOLF DOG.

In other words, a conceptual representation cannot be a merely perceptual, particular and specific representation that corresponds—using Dretske's terminology—to some kind of analogic representation formed when we look e.g. at the dog of the neighbors digging a hole in the garden. For this reason, to serve as a conceptual representation, the proxytype must be a general and non-contextual representation, which can include more specific and contextual representations as its parts. More precisely, we can say that a person has the concept of DOG, if s/he has a general proxytype of DOG, which includes as its parts the general concepts of SLED DOG, of DOBERMAN PINSCHER, of WOLF DOG etc. Once we concede this, we can follow the line of Prinz's argument and hypothesize that proxytypes can also be much more specific and can assume different forms in different occasions (e.g. when we think of a working dog we usually have specific kinds of dogs in specific contexts in mind).

Prinz admits (somehow[26]) the need for general proxytypes and for this reason introduces the notion of a "default proxytype": "A default proxytype is the representation that one would token if one were asked to consider a category without being given a context. When no context is specified, we use representations of the kind that figure most frequently in our interactions with category members. I speculate that default proxytypes are relatively stable, widely shared, and frequently responsible for guiding ordinary category-directed behavior" (Prinz 2002: 154). And further: "I hypothesized that default proxytypes are used most of the time" (Prinz 2002: 157).

As the name suggests, proxytypes are conceived as analogous to prototypes: they are constituted by features that are very frequently exhibited by the objects included in the class designed by the proxytype. According to Prinz they are highly intersubjectively stable and only weakly influenced by subjective theoretical knowledge about the designated objects. However, unlike prototypes they are not conceived in terms of unstructured and verbally descriptive feature lists [as, for instance, BIRD = animate, feathers, wings, head, legs, eyes, lays eggs, flies, nests, eats worms and flies etc.] (for a description and a critique of the view that describes concepts in terms of a 'feature list' see Sect. 4 of this chapter).

On the contrary, the proxytype is meant as a generalization of the object representation which describes in an analogic, schematic and parametric manner the perceptual characteristics of the instances belonging to a certain concept. For instance, with respect to visual information, the proxytype describes in an analogic,

[26]In fact, Prinz frames this issue in different terms from those we highlight here. His main concern is the necessity for justifying the public nature of concepts: i.e. the fact that they are largely shared by people, even though everyone had different particular perceptual experiences: see Prinz (2002: 152ff).

schematic and parametric manner the geometrical and topological characteristics of objects. It describes, among other things:

- the parts that the objects belonging to a certain concept are made up of (e.g. dogs have four legs, a head, a tail, two ears, etc.);
- the points of attachment of the parts (e.g. in the case of dogs the legs are attached to the torso at specific points, so are the head, the tail, etc.);
- the variation parameters for the size of the objects belonging to a certain concept (dogs and the parts dogs are made of can vary between specific minimum and maximal sizes);
- the variation parameters for the relative sizes of the parts (dog-parts have e.g. specific relative proportions for which the possible variation cannot exceed specific limits);
- the angles of attachment and their possible parametrical variations (in the case of dogs, they specify e.g. the possible sizes of the angles of attachment of legs, tail, head to the torso);
- the topological variations of the objects belonging to a certain concept (e.g. the possible differences of the overall shape and form of dogs);
- the possible movement of the objects belonging to a certain concept (e.g. all the movements that dogs can perform).

Since the proxytypes are structured and describe the composition of the objects belonging to a certain concept in a way that is both schematic and analogical (geometrical and topological as far as vision is concerned), they allow us to compare the information we perceive in a certain moment—the object model we presently perceive—with the general structure common to all the instances of that concept (e.g. DOG) which is stored in long term memory. This kind of comparison is not vulnerable to the objection introduced in the previous section by Murphy, who says that "A pile of bird features does not make a bird—the parts need to be tied together in just the right way" (Murphy 2004: 48). As e.g. also Prinz points out:

> Prototypes are often criticized for their failure to represent relations between the features that compose them. They do not represent simple spatial relations (e.g., the fact that the beak of a bird protrudes from a different part than its wings) or more complex explanatory relations (e.g., the fact that having wings is responsible for being capable of flight). In contrast, proxytypes typically encode a considerable amount of relational information. The spatial relation between beaks and wings is inevitably represented in bird proxytypes as a consequence of the fact that they are built up from perceptual representations, such as visual object models. The relationship between having wings and being capable of flight is also inevitably represented because being capable of flight is represented by the sequence of wing movements (Prinz 2002: 156).

Since proxytypes summarize structural information of perceptual origin on the appearance of the objects, they are apt to describe categorization (only or primarily) in the case of instances that belong to the same class by virtue of their physical similarity—i.e. in virtue of the fact that they share similar observable characteristics. If we consider conceptual hierarchies of decreasing or increasing generality like e.g. [animals—mammals–dogs–poodles] and [vehicles—cars–station wagons], it is clear that the proxytype can explain the categorization (only or primarily) at low generality levels: in the proposed example, it can explain how dogs, poodles, cars and station wagons are classified together, but not how we categorize animals, mammals and vehicles, whose instances have very different exterior appearances. In the proposed hierarchies the classes of cars and dogs represent the highest generality level at which the objects belonging to them are similar in shape. Research on concepts calls this level 'basic level' (for a more detailed discussion on conceptual hierarchies and basic level concepts see the next chapter). *Proxytype theory offers a good explanation for categorization as far as basic and lower than basic level concepts are concerned.* Higher-level concepts like animal or vehicle could be explained by assuming that they consist of a collection of lower-level proxytypes.

These considerations allow us to make some concluding remarks on the theoretical applications of proxytypes. Prinz's proxytype is a schematic representation of a perceptual kind that allows for the recognition and identification of objects belonging to a certain concept. Proxytypes allow us to explain what we called referential competence with respect to concepts at a medium or low level of generality by referring to concrete objects that can be perceived through the senses. Proxytypes easily and intuitively remedy the problems connected with feature lists as far as these basic concepts are concerned. So, even though the notion proxytype has a limited application and it offers a solution to the grounding problem only for concepts referring to observable objects in the external world, this is in any case suitable for our theory, since it offers a cognitive explanation for referential competence at least as far as these concepts are concerned.

However, since proxytypes do not explain referential competence where other kinds of concepts are concerned, in the following discussion (starting from the next section) this notion will be blended with a more general one—the notion of '*perceptual schema*'—on the basis of which we will try to give a wider explanation of referential competence that is able to account for referential competence also as far as other conceptual classes are concerned, especially those referring to abstract states, internal physical states and emotions. Differing from Prinz, the position we will put forward does not aim to show that conceptual representations are perceptual representations *tout court*. We will rather try to explain conceptualization and the semantic competence that relies on it on the basis of two different factors: we will suggest that inferential competence requires the availability of linguistic information, while referential competence relies on perceptual information (i.e. on a representation called a 'perceptual schema'). Therefore, the position we will argue for hypothesizes that a concept is composed of two kinds of information: its core will consist of perceptual information organized into a perceptual schema, while its periphery will consist of linguistic information.

6 Schematic Image and Perceptual Schema

The proxytype is a *perceptual scheme* that allows us to identify observable objects: it is a non-propositional representation, composed of analogic, dynamic and modal information of perceptual origin. Its schematic character relies on the fact that it constitutes a structure—a skeleton—that applies to different instances and that allows us to identify and to recognize different perceived instances. This notion is extremely useful for explaining categorization as far as concepts referring to concrete terms are concerned. However, the question arises whether other conceptual classes could also be perceptually grounded and rely on something like a perceptual scheme of this kind. The literature that discusses the problem of embodiment and grounded cognition in categorization has often suggested that the perceptual foundation of concepts must be much broader and that most concepts, if not all of them, must be based on and originate from perceptual information. The idea some authors appeal to in order to support this position comes primarily from cognitive linguistics and can be traced back mainly to Johnson (1987) and Lakoff (1987) They suggest that abstract concepts also rely on a perceptual representation which they call a 'schematic image'.[27]

The theories that make recourse to the notion of image have a long and complex story both in philosophy ad in psychology (philosophy has investigated the nature and the function of imagination as a faculty that mediates between sensations and thought (see e.g. Johnson 1987: 139–172); while psychology in its recent history has focused on imagery understood as the processes leading to the development and the use of perceptual mental images (see mainly Paivio 1971, 1986; Neisser 1976; Kosslyn 1973, 1980).[28] Traditionally the common denominator of the theories that make use of the notion of image is to assume that "knowledge has a strong perceptual character" (Barsalou 1999: 578) and that "the underlying structure of many kinds of thinking and understanding is not propositional in nature, but consists instead of manipulating spatial models" (Mandler 1998: 263).

Lakoff maintains e.g. that "[...] experience is structured in a significant way prior to, and independent of, any concepts. Existing concepts may impose further structuring on what we experience, but basic experiential structures are present regardless of any such imposition of concepts" (Lakoff 1987: 272). It is to account for this prelinguistic, perceptual organization that Johnson introduces the notion of 'image' in association with those of 'schema'. His notion of image adopts some

---

[27]The notion of schematic image has been discussed in several works which provide an overview of the issue of embodiment and grounded cognition from a cognitive point of view (see e.g. Gärdenfors 2007; Gibbs 2005; Spivey et al. 2005; Sinha and Jensen de López 2000). The bibliographical sources to which this notion is traced back are however quite limited and repetitive: in addition to Lakoff and Johnson these studies mention a number of other researchers working in cognitive linguistics, who embrace the same prospective, such as Langacker (mainly Langacker 1987) and Fauconnier (1994).

[28]For an overview that—although not very recent—classifies theories on imagery on the basis of the idea of perception they embrace: see e.g. Finke (1985).

elements both from the philosophical and the psychological tradition. From the philosophical—and specifically from the Kantian—tradition, Johnson takes the idea that we need to assume a capacity of imagination, i.e. a "basic image-schematic capacity for ordering our experience" (Johnson 1987: xx). This can be understood as a schematizing activity that allows humans to organize perception into a conceptual form; more specifically, according to Johnson: "into meaningful unities that we can comprehend. *Imagination generates much of the connecting structure by which we have a coherent, significant experience, cognition and language*" (Johnson 1987: 165). Differently from Kant, however, Johnson maintains that imagination cannot be conceived as something strictly separated from understating and sensation. In his opinion there is no unbridgeable gap between sensations and understanding and for this reason imagination can be a common part of both—or better: the central part of both (Johnson 1987: 166ff).

From this same philosophical tradition, he draws inspiration for developing his idea of schema:

> I want to propose a meaning for the term 'schema' that differs in important respects from what has come to be the standard meaning of the term in recent cognitive science. My use of the term derives from its original use as it was first elaborated by Immanuel Kant. He understood schemata as nonpropositional structures of imagination. Today, by contrast, schemata are typically thought of as general knowledge structures, ranging from conceptual networks to scripted activities to narrative structures and even to theoretical frameworks (Johnson 1987: 19).

Therefore, Johnson's schema must be understood in terms of a schematizing activity meant as a procedure for constructing images that allow us to organize and give sense to our experience in a nonpropositional, but perceptual manner.

> I am identifying the schema as a continuous structure of an organizing activity. Yet, even though schemata are definite structures, they are dynamic patterns rather than fixed and static images, as their visual diagrams represent them. They are dynamic in two important respects. (1) Schemata are structures of an activity by which we organize our experience in ways that we can comprehend. They are a primary means by which we construct or constitute order and are not mere passive receptacles into which experience is poured. (2) Unlike templates, schemata are flexible in that they can take on any number of specific instantiations in varying contexts. It is somewhat misleading to say that an image schema gets 'filled in' by concrete perceptual details; rather, it must be relatively malleable, so that it can be modified to fit many similar, but different, situations that manifest a recurring underlying structure (Johnson 1987: 29–30).

Johnson and the other authors who embrace the same notion of image adopt from the psychological tradition the idea that images are embodied imaginative structures. For instance, they consider images to be *experiential gestalts* that make use of information taken from different sensory modalities (not only vision), including primarily information drawn from our sensorimotor activity. As e.g. Jean Mandler points out: "I stress that image-schemas are not images, because this is often mistakenly believed to be the case. This, in fact, is where the debate between the imagists and the propositionalists that took place in the late 1970s went wrong" (Mandler 2004: 79). The examples of image schema discussed by Johnson are various and not necessarily homogeneous.[29] They concern, among other things, concepts like those of FORCE, MOTION, CONTAINMENT, PATH, BOUNDEDNESS, DIFFERENTIATION etc. To have a more concrete picture of how Johnson thinks about image schemas, let us consider e.g. the example of force (Johnson 1987: 42ff; on this see also Gärdenfors 2007). Originally, we experience force through bodily interaction, when we are pushed or we push something/somebody in some direction. In our experience force involves a movement from one place to another along a specific path. Moreover, we experience forces applied to our body as characterized by a certain degree of intensity and as involving a causal sequence: the force generates a compulsion (i.e. a state of being compelled) in what it is applied to. The effect of the compulsion depends of the intensity both of the force and of the blockage (the resistance to force). The understanding of the structure *force → compulsion* leads us to organize experience in terms of relationships mediated by forces and compulsions generated by forces (against which it is possible or impossible to make a stand).

More specifically, this complex experience leads us to form a schematic image of force that make us understand phenomena according to the dynamic of forces: we understand what force is and how force works and, on this basis, how the things in the world are organized and are related to each other from the point of view of the forces that act on them. This physical (or more specifically bodily) understanding of force is also the starting point for building the more abstract concept of force we use for example to understand other 'immaterial' phenomena that can be metaphorically understood as forces: sexual attraction can be seen e.g. a compulsory force that pushes one toward the object of desire (for a more detailed explanation of this example see e.g. Johnson 1987: 1–40).[30]

According to Johnson, the image schema of force is mentally represented in a continuous and dynamic form as a complex perceptual image that can be described as a vector with a directionality which follows a specific path and has an origin, a source and a target that undergoes a compulsion. This is conceived as "a dynamic pattern that functions somewhat like the abstract structure of an image, and thereby connects up a vast range of different experiences that manifest this same recurring structure" (Johnson 1987: 2). A schematic image is considered to be mainly composed of visual, tactile, kinesthetic and proprioceptive information related to the concrete bodily experience of applying and undergoing a force. "The experiential (image-schematic) gestalts for force that I have just described are actual repeatable structures of experience that emerge from forceful interactions in our

[29]As we will specify in the next chapter, some of Johnson's examples concern very general organizing structures like e.g. that of 'containment' that can be classified as categorical, while others are more specific and relate to particular abstract concepts (in fact, they explain how we come to understand these abstract concepts: see Chap. 4, Sect. 2 and Chap. 6, Sect. 3.

[30]On the role assigned to metaphors see e.g. Lakoff and Johnson (1980). The issue of the cognitive theories of metaphor and of the role metaphors assumed by some theories of conceptual understanding developed in the field of embodied cognition will be discussed in Chap. 6, Sect. 3 of this work.

world. These patterns exist for us prelinguistically, though they can be considerably refined and elaborated as a result of the acquisition of language and the conceptual system that language makes possible" (Johnson 1987: 48).

Even though the description of the image given by Johnson in terms of a vector and the figure he uses to illustrate it (an arrow with a specific orientation) might suggest a visual picture, it is actually of primary importance not to interpret schematic images in this way. A picture is a static object which visually represents a particular thing in a specific position. A schematic image is built starting from information coming from different modalities (i.e. bodily, kinesthetic, proprioceptive and perceptive information) and must serve as an interpretative schema for multiple concrete phenomena which differ in form and content. As e.g. Mandler states: "It is correct that image schemas are abstracted from perceiving objects in space, but they eliminate figural details, including many details of movement. They are not temporary constructions, as are images, but permanent representations of meaning. And although like images they are analog representations, they are not visual ones" (Mandler 2004: 80–81).

Unlike e.g. proxytypes—which can be easily conveyed graphically through composite schematic figures like Marr's and Nishihara's 3D model (Marr and Nishihara 1978)—an image schema like the ones that Johnson and Lakoff hypothesize must support the understanding of abstract concepts e.g. that of FORCE are difficult to depict through a drawing and to envisage concretely. Any attempt to give a symbolic description runs the risk of being counter-productive since it recalls a gestalt structure of a visual kind. The best way to give some concreteness to the image schema is by focusing on its function. Concepts cannot be understood uniquely on the basis of linguistic information of a propositional kind, but they need —at least in part—to be of perceptual origin and to rely on perceptual information. If they didn't, we would neither be able to referentially connect these concepts with the world, nor to learn a language on the basis of prelinguistic organizing structures of experience.

It was precisely this need to explain conceptual learning on the basis of a perceptual, prelinguistic organization of experience that led some researchers working in the field of developmental psychology to assume the existence of perceptual and nonlinguistic representations analogous to, if not identical with, image schemas. As e.g. Jean Mandler maintains:

> The approach to prelinguistic conceptual representation that I have developed was inspired by the work of cognitive linguists, in particular Mark Johnson (1987) and George Lakoff (1987), but also Gilles Fauconnier (1994), Ronald Langacker (1987), and Leonard Talmy (1988). These theorists, although not developmentalists, are interested in the underlying basis of the concepts expressed in language. This quest has a natural affinity with development. Some of the concepts expressed in language must be preverbal, because it is these on which language acquisition rests. A basic claim of cognitive linguistics is that one of the foundations of the conceptualizing capacity (in or out of language) is the ability to form image-schemas. Image-schemas are spatial representations that express primitive or fundamental meanings. Primitive in this sense means foundational, not that image-schemas are atomic, unitary, or without structure (Mandler 2004: 78).

The image schemas Mandler speaks of seem to be structures combining Lakoff's and Johnson's notion with Prinz's proxytypes, since they are hypothesized to be used to develop various kinds of concepts both concrete and abstract. According to Mandler, the perceptual information image schemes are composed of "is derived in the usual case by vision, but can be provided by touch, audition, and even one's own movements" (Mandler 2004: 79). Even though image schemas are derived from perception, they are permanent, analogue representations that describe schematically the spatial and movement properties of classes of objects.

Their function is both to allow the learning of a conceptual system and therefore of a natural language and to perceptually ground that conceptual system:

> Thus, in this view, the initial conceptual representations consist of transformations of perceptual information in analogical forms, rather than the activation of built-in propositional language of thought […]. This claim that early concepts are transformations of perceptual information is partly a claim about where meaning comes from. One problem with traditional symbolic systems is what has been called the symbol grounding problem […]. How do symbols in a physical symbol system get their meaning? […] One approach to the grounding problem is to have symbols intrinsically carry meaning, as in the case in the various analog systems that have been proposed. Image-schemas by their very structure express meaningful relations (Mandler 1998: 264).

And further:

> That is the nontrivial issue here, namely, how to solve the symbol grounding problem. […] The solution, I suggest, is that whether vectors or abstract analog structures are used, meaning is grounded by being extracted from perceptual information and directly represented by its format, not by assigning perceptual information to an otherwise arbitrary symbol. It is soon enough to deal with arbitrary symbols when language begins to be learned! (Mandler 2004: 82–83)

Other authors who embrace a different position from Mandler supporting a more strongly nativist view of mental structures (Mandler follows an empiricist line of research) share the idea that, in order to explain how it is possible to acquire a linguistic conceptual system, we need to assume the existence of prelinguistic representations. Susan Carey distinguishes e.g. among three levels of representations consisting in (a) a sensory/perceptual representational system, (b) theoretical conceptual knowledge and (c) "core cognition", which is based on Elizabeth Spelke's notion of "core knowledge" (Carey 2009: 3–65; Carey and Spelke 1994: 169–200; Spelke 2000; Spelke and Kinzler 2007). Conceptual representations are identified with linguistic representations and are characterized by a strong inferential component that is influenced by our propositional knowledge about objects, while perceptual representations are described as iconic and analog representations of the features of immediate experience which play only a weak inferential role. So-called "core cognition" describes a level of information processing that takes place between these two: this level is shared by humans and nonhuman animals, and "is the developmental foundation of human conceptual understanding" (Carey 2009: 11). Representations that are at the level of core cognition are more complex than sensory/perceptual representations and play a richer inferential role in thought, but—differently from conceptual representations—the information they include is not fully explicit and is not influenced by the propositional knowledge of the subject (i.e., according to Carey, by the intuitive theories that articulate people's knowledge of the world, see Carey 2009: 3–65).

Consistent with these studies, the perspective we embrace in this work also shares the idea that in order to explain both how conceptual development starts and how concepts are grounded in reality and support our referential capacities we need to assume the existence of something like core knowledge—i.e. we need to assume that conceptual representations are also characterized by a non-propositional perceptual counterpart which supports our referential competence. This perceptual counterpart might be understood either as a (more fundamental) internal component of the concepts, as the word 'core' suggests, or as an autonomous perceptual representation connected with the conceptual representations. In the first case, the conceptual representation will be composed of both perceptual information (at its core) and of propositional information (at its periphery), while in the second, our conceptual capacity will be supported by two different kinds of representation linked to each other: the one perceptual and the other propositional.

Going back to the question introduced in Sect. 4 of this chapter, we can now specify that the perceptual component of the core explains where the features that compose the feature lists which so many studies of categorization rely on comes from. With respect to a concept like e.g. BIRD, the perceptual component/counterpart of the conceptual representation describes how people know that birds have feathers, wings, head, legs, eyes, lays eggs, flies, etc. (the related issue of how people learn that birds are animate will be discussed in the next chapter). The hypothesis we are considering suggests that this information is not represented/stored in linguistic form, but in perceptual form: i.e. in a schematic format analogous to that of schematic images, or more specifically, in this case to proxytypes (since birds are concrete, observable objects of the external world). It is information codified analogically and spatially concerning the geometrical, topological and movement properties of the object. Other information on the fact e.g. that birds do not breastfeed, but lay eggs, or that some migrate, or that some species of birds produce saliva to build their nests can on the contrary be acquired linguistically and constitutes the information which the inferential competence exhibited by individuals is based on.

For the aim of this work it is not fundamental to determine whether a perceptual representation is a component (i.e. the internal core) of the conceptual representation or whether the two are separate representations linked to each other, even though we often rely on the idea of a core representation suggesting a preference for the first position. The main issue that remains to be discussed relative to this point is rather whether other concepts in addition to concrete ones include a perceptual core (or are linked to a perceptual representation) and, if so, what kind of information this perceptual core/representation consists of. The thesis we will argue for is that perceptual representations are not only fundamental for the development of concrete concepts and for the mastery of words related to concrete concepts but also for the development of both categories (i.e., as we will see in the next chapter, of high level organizing structures) and other kinds of concepts. Our analysis will try to show that referential mastery of words referring to internal states of a bodily and emotional kind and to some specific abstract concepts is based on concepts characterized by a perceptual core (or by a link to perceptual information).

To define this core, we will however not use Johnson and Lakoff's term 'image schema' which we consider misleading in respect to our position since on the one hand of the term image evokes a picture, and on the other it recalls the research tradition of imagery, which we refer to only for in some minor respects (see Chap. 6, Sect. 4). We will instead use the term '*perceptual schema*'. The word 'perceptual' recalls the perceptual nature of the information that composes the core concept. This notion of perception is not meant to include external senses only, but also bodily information and in particular 'proprioception' as it has been defined previously (see Chap. 2). The word 'schema' assumes a meaning similar to that intended by Lakoff and Johnson, and indicates the way in which perceptual information is organized in the perceptual representation or in the perceptual core. The perceptual information of the core must accomplish a quasi-conceptual function in the sense that it is needed to recognize and to classify different instances. For this reason, it cannot be disorganized, static or particular, but must form a schema—i.e. a pattern, a map or a model —that can always be compared to new stimuli and lead to their recognition and classification. In other words, the information this perceptual core is composed of must be arranged in the form of an organizing structure in order to accomplish the function of a schema for the recognition and classification of instances.

If we go back to the problem of the semantic competence we started with in this chapter, the hypothesis we argue for suggests that the perceptual schema is what allows us to have referential competence. The fact that this schema is perceptual and therefore independent from any linguistic information assures that referential competence will be largely independent from inferential competence and provides an argument in favor of the idea that linguistic mastery is not in any case a necessary condition for referential competence, in the sense that even without language it is possible to group together some of the instances one perceives (through the external senses or, as we will see, through proprioception) on a non-linguistic basis.

7 The Issue of Referential Competence

With the exception of the studies that rely on Johnson and Lakoff's view, the idea that cognition is grounded on perceptual representations that allow for recognition and categorization of instances is usually applied only to concrete concepts which refer to observable objects in the external world and have spatial (geometrical and topological) properties. In this sense, the notion of perceptual representation that is implicit in most studies that address the issue of the relationship between perception and conception (for a review on grounded cognition see Chap. 2, Sect. 1) is analogous to the notion of proxytype introduced by Prinz. This might lead one to think that only concrete concepts that refer to material objects located in the external world, which are intersubjectively observable and have spatial properties should have a perceptual core of this kind. Therefore, concepts like BIRD or APPLE would have such a core, while concepts such as ANGER, QUARK or FREEDOM (emotional and abstract concepts) would not have any core of this kind but consist merely of inferential conceptual knowledge.

From this point of view, it would follow that it is only possible to form perceptual representations starting from concrete and observable objects and further, that referential competence develops before and remain relatively independent from inferential competence with respect to the class of concrete concepts. If someone can form, on the basis of his/her perceptual experience, a schematic representation for e.g. birds like the one hypothesized by Prinz, s/he will be able to identify the birds in the external world in a non-linguistic fashion and to differentiate them from other types of objects on the basis of their spatial properties (overall shape, parts, relationships between the parts, etc.). In fact, the capacity to group birds together will be independent from and prior to language acquisition. If the subject does know a language, it is plausible to assume that his capacity to recognize birds in the world will still be largely independent from and poorly influenced by his/her inferential knowledge on birds.[31] However, in this respect the issue arises as to what would happens in the case of concepts referring to objects that do not have intersubjectively observable, physical and spatial properties. In fact, Prinz's notion of proxytype as a possible description of a perceptual core concept only applies to concepts which do indeed refer to intersubjectively observable objects with spatial properties.

The suggestion implicit in Marconi's theory in relation to this issue is that "the distinction between the two aspects of semantic competence, inferential and referential, is not relevant to all words in the lexicon of a natural language" (Marconi 1997: 64): the idea of referential competence as relatively independent from inferential competence only applies to some particular cases (concrete concepts), while in other cases inferential competence entirely determines our referential capacities. Marconi embraces Harman's point of view that: "it is possible that there are certain sorts of theoretical terms, like 'quark', that play no role in perception at all, so that the content of the concepts they express is determined entirely by inferential role" (Harman 1987: 61). This position refers to an idea of semantics according to which "our ability to apply words may be based on the ability to apply other, more common words" that can be "cognitively basic" (Marconi 1997: 62). In fact, Marconi's implicit suggestion is that all kinds of concepts not referring to concrete, physical and observable object with spatial properties are made uniquely or primarily of inferential information, which is driven by the language and depends on the way words are defined in this language and on the way individual speakers

[31]It is also plausible to assume—consistently with Carey (2009)—that a perceptual representation would also allow us to make some non-linguistic inferences.

have learned these definitions (i.e. from the specific inferential competence they have developed).

This hypothesis describes a position which is largely shared by both philosophical and cognitive research, and which is shared also by the literature on the so-called grounding problem mentioned before (Chap. 2, Sect. 1) and by the classical perspective on language and linguistic structure of neopositivistic origins. An example in this direction is e.g. the idea of Harnad (1990) that—in order to solve the symbol grounding problem—it is sufficient to show that a subset of representations (i.e. those that have an observable referent) are empirically grounded, while the others can be conceived as indirectly grounded, since they are constructs built on the grounded ones. As it has been suggested also by Alan Cruse (2000) with reference to the philosophical tradition:

> A standard picture of meaning within the philosophy of language identifies a set of words, known as the observation vocabulary, whose meaning are fixed by their relations with observable properties of the environment. The meanings of words not belonging to this set are fixed by a network of inferential or other relations to the meanings of other words, including those belonging to the observation vocabulary. We can take observation vocabulary items to be the more basic. A general assumption is that the concrete/observable/basic terms will be the first learned, probably the first to arise in the evolution of human language, the most accessible in psycholinguistic terms, and the most likely to be points of convergence between widely different languages, and so on (Cruse 2000: 52).

The view that Cruse characterizes as the standard picture of language draws a sharp line of demarcation between the terms that have a concrete referent, observable in the external word, and the terms that do not have a referent of this kind and further hypothesizes that language has a constructive structure, in virtue of which all terms that do not belong to the observation vocabulary are inferential constructs. This idea is shared by many studies that address the issue of the origin of concepts from a cognitive point of view. As e.g. Susan Carey also states in some preliminary observations leading to her hypothesis on the origin of concepts: "While there certainly is also a causal story to be told about how the content of the concept *electron* is determined, it will have a very different flavor form the explanation for how the content of round is determined. I shall argue […] that inferential role plays a part in the causal process through which conceptual representations [i.e. theory laden representations] pick out their referents" (Carey 2009: 8).

This constructive view of language and of linguistic competence suggests that—in order to master words that do *not* have observable, physical and concrete, referents in the external world, people don't need any referential competence distinct from inferential competence. In such cases full lexical semantic competence corresponds mainly to inferential competence, since the knowledge we have of these words is in fact primarily inferential. When we identify e.g. an action as 'honest' or a condition as 'peace', we do so in virtue of the fact that we have specific inferential knowledge about what makes an action 'honest' and what 'peace' is. This thesis underlies a position that can be called the *principle of inferential compensation* suggesting that in the case of concepts that do not denote anything concrete, spatial and material, directly observable through the senses in the external world, our propositional knowledge provides (compensates) for the lack of perceptual information and the categorization is carried out on the basis of this knowledge. We cannot observe anything in the world that directly corresponds e.g. to 'honesty' and 'peace'. We inferentially define 'honesty' and 'peace' and we identify in the world something as honest or as peace on the basis of this definition, depending on whether the considered action fulfills the definition of 'honest' or on whether the considered situation fulfills the definition of peace.

The view we just sketched raises a serious problem for research on concepts and on the constitution of concepts: are concepts that do not denote spatial and observable material objects really composed of inferential information only or are they also made up—at least in part—of some kind of perceptual or quasi-perceptual information, even though we are not aware of it and even though it is a different form of perceptual information with respect to those that we assumed to form the proxytype?

This issue concerns different conceptual classes that type instances which are also very different from each other. Without demanding a complete taxonomy, we can certainly consider the class of theoretical (abstract) concepts referring to instances of the external world that can be observed only through an inferentially guided procedure like e.g. CELL or ELECTRON, QUARK, FREEDOM or DEMOCRACY and general categories we use to organize our experience from a global point of view like ANIMATE/INANIMATE, LIVING/NON-LIVING etc. But we can also raise this concern in relation to concepts referring to internal states of subjects like bodily states (e.g. PAIN, THIRST, HUNGER) and various kinds of emotions (e.g. FEAR, ANGER, LOVE, FRIENDSHIP, ENVY). Theoretical concepts, categories and internal states cannot be represented through a perceptual core made of spatial information like the one that supports our categorization of e.g. cats. This does not mean however that these must be concepts made entirely of inferential information and that our referential competence with respect to them must entirely depend on and be fully mediated by our inferential competence. The analysis we will carry on in the next chapters aims at showing that these concepts and the capacity we have to categorize the world according to them do not rely exclusively on inferential information. In these chapters we will try to show that—also with respect to these 'non-concrete concepts'—people exhibit a specifically referential competence that is based on quasi-perceptual information, which allows them, among other things, to fully master the corresponding words.

The thesis we will put forward does not suggest that inferential competence does not play any role with respect to the constitution of these concepts and with respect to their mastery. However, we will claim that the concepts that classify instances which lack spatial properties are also not only composed of inferential information, but also of referential information and are based on quasi-perceptual representations which are essential in order to identify the instances belonging to them. Also in the case of these concepts, referential competence comes first, i.e. it is genetically primary with respect to inferential competence, and is (more or less—depending on the case) relatively independent from it.

In the case of observable, spatial instances in the external world, referential competence is enabled by a perceptual representation made of geometrical and topological information which is extracted primarily thorough vision. The hypothesis we put forward in the next chapters suggests that, in the case of non-observable and non-spatial instances like the one belonging to the mentioned classes—mainly: general categories, theoretical (abstract) concepts and concepts referring to internal states—referential competence *is enabled by information generated by corporeity*, i.e. by *proprioceptive information* as it was previously defined (Chap. 2). The analysis we carried out in this chapter does not challenge the idea—accepted almost unanimously by the studies on the conceptual architecture—that our conceptual system is constructive and that our capacity to understand more complex concepts (and to master the corresponding words) is based on that of understanding more basic concepts, which are characterized by a stronger referential (i.e. perceptual) component.

The presupposition that characterizes this work is rather that—before being assembled—the parts (i.e. the concepts our constructions are built of) must be constituted. We maintain that 'proprioception' is a necessary means—an implicit source of information that operates without (direct) awareness—for the constitution of many (in not all) the contents that people uses in the architecture of their knowledge. The aim of our analysis is foundational: our goal is to investigate the composition of specific conceptual classes in order to identify their constituents, i.e. the kinds of information they are made of. In the following chapters we will try to show that we have referential competence also with respect to conceptual classes like general categories, theoretical (abstract) concepts and concepts referring to internal states which do not denote observable, spatial instances in the external world and that in these cases referential competence relies on proprioceptive (propriosensitive) information.

References

Anggoro, F. L., Waxman, S. R., & Medin, D. L. (2008). Naming practices and the acquisition of key biological concepts. Evidence from English and Indonesian. *Psychological Science, 19*(4), 314–319.

Armstrong, S. L., Gleitman, L. R., & Gleitman, H. (1983). What some concepts might not be. *Cognition, 13*(3), 263–308.

Baghramian, M., & Carter, J. A. (2016). Relativism. In E. N. Zalta (Ed.) *The stanford encyclopedia of philosophy*. https://plato.stanford.edu/archives/win2016/entries/relativism/.

Barsalou, L. W. (1987). The instability of graded structure: Implications for the nature of concepts. In U. Neisser (Ed.), *Concepts and conceptual development: Ecological and intellectual factors in categorization* (pp. 101–140). Cambridge: Cambridge University Press.

Barsalou, L. W. (1992). Frames, concepts and conceptual fields. In A. Lehrer & E. F. Kittay (Eds.), *Frames, fields, and contrasts* (pp. 21–74). Hillsdale, NJ: Lawrence Erlbaum.

Barsalou, L. W. (1999). Perceptual symbol systems. *Behavioral and Brain Sciences, 22*(4), 577–660.

Barsalou, L. W. (2012). The human conceptual system. In M. J. Spivey, K. McRae, & M. F. Joanisse (Eds.), *The Cambridge handbook of psycholinguistics* (pp. 239–258). Cambridge: Cambridge University Press.

Bermúdez, J. L. (2003). *Thinking without words*. Oxford: Oxford University Press.

Biedermann, I. (1987). Recognition-by-components: A theory of human image understanding. *Psychological Review, 94*(2), 115–147.

Biedermann, I. (1990). Higher-level-vision. In N. D. Osherson, S. Kosslyn, & J. Hollerbach (Eds.), *An invitation to cognitive science: Visual cognition and action* (pp. 41–72). Cambridge, MA: MIT Press.

Block, N. (1986). Advertisement for a semantics for psychology. In P. A. Rench, T. E. Uehling, & H. K. Wettstein (Eds.), *Midwest studies in philosophy. Volume 10: Studies in the philosophy of mind* (pp. 615–678). Minneapolis: Minnesota University Press.

Burge, T. (1979). Individualism and the mental. In P. A. French, T. E. Uehling, & H. K. Wettstein (Eds.), *Midwest studies in philosophy. Volume 4: Studies in metaphysics* (pp. 73–122). Minneapolis: University of Minnesota Press.

Carey, S. (1985). *Conceptual change in childhood*. Cambridge, MA: MIT Press.

Carey, S. (1991). Knowledge acquisition: Enrichment or conceptual change? In S. Carey & R. Gelman (Eds.), *The epigenesis of mind: Essays on biology and cognition* (pp. 257–291). Hillsdale, NJ: Lawrence Erlbaum.

Carey, S. (2009). *The origin of concepts*. Oxford: Oxford University Press.

Carey, S., & Spelke, E. S. (1994). Domain-specific knowledge and conceptual change. In L. Hirshfeld & S. Gelman (Eds.), *Mapping the mind: Domain specificity in cognition and culture* (pp. 169–200). New York: Cambridge University Press.

Carruthers, P., & Boucher, J. (1998). *Language and thought. Interdisciplinary themes*. Cambridge: Cambridge University Press.

Chomsky, N. (1986). *Knowledge of language. Its nature, origin and use*. London: Praeger.

Clark, E. V. (1993). *The lexicon in acquisition*. Cambridge, New York: Cambridge University Press.

Cruse, A. (2000). *Meaning in language. An introduction to semantics and pragmatics*. Oxford: Oxford University Press.

Deacon, T. (1997). *The symbolic species: The coevolution of language and the brain*. New York: Norton.

Dellantonio, S., & Pastore, L. (2006). What do concepts consist of? The role of geometric and proprioceptive information in categorization. In P. Hanna, A. McEvoy, & P. Voutsina (Eds.), *An anthology of philosophical studies* (pp. 91–102). Athens: ATINER Press.

Dellantonio, S., & Pastore, L. (2009). Concetti e rappresentazioni semantiche: Un'ipotesi genetica. In L. Pastore & S. Dellantonio (Eds.), *Percezione, rappresentazione e coscienza. Questioni filosofiche e problemi epistemologici* (pp. 233–275). Pisa: ETS.

Dellantonio, S. (2007). *Die interne Dimension der Bedeutung. Externalismus, Internalismus und semantische Kompetenz*. Frankfurt am Main: Peter Lang.

Dretske, F. (1999). *Knowledge and the flow of information*. Cambridge, MA: CSLI Publications.

Fauconnier, G. (1994). *Mental spaces: Aspects of meaning construction in natural language*. Cambridge: Cambridge University Press.

Finke, R. A. (1985). Theories relating mental imagery to perception. *Psychological Bulletin, 98*(2), 236–259.

Fodor, J. A. (1975). *The language of thought*. New York: Thomas Y. Crowell Company.

Fodor, J. A. (1983). *The modularity of mind*. Cambridge, MA: MIT Press.

Fodor, J. A. (1987). *Psychosemantics: The problem of meaning in cognitive science*. Cambridge, MA: MIT Press.

Fodor, J. A. (1994). *The ELM and the expert*. Cambridge, MA: MIT Press.

Fodor, J. A. (1998). *Concepts. Where cognitive science went wrong*. Oxford: Oxford University Press.

Fodor, J. A. (2003). *Hume variations*. Oxford: Oxford University Press.

Fodor, J. A. (2008). *LOT 2. The language of thought revisited*. Oxford: Clarendon Press.

Gärdenfors, P. (2007). Cognitive semantics and image schemas with embodied forces. In J. M. Krois, M. Rosengren, A. Steidele, & D. Westerkamp (Eds.), *Embodiment in cognition and culture* (pp. 57–76). Amsterdam, Philadelphia: John Benjamins.
Gibbs, R. W. (2005). Embodiment in metaphorical imagination. In D. Pecher & R. A. Zwaan (Eds.), *Grounding cognition. The role of perception and action in memory, language, and thinking* (pp. 65–92). Cambridge: Cambridge University Press.
Gibson, J. J. (1977). The theory of affordances. In R. Shaw & J. Bransfor (Eds.), *Perceiving, acting, and knowing: Toward an ecological psychology* (pp. 127–143). Hillsdale, NJ: Lawrence Erlbaum.
Goldman, A. I. (1993). *Philosophical applications of cognitive science*. Boulder, CO: Westview Press.
Goss, A. E. (1961). Verbal mediating responses and concept formation. *Psychological Review, 68,* 248–274.
Greenberg, M., & Harman, G. (2006). Conceptual role semantics. In E. Lepore & B. Smith (Eds.), *The Oxford handbook of philosophy of language* (pp. 295–323). Oxford: Clarendon Press.
Harman, G. (1987). (Nonsolipsistic) conceptual role semantics. In E. Lepore & B. Loewer (Eds.), *New directions in semantics* (pp. 55–81). London: Academic Press.
Harnad, S. (1990). The symbol grounding problem. *Physica D: Nonlinear Phenomena, 42,* 335–346.
Johnson, M. (1987). *The body in the mind: The bodily basis of meaning, imagination, and reason*. Chicago: Chicago University Press.
Jackendoff, R. (1983). *Semantics and cognition*. Cambridge, MA: MIT Press.
Jackendoff, R. ([1987] 1994). *Consciousness and the computational mind* (4th ed.). Cambridge, MA: MIT Press.
Jackendoff, R. (1990). *Semantic structures*. Cambridge, MA: MIT Press.
Jackendoff, R. (1992). *Languages of the mind. Essays on mental representation*. Cambridge, MA: MIT Press.
Jackendoff, R. (1996). How language helps us to think. *Pragmatics & Cognition, 4*(1), 1–34.
Johnson-Laird, P. N. (1983). *Mental models: Towards a cognitive science of language, inference, and consciousness*. Cambridge: Cambridge University Press.
Kluender, K. R. (1994). Speech perception as a tractable problem in cognitive science. In M. A. Gernsbacher (Ed.), *Handbook of psycholinguistics* (pp. 173–217). San Diego (CA): Academic Press.
Kluender, K. R. & Kiefte, M. (2006). Speech perception within a biologically realistic information-theoretic framework. In M.J. Traxler & M.A. Gernsbacher (Eds.), *Handbook of psycholinguistics* (2nd ed., pp. 153–199). Amsterdam, Boston: Elsevier.
Kosslyn, S. M. (1973). Scanning visual images: Some structural implications. *Perception and Psychophysics, 14*(1), 90–94.
Kosslyn, S. M. (1980). *Image and mind*. Cambridge, MA: Harvard University Press.
Kripke, S. (1979). A puzzle about beliefs. In A. Margalit (Ed.), *Meaning and use* (pp. 239–283). Dordrecht/Berlin: Springer.
Lakoff, G. (1987). *Women, fire and dangerous things: What categories reveal about the mind*. Chicago: University of Chicago Press.
Lakoff, G., & Johnson, M. (1980). *Metaphors we live by*. Chicago: Chicago University Press.
Langacker, R. W. (1987). *Foundations of cognitive grammar. Volume 1: Theoretical prerequisites*. Stanford: Stanford University Press.
Lappin, S. (2005). Semantics. In R. Mitkov (Ed.), *The Oxford handbook of computational linguistics* (pp. 91–111). Oxford: Oxford University Press.
Laurence, S., & Margolis, E. (1999). Concepts and cognitive science. In E. Margolis & S. Laurence (Eds.), *Concepts. Core readings* (pp. 3–81). Cambridge, MA: MIT Press.
Loar, B. (1985). Social content and psychological content. In R. H. Grimm & D. D. Merrill (Eds.), *Contents of thought* (pp. 99–110). Tucson: Arizona University Press.

Mandler, J. M. (1998). Representation. In D. Kuhn & R.S. Siegler (Eds.), *Handbook of child psychology. Volume 2: Cognition, perception, and language* (5th ed., pp. 255–308). New York: Wiley.

Mandler, J. M. (2004). *The foundations of mind. Origin of conceptual thought*. Oxford: Oxford University Press.

Marconi, D. (1987). Two aspects of lexical competence. *Lingua e Stile, 12*(3), 385–395.

Marconi, D. (1991). Understanding and reference. *Sémiotiques., 1*(1), 9–25.

Marconi, D. (1995). On the structure of lexical competence. *Proceedings of the Aristotelian Society. New series, 95*, 131–150.

Marconi, D. (1997). *Lexical competence*. Cambridge, MA: MIT Press.

Margolis, E., & Laurence, S. (Eds.). (1999). *Concepts. Core readings*. Cambridge, MA: MIT Press.

Marr, D. (1982). *Vision: A computational investigation into the human representation and processing of visual information*. San Francisco (CA): W.H. Freeman.

Marr, D., & Nishihara, K. (1978). Representation and recognition of spatial organization of three-dimensional shapes. *Proceedings of the Royal Society of London, Series B: Biological Sciences, 200*(1140), 269–294.

Medin, D. L., & Schaffer, M. M. (1978). Context theory of classification learning. *Psychological Review, 85*(3), 207–238.

Miller, G. A. (1991). *The science of words*. New York: W.H. Freeman.

Miller, G. A., & Fellbaum, C. (1991). Semantic networks of English. *Cognition, 41*(1–3), 197–229.

Moravcsik, J. M. E. (1981). How do words get their meanings? *The Journal of Philosophy, 78*(1), 5–24.

Murphy, G. L. (2004). *The big book of concepts*. Cambridge, MA: MIT Press.

Murphy, G. L., & Medin, D. (1985). The role of theories in conceptual coherence. *Psychological Review, 92*(3), 289–316.

Neisser, U. (1976). *Cognition and reality. Principles and implications of cognitive psychology*. San Francisco (CA): Freeman & Co.

Pagin, P. (2006). Meaning holism. In E. Lepore & B. Smith (Eds.), *The Oxford handbook of philosophy of language* (pp. 213–232). Oxford: Clarendon Press.

Paivio, A. (1971). *Imagery and verbal processes*. New York: Rinehart & Winston.

Paivio, A. (1986). *Mental representations. A dual coding approach*. Oxford: Oxford University Press.

Pinker, S., & Jackendoff, R. (2005). The faculty of language: What's special about it? *Cognition, 95*(2), 201–236.

Pinker, S. (2007). *The language instinct. How the mind creates language*. New York: Harper Perennial Modern.

Prinz, J. J. (2002). *Furnishing the mind. Concepts and their perceptual basis*. Cambridge, MA: MIT Press.

Putnam, H. (1975). The meaning of 'meaning'. In H. Putnam (Ed.), *Philosophical papers. Volume 2: Mind, language and reality* (pp. 215–271). Cambridge: Cambridge University Press.

Rosch, E. H., Mervis, C. B., Gray, W. D., Johnson, D. M., & Boyes-Braem, P. (1976). Basic objects in natural categories. *Cognitive Psychology, 8*(3), 382–439.

Sapir, E. ([1921] 2004). *Language. An introduction to the study of speech*. New York: Dover Publications.

Searle, J. R. (1980). Minds, brains, and programs. *Behavioral and Brain Sciences, 3*(3), 417–457.

Sinha, C., & Jensen de López, K. (2000). Language, culture and the embodiment of spatial cognition. *Cognitive Linguistics, 11*(1–2), 17–41.

Smith, E. E., & Medin, D. L. (1981). *Categories and concepts*. Cambridge, MA: Harvard University Press.

Spelke, E. S. (2000). Core knowledge. *American Psychologist, 55*(11), 1233–1243.

Spelke, E. S., & Kinzler, K. D. (2007). Core knowledge. *Developmental Science, 10*(1), 89–96.

Spivey, M. J., Richardson, D. C., & Gonzales-Marques, M. (2005). On the perceptual-motor and image-schematic infrastructure of language. In D. Pecher & R. A. Zwaan (Eds.), *Grounding cognition. The role of perception and action in memory, language, and thinking* (pp. 246–281). Cambridge: Cambridge University Press.

Talmy, L. (1988). Force dynamics in language and cognition. *Cognitive Science, 12*(1), 49–100.

Tomasello, M. (1999). *The cultural origins of human cognition*. Cambridge, MA: Harvard University Press.

Tomasello, M. (2001). Perceiving intentions and learning words in the second year of life. In M. Bowerman & S. C. Levinson (Eds.), *Language acquisition and conceptual development* (pp. 132–258). Cambridge: Cambridge University Press.

Ullman, S. (2000). *High-level vision. Object recognition and visual recognition*. Cambridge, MA: MIT Press.

Waxman, S. R. (1999). The dubbing ceremony revisited: Object naming and categorization in infancy and early childhood. In D. L. Medin & S. Atran (Eds.), *Folkbiology* (pp. 233–284). Cambridge, MA: MIT Press.

Waxman, S. R., & Lidz, J. L. (2006). Early word learning. In D. Kuhn & R. Siegler (Eds.), *Handbook of child psychology* (Vol. 2, 6th ed., pp. 299–335). New York: Wiley.

Waxman, S. R., & Markow, D. B. (1995). Words as invitations to form categories: Evidence from 12- to 13-month-old infants. *Cognitive Psychology, 29*(3), 257–302.

Weiskopf, D. A. (2008). The origins of concepts. *Philosophical Studies, 140*(3), 259–384.

Whorf, L. B. ([1956] 2012). In J. B. Carroll (Ed.), *Language, thought and reality: Selected writings of Benjamin Lee Whorf*. Cambridge, MA, New York, London: MIT Press.

Chapter 4
In the Beginning There Were Categories

The Bodily Origin of Prelinguistic Categorical Organization: The Example of Folkbiological Taxonomies

Abstract Studies of categorization in psychology and the cognitive sciences have made use of the notions of 'category' and 'concept' without precisely defining what is meant by either; in fact, often these terms have been used as synonyms, making it difficult to address specific issues related to conceptual development. This chapter begins by discussing the definitions of, as well as the distinctions between, 'categories' and 'concepts' in the classical philosophical tradition (Aristotle, Kant and Husserl). We introduce our view of categories with reference to Husserl. Categorization is defined as the way in which our experience is originally (pre-linguistically) organized in a passive and fully unconscious manner on the basis of universal structuring principles. Conceptualization is explained as a later process in which the earlier categorical macro-classes are further subdivided into more specific and detailed sets; this later process also relies on linguistic learning and exposure to culture. On the basis of Ray Jackendoff, Jean Mandler, George Lakoff and Mark Johnson's work, we hypothesize that categorization is a two-step process that begins with the formation of homogeneous sets of entities partitioned into regions that describe the ontological boundaries of the objects that humans perceive (categories) and continues at a later stage with the development of more specific classifications (concepts). In this chapter, we mainly address three related issues: Why should we assume that there is categorical organization which precedes the development of a conceptual system? How do categories and concepts relate to each other? Shall we hypothesize that categories are innate or that they are formed before concepts on the basis of information and organizational structure available at a very early developmental stage? We show that categorical partitions are necessary for categorization and, following Mandler, that the general categories we form at an early age do not match our adult superordinate concepts. As for the third issue, we argue that there might be no need to assume that categories are innately present in the human mind, since their formation can be explained—at least in certain cases—by basic mechanisms that work on body (propriosensitive) information. This hypothesis will not be discussed in general, but in relation to a particularly relevant example of categorical partition, i.e. the folk-biological dichotomy between ANIMATE/INANIMATE. This is compared with other dichotomies that derive from it, but are not directly categorical such as LIVING/NON-LIVING and BIOLOGICAL/NON-BIOLOGICAL.

S. Dellantonio and L. Pastore, *Internal Perception*, Studies in Applied Philosophy, Epistemology and Rational Ethics 40, DOI 10.1007/978-3-662-55763-1_4

1 Categories—Towards a Definition (I): A Brief Overview of the Classical Philosophical Tradition

Studies on categorization carried out in the field of psychology and the cognitive sciences have made use of the notions of 'category' and 'concept' without precisely defining what is meant by either. It is e.g. Jean Mandler who underscores how: "With a few notable exceptions [...] terms such as concept and category usually go undefined in the literature, even though psychologists use these expressions all the time" (Mandler 2004: 4). This lack of precise definition is accompanied by the fact that the two notions have often been used as if they were more or less synonymous. This might be ascribed to the fact that many studies rely on a view of concepts and of categories which implicitly assumes these notions are two sides of the same coin. As e.g. Douglas Medin and Lance Rips point out: "Cognitive scientists generally agree that a concept is a mental representation that picks out a set of entities, or a category. That is, concepts *refer*, and what they refer to are categories. It is also commonly assumed that category membership is not arbitrary but rather a principled matter. What goes into a category belongs there by virtue of some law-like regularities" (Medin and Rips 2005: 37). In substance, this definition suggests that concepts are mental entities referring to sets of objects in the external world: these sets are called categories. The concept of dog for example is a mental entity which refers to the set—i.e. the category—of dogs (Fig. 1).

An analogous definition is given by Robert Goldstone and Alan Kersten in a work that aims at summarizing the main lines of the debate on categorization:

> It is common to distinguish between a concept and a category. A concept refers to a mentally possessed idea or notion, whereas a category refers to a set of entities that are grouped together. The concept dog is whatever psychological state signifies thoughts of dogs. The category dog consists of all the entities in the real world that are appropriately categorized as dogs. The question of whether concepts determine categories or vice versa is an important foundational controversy (Goldstone and Kersten 2003: 600).

These definitions are problematic for many aspects. First of all, they might lead to an ontological interpretation of the notion of category as the set of entities in the real world that 'belong together'. This is highly disputable from the point of view of a theory of knowledge, since it implies that things 'naturally' belong to specific

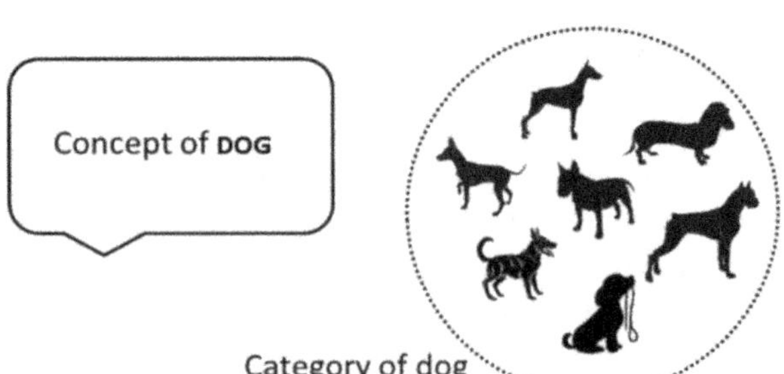

Fig. 1 DOG: mental concept versus category understood as the set of all real dogs

groupings which existed before and independently of the way in which we come to know them. The justification of these groupings would therefore require ontological or metaphysical, instead of epistemological, criteria. Secondly, these definitions of 'concepts' and 'categories' overlook the specificity of the classic philosophical notion of category defined mainly by a line of thought that goes from Aristotle to Kant and to Husserl. Among other things, neglect of the philosophical tradition makes it difficult to address specific issues about categories and concepts that are essential in contemporary cognitive research on these topics. For this reason, it is useful to go back to the classical tradition, at least briefly; to show how categories were originally conceptualized as well as the analogies and differences between the classic views and the ones described at the beginning of this section. In this way we aim to provide a revised definition of category and an explanation of what categories are and what they are used for: in fact, this will be the issue at stake in this chapter.

According to Aristotle's definition, *categories*[1] describe the largest *kinds* (or *genera*) *of properties that can be predicated of something existing, of an entity* (Aristotle 1963). A category is the highest *genus* that unifies all possible predications that can be referred to the same aspect of reality. Taken together, they express the largest groups (the highest *genera*) of 'properties' (of 'aspects') existing in reality. For this reason, Aristotle's theory of categories has a twofold function: on the one hand it allows us to linguistically describe the existing entities (under the presumption that the description given is exhaustive) and on the other it reveals (it allows us to know) the structure of reality as it is.

Since the properties that are said of something (are predicable) must refer to something existing, in the system of categories it is possible to trace a fundamental partition between *substance* and *accidents*. The substance defines the existing entities: it identifies at the highest level of concreteness what exists and describes the what-it-is of the existing entities; as such, substance fulfils the function of the *substratum* of all existing things—it is the carrier of all predications and their fundamental referent. The accidents are, so to speak, the 'qualitative aspects' that can be predicated of the substance: among them some belong to the substance in an essential manner while others belong in a merely contingent manner (some are essential while others are contingent properties of the substance). Categories fulfil the functions of specifying through progressive explications the properties of what exists (of the substance), and identifying and describing it as it is.

The Latin word *categoria* comes from the Greek verb *kategoréo* ('I accuse of', 'I show publicly"), which in turns derives from the verb *agoréno* (I say, I express) and from the prefix *katà* (against). Since they are predicable of an existing entity (i.e. substance—e.g. of a horse), categories individuate the necessary properties for its description and for its being known: e.g. its quantity (one), its quality (e.g., its color), its relations (is it bigger or smaller than some other thing), its place (where it

[1]Aristotle speaks of "kinds of predication" (*genê tôn katêgoriôn*). The word 'category' is introduced by the Romans.

is), the date (the time we are considering), but also other aspects like its posture (e.g. is it lying down or is it standing), its state (has a saddle on; has horseshoes, etc.), its action (it gallops, it runs), and its passion (it is injured).

This way of expressing the properties of substance is not meant merely as a linguistic description. Aristotle maintains that it reveals the way in which reality itself is structured and can be known: as the highest genera of all possible predications of the existing entities, the categories are also the fundamental partitions which reality is organized into. Therefore, in this view categories are characterized by a particular form of objectivity, which is granted by the ontological constitution of the world. They are not forms of organization of our knowledge and of our language—i.e. they are not generated by our mind and by the way in which our thought is organized; on the contrary, our thought and our language must be organized according to them in order to effectively know and describe the reality of the world. Categories are constitutively independent from our perception of the world since they depend on the way in which the world is organized and presents itself to our perception. As e.g. Robin Smith explains: "A given thing can be classified under a series of progressively wider genera: Socrates is a human, a mammal, an animal, a living being. The categories are the highest such genera. Each falls under no other genus, and each is completely separate from the others" (Smith 2012: §7.3).

Aristotle thinks that everything there is can be described and cataloged using this hierarchical system of categories, *genera* and *species* (Fig. 2).

Contemporary cognitive research (in the main)[2] is no longer concerned with the issue of whether there are highest genera and what they might be: from this point of view the definition of category proposed at the beginning of this section has nothing to do with that of Aristotle. By contrast, Aristotle's idea that our system of knowledge is hierarchically organized is compatible with a well-accredited hypothesis in contemporary cognitive research, according to which our conceptual knowledge is taxonomically organized and ranges from very specific (i.e. subordinate) concepts to very general (i.e. superordinate) concepts.

While in Aristotle's terminology only the highest superordinate concepts should be called categories, cognitive research speaks indifferently of categories and concepts at all levels of the hierarchy. Nevertheless, in spite of this difference, Aristotle's theory matches an idea that, as we mentioned before, could possibly be implied in the definition of category given at the beginning of this section on the basis of the current cognitive literature. In fact, according to Aristotle, the taxonomy underlying the category system does not describe only the way we speak or we think about reality, but also the way reality is; i.e. it corresponds to the way reality is ontologically organized.

However, this ontological characterization of categories and this notion of our system of knowledge was challenged by Immanuel Kant along with the idea of a correspondence between word categories and the way the world is ontologically

[2]We consider some exceptions later in this chapter.

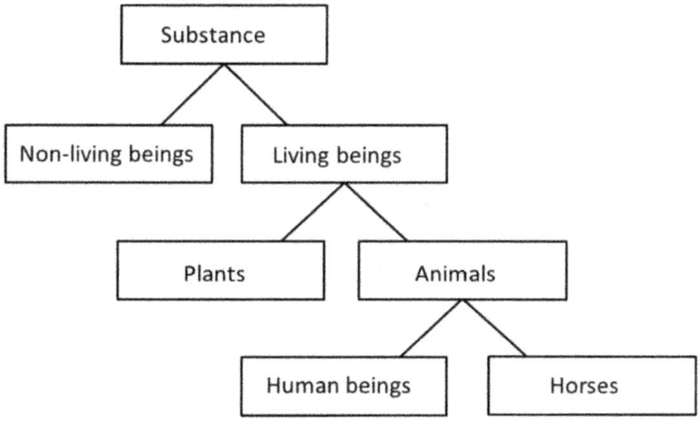

Fig. 2 Aristotle's hierarchical system of categories

organized. Kant's so-called 'Copernican Revolution' suggests exactly the opposite: that knowledge is not obtained by a process in which cognition conforms to objects, but, on the contrary, is obtained by a process in which the object (as an object of the senses) conforms to the constitution of our faculty of intuition—i.e. (in more contemporary terms) to the way we perceive the world (Kant [1787] 2003: 20ff). A view of this kind reverses the previous objectivist conception of the relationship between ontology and categories: categories are no longer seen as objective partitions of reality, but they are conceived as structures through which we organize our experience of the world. Furthermore, the 'place' of categories is the mind: categories are interpreted as mental functions that allow the constitution of our experience and of our knowledge; i.e. that allow us to form a *unified* and *coherent* representation of the world.

In order to understand how Kant defines categories, we need to briefly consider some relevant aspects of his theory of knowledge. In Kant's view our knowledge is the result of the continuous and steady cooperation between two distinct faculties: sensibility—which sensory data (*Empfindungen*) come from and which supplies the raw material of perceptual experience—and understanding (*Verstand*)—through which the material supplied by sensibility is conceptualized, thought and known (Kant [1787] 2003: 61–62; on this see also Aquila 1983: 33–82). Kant thinks of sensibility as a passive and receptive faculty which merely catches the constant flow of sensory stimuli that—through the channels of sensibility—affect (i.e. lead to modification of) the subject. The stimulation offered by sensibility to the subject is disorganized, chaotic and unstructured. However, in spite of this the world is perceived and known by subjects as something always coherent, unified and structured. This is possible because the raw material provided to the subject by sensibility must be endowed with structure through the action of understanding, which is described as a faculty that *actively* and *spontaneously* operates on sensory data, structuring them by means of concepts. The understanding is the seat of concepts which are described as functions—i.e. ordering strictures or rules imposed by the mind on the data coming from sensibility—which bestow coherence and structure on our experience and knowledge. In Kant's view, experience, thought and knowledge are considered to be representational activities of the human mind: the fact that we perceive and see the world in the organized way we do is a result of the working of our understanding and its concepts, which process the raw material supplied by sensibility into representations (*Vorstellungen*), i.e. into contents of thoughts which are accessible to our awareness.

According to Kant there are however different kinds of concepts which can be distinguished from each other on the basis of their content. Kant calls *empirical* the concepts that include sensory elements and that have therefore been formed on the basis of the flow of experience, while he calls *a priori* or *pure* the concepts whose existence do not depend on experience or sensibility, but are structures of the mind. Pure concepts are present with their own structure in the understanding before and independently of any real or possible experience. Empirical concepts are the results of the application to experience of the a priori structure of pure concepts. Since concepts are ordering structures imposed on experience, concepts can carry out their function in full only when they have empirical content. As Kant wrote: "Thoughts without intuitions are empty, intuitions without concepts are blind" (Kant [1787] 2003: 93).

In Kant's view, *categories are pure concepts—i.e. they are ordering structures whose form do not depend on experience*. Categories are therefore formal structures that carry out the duty of making sensible experience thinkable, organized and coherent: "categories are due to the nature of the mind and are imposed by the mind on the objects which it knows" (Paton 1936: 258). They describe the ways in which the human mind represents reality. Categories are a necessary condition of thought and of our discursive abilities (Waxman 1991: 118–155). As Kant states: "[T]he objective validity of the categories as a priori concepts rests, therefore, on the fact that, so far as the form of thought is concerned, through them alone does experience become possible. They relate of necessity and *a priori* to objects of experience, for the reason that only by means of them can any object whatsoever of experience be thought" (Kant [1787] 2003: 126).

Relying on Aristotle's theory of categories as predicable, Kant identifies four clusters of categories (quantity, quality, relation and modality—each of which is internally tripartite). The specific function categories carry out is to bestow logical 'form' and 'structure' on the elements that constitute the content of our understanding activity (i.e. of our thought), which we are aware of. The sensory stimuli are organized by means of the pure concepts of the understanding (categories) in a coherent and systematic representation: i.e. as objects which can be consciously perceived, thought and known.

According to Kant categorization is however only the first step in the constitution of our representation of the world. The second step, subsequent to categorization, is *conceptualization,* i.e. the development of concepts which are *not pure*, but provided with *empirical* content that serve the purpose of organizing and classifying our experience and to structure our knowledge. It should be however specified that this distinction between categorization and conceptualization can be made only as a merely logical abstraction, since the percepts and the objects of our thought and of our knowledge arise to our awareness always already categorized *and* conceptualized, so that we cannot be aware that they are the product of two different and consecutive operations.

To put Kant's view in contemporary terms, the word 'concepts' (in the sense of empirical concepts) indicates all those representations which we are aware of and that refer to objects that can be perceived and thought of *by more than one person* (i.e. intersubjectively). This characterization of concepts is introduced by Kant in order to differentiate the object of a concept from a mere sensation, which is also a conscious representation whose content is entirely 'private' (Kant [1787] 2003: 314). This is the reason why our shared experience and therefore our knowledge are not based on stimuli or sensations, but on concepts which are meant as functions that subsume and organize the manifold of our numerous empirical representations. While categories (the pure concepts of the understanding) articulate only the formal conditions to make conceptualization—i.e. the acquisition of contents—possible, concepts (empirical concepts) have content driven by the senses; therefore, they can also have a referent in the world and perform the function of partitioning and classifying things in the world in an always more detailed and specific manner.

This brief overview of Kant's position clearly shows that Kant introduces a very different view of categories with respect to that of Aristotle. While Aristotle conceives of categories in an objectivist and realistic sense (categories reveal the way reality is structured), according to Kant categories are a function of the mind: they are different from concepts only because they lack empirical content and the structuring they carry out is preliminary to conceptualization. This view introduces an approach to the theory of knowledge based on the study of the elements and of the structures the human mind is equipped with: based on representations, concepts and categories as mental entities (about the link between Kantian philosophy and contemporary cognitive research see e.g. Gardner 1985: 56ff). From this point of view, Kant's view is very similar to some conceptions proposed in contemporary cognitive research which we will consider in the next section. The common underlying idea is mainly that (a) categories are elements internal to the human mind (b) which serve the purpose of conferring structure on stimuli driven by the external world.

Further developments in post-Kantian philosophy exhibit an interesting variation on Kant's position introducing a more elaborated distinction between categories and concepts that still explain conceptualization as a later step in the structuring of our experience, preceded by other more basic processes which organize what we perceive. The position we will consider in this regard is one developed in the first half of the 20th century by the Austrian philosopher Edmund Husserl, which gave rise to the so called phenomenological tradition.

For Husserl as for Kant, the notion of 'category' identifies a basic element in the constitution of our experience and knowledge of the world; for Husserl as for Kant, categories form part of the system of concepts, i.e. they are special concepts which play specific functions. Categories precede concept formation and thus we can claim that also in Husserl's case categorization precedes conceptualization. Differing from Kant, however, for Husserl categories are not purely formal elements, i.e. they are not only purely logical functions we use to structure the world into entities (objects) ready to be thought and connected to each other.

Husserl distinguishes between formal and material categories.[3] Formal categories bestow logical structure on the world; they are the precondition for developing a language that enables us to speak of the world and its objects in a structured and coherent way. Material categories—which Husserl also calls *regions*—define the most general types or *genera* of objects we can find in the world experience, and speak about. Formal categories have a purely logical and formal nature and are therefore neutral, in the sense that they can be applied to any possible content, i.e. to any thinkable entity: formal categories define the essence of the content they are applied to which is made up of specific properties like number, quality, the way they might appear, their relationships with other objects, the structure they have, their dependence/independence from other objects or properties etc. Material (or regional) categories describe instead the largest fields to which something existing and thinkable might belong. Husserl identifies three macro-classes of regions—*nature*, *consciousness* and *culture*—which are in turn internally divided into subsets of objects. The *region of nature* is organized into classes like living things, non-living things, plants, animals, humans, etc.; the *region of consciousness* is made up of the different modalities of the acts of thoughts (ego, intentionality, experience, remembering, perceiving); while the *region of culture* is organized in classes like societies, values, different kinds of artifacts, the contents of scientific knowledge, etc.).

Both formal and material categories contribute to the constitution of our concrete experience and the objects of our thought. However, all the aspects concerning the content of our thought—concerning the 'matter' that 'fills up' (i.e. gives a concrete reference to) the logical forms through which we think and we speak—are due to the structure of material categories. Roughly speaking, *the structure of material categories determines the possible contents of the conceptual forms we use when we speak of our experience and when we investigate the world in order to develop some knowledge of it.* Formal and material categories have different paths of development. Both contribute to determining the structure of our knowledge, but material categories start operating at a very early stage in the constitution of our

---

[3]Categories are used to organize both our material and our ideal experience. Since our knowledge can either be about ideal entities (like numbers, geometrical forms, merely linguistic contents, etc.) or about material things (objects, perceptions, memories, people etc.), Husserl distinguishes between formal and material categories. Formal categories individuate the logical characteristics that allow the constitution of purely logical and linguistic objects: they allow e.g. the constitution of well-formed syntactical structures that make our language coherent and compositional. Material categories are, on the contrary, related to meanings and reference: they lead to what Husserl calls 'material fulfilment' of what has been constituted on the basis of the formal categories. This is of course a material fulfilment which can be further developed on the basis of conceptualization. On Husserl's terminology see Drummond (2008). On Husserl's definition of categories and their distinction into formal and material categories see Smith (1995, 2004: 176ff and 242ff).

experience: they work primarily on pre-linguistic—sensory and bodily—information; and are partially independent of higher modalities of the functioning of the mind. These are the reasons why the topic we specifically address in this section is material categories.

Kant hypothesized that—even though categorization and conceptualization can be logically distinguished as far as their function and their sequence in time are concerned—the contents that determine what we are aware of, and therefore how we see the world and what we know of it, are always the product of the working of both. Differing from Kant, Husserl thinks that our experience begins to be organized *before* the intervention of our logical and linguistic activity (in Kantian terms: before the faculty of understanding comes into play) and that we are aware of the results of this early, non-linguistic form of organization which therefore does play a role with respect to the way we see the world and how we develop our knowledge of it. In fact, according to Husserl, sensibility does not merely provide chaotic and disorganized stimulations, since already at the level of sensibility experience begins to be organized and structured. This structuring is carried out in a purely *passive* manner, in the sense that the structuring processes are automatic and the subject is completely unaware of them. The principles they rely on depend on the working of sensibility itself. In this way, sensibility provides contents that are already structured and formed, even if in a still vague way, to the 'highest' level structuring activities of understanding. This higher structuring can therefore build on the earlier organization making it more explicit and detailed and forming linguistic structures appropriate to its description.

According to Husserl this first form of categorical organization is the product of the self-organization of sensibility. Material categories describe the highest genera of concrete phenomena we can experience—or, in other words, the larger classes into which our experience is spontaneously organized. As we already mentioned, the notion of (material) category—as Husserl defines it—is closely related to that of *region* (see Husserl [1913] 1982: 28ff). Husserl maintains that categories are *regions* in the sense that they describe the original and fundamental partition of our experience into broad realms of objects which we are subjectively aware of. From the point of view of our experience in its concreteness, the regions delimit possible ontologies (sets of objects and perceptual phenomena) which describe the most fundamental partitions that characterize our direct and naïve experience of the world and of ourselves. In our awareness we experience the world as already 'fractioned' in macro-classes (i.e. regions) of phenomena and we cannot experience it otherwise.

As the geographical metaphor recalled by the word already suggests, the regions describe elements of a *morphological* nature (in fact, Husserl also speaks of 'morphological concepts': see e.g. Husserl [1929] 1977: 166ff), i.e. elements related to the form and the structure of a territory which depend in turn on how this territory is, independently of who observes or describes it. In Husserl's metaphor the limits of the regions depend on the natural partitions internal to the territory; i.e. on the 'texture' of the territory itself. As far as experience is concerned, this suggests that regions are something constituted inside experience on the basis of its 'texture' before the active and explicit intervention of consciousness. According to Husserl our experience is initially structured into regions; each region defines or, more specifically, expresses the essential properties that characterize the objects which the region is composed of. Regions describe therefore the fundamental partitions of all the phenomena forming our experience—they also define the ontology of our experience. For this reason, since categories are the constituents of the primary structuring of our way of experiencing the world as divided into regions, Husserl calls them also 'objective regional concepts' (see e.g. Husserl [1913] 1980: 22).

To sum up, regions as (material) categories are the product of the self-organizing structures of sensory experience; they are part of the awareness and knowledge subjects have, but they have their origin before and independently of the activity of the subject. This notion of category is analogous to Kant's definition in the sense that for both philosophers categories are the basic means for structuring our knowledge. However, Husserl's characterization of this notion does not share Kant's intellectualism: in fact, according to Husserl categories are not purely intellectual structures that are imposed on sensibility to produce experience. On the contrary, categories are part of the material dimension of experience and pertain specifically to sensibility and its organizing principles. Husserl distances himself from Kant also with respect to the way in which categories need to be identified is concerned. According to Husserl, categories cannot be identified independently from experience: they are not 'postulates' or elements that are not included in the complex of our perception, but they can be identified only following clues from our sensory experience: we cannot deduce categories, but rather can find them ourselves and step by step we can grasp them ourselves (see Husserl [1913] 1980: 22).

In this way, Kant's intellectualistic and radically mentalistic view on categories is further developed by Husserl in the direction of a critical reconsideration of Artistotle's position. According to Husserl categorical organization is not separated from material experience. However, Husserl's position is not merely objectivistic as is Aristotle's conception because in Husserl's view it is material categories which partition the world into the regions experienced by subjects by means of the structuring produced by the material categories That is, categorical organization depends on the subjects' capacities to organize the sensory information they receive.

The idea of category we will develop in this chapter mainly relies on Husserl's position, even though it also makes use, at least in part, of Kant's definition. In fact, Kant's position is essential for understanding some mentalistic views of categories developed in the field of contemporary cognitive research (e.g. that of Ray Jackendoff, presented in the next section). The first point we will make is that the notions of category and of concepts are not interchangeable, but describe different things. We will rely on and (freely) reinterpret the idea of categories proposed by Husserl in order to show that categories are *the material condition of possibility for the development of concepts*: in our view the development of categories is a necessary precondition for the formation of concepts. Categorization will be described as the passive and fully unconscious process through which our experience is originally organized in a prelinguistic manner on the basis of universal structuring principles that depend on the way humans get and organize their perceptual information. On the contrary, conceptualization will be explained as a later process in which the earlier categorical macro-classes are further subdivided into more particular and detailed sets also on the basis of linguistic learning and exposure to a culture.

By the term 'category' we mean therefore macro-classes of highly general kinds that are produced by a spontaneous and automatic structuring of our experience. This structuring leads to the formation of homogeneous sets of entities partitioned into regions that describe the ontological boundaries of the objects that humans can perceive, represent and therefore come to know and conceptualize in a more particular manner. Since, according to this interpretation, categories are the basic and primary constructs of our experience, they are considered to precede conceptualization and are a necessary preliminary step for it.

2 Categories—Towards a Definition (II): Contemporary Cognitive Research

Even though cognitive research has largely overlooked the classical problem of categories as the highest genera, some authors do take into consideration the classical issue of categories and state that the notion of category as it has been defined by the classical philosophical tradition has in fact an important cognitive application. One of most important authors who deals with this problem is Ray Jackendoff, who speaks of the major ontological categories as elements that "characterize the distinction among the major classes of #entities# that we act as though the #world# contains" (Jackendoff 1983: 51). From an epistemological point of view Jackendoff embraces a position which he declares to be based on Kant's view (Jackendoff 1983: 29). This leads him to distinguish between two different ways in which references—i.e. the entities of the external world—can be conceived. On the one hand, we can consider the world as it is, independently on how people conceive it: in this case Jackendoff speaks of the "real world". On the other hand, we can be interested in the knowledge we have of the world, i.e. of "the world as unconsciously organized by the mind" (Jackendoff 1983: 29) and in particular the conceptual structure of this knowledge (Jackendoff 1983: 31): in this case Jackendoff speaks of the "projected world". Conceptual structure is—as the name suggests—a structure that organizes perceptual information (Jackendoff focuses especially on vision) in linguistic form (Jackendoff 1983: 6). The markings # # indicate the projected-world entities, i.e. the #entities# as they are experienced after they are processed by the conceptual structure (Jackendoff 1983: 31). On the basis of this clarification, categories as Jackendoff conceives them are the highest genera our conceptual knowledge is organized into, i.e. they are the highest genera in which the #entities# of our projected world are partitioned.

The notion of "projection" describes the process through which sensory information is represented in a structured way. According to Jackendoff this process requires, first of all, that information be categorized from an ontological point of view. In order to represent certain information as e.g. #vase# must first be categorized as OBJECT (instead of e.g. PLACE, EVENT, ACTION or something else). Furthermore, the information [OBJECT] must itself become part of the representation of the vase: it must become a conceptual constituent of the concept of VASE. The information that a vase is an object becomes therefore an indispensable constituent of our concept of VASE.[4]

In other words, in order to conceptualize something, it must first of all be located in a specific "ontological region" (in Husserl's sense), i.e. people need to understand which kind of #entity# it is: whether it is e.g. a [THING], a [PLACE], an [ACTION], an [EVENT], or a [MANNER], etc. This position has important consequences as far as conceptual structure in concerned. As Jackendoff states: "To support the projection of this variety of #entities#, conceptual structure must contain, in addition to [THINGS], constituents whose major features are [PLACE], [DIRECTION], [ACTION], [EVENT], [MANNER], and [AMOUNT]" (Jackendoff 1983: 50). A few lines later Jackendoff points out that in his opinion a larger number of categories is needed like e.g. those of *SOUND*, *SMELL* and *TIME* (Jackendoff 1983: 51). In later works he also mentions *STATE*, *PATH*, *PROPERTY* (Jackendoff 1995: 34) and *DISTANCES* (Jackendoff 2002: 317). Information about the ontological nature of the perceived entity is information about its kind: information regarding whether what it is being perceived is a [THING], a [PLACE], an [ACTION], etc.—becomes part of the conceptualization of the entity and an informational constituent of the corresponding concept: e.g. a #vase# is a [THING], #Sicily# is a [PLACE], #running# is an [ACTION] etc.). According to Jackendoff, the attempt to reduce the number of categories in search of something like a *summum genus* or of a few *summa genera* is not the right way to go. On the contrary, all the categories that are needed to explain our conceptual capacities and the way in which we use language to refer to objects in the external world must be included in our theory.[5]

---

[4]In line with ordinary usage in the field of cognitive research, in his works Jackendoff uses 'concepts' and 'categories' mainly as interchangeable words. When he focusses on the issue of categories in the specific sense we discussed in this section, he speaks explicitly of *ontological* categories. Graphically, ontological categories are indicated with small capitals (THING: see e.g. Jackendoff 1983) or at least with an initial capital letter (Thing: see e.g. Jackendoff 1995, 2002). The square brackets are used to indicate categories as conceptual constituents, i.e. as pieces of information that contribute to the content—in the determination—of a concept. Since in this work we use small capitals for concepts and categories, to distinguish our use from Jackendoff's we indicate ontological categories in italics.

[5]To be precise, Jackendoff's argument is more technical and even reminiscent of the justifications proposed first by Aristotle and then by Kant for their respective tables of categories. Indeed, Jackendoff introduces the notion of 'pragmatic anaphora' to indicate all anaphoric expressions including pragmatically controlled pronouns (like e.g. 'I bought that yesterday.'). To understand these sentences a speaker must pick out the intended referent. According to Jackendoff, if we were able to enumerate all the kinds of pragmatic anaphora speakers are able to understand, we would obtain a complete list of the referents that speakers are able to identify. See Jackendoff 1983, §3.2. A further development of this view can be found in Jackendoff [1987] 1994, §8.4 and in Jackendoff 1995, §2.6.

A second, different line of study in cognitive research that gives a similar account of categories is found in the field of child psychology. One of the most important authors in this respect is Jean Mandler (together with her group). Like other authors in the field of cognitive science, Mandler uses the terms 'concepts' and 'categories' interchangeably, however, within what we have defined as categories, she makes a distinction between "global concepts" and "global categories". According to Mandler, the "conceptual system is categorical from its inception, with global categories being formed first, followed by subdivision of these broader categories" (Mandler 2002: 315).

Global categories are considered to be, at least in some respects, similar to the so called superordinate categories, even though it would be a mistake to use the equivalent term since global categories differ (i.e. largely by being vaguer and undifferentiated) from the adult's superordinate categories (see e.g. Mandler 2000: 8). As Mandler and colleagues state: "Although we will sometimes use the term 'superordinate' when referring to existing research, we prefer the terms 'global' or 'domain level' to characterize such categories, to make clear that young children can conceptualize broad domains without having specified hierarchical subclass relationships within them" (Mandler et al. 1991: 264).

According to Mandler, global concepts are preverbal concepts by means of which infants make broad differentiations between kinds of things. They are the first conceptual knowledge infants develop and constitute the precondition for them to carry out more specific differentiations and to develop a conceptual system. Since global categories are broader than concepts, Mandler argues that conceptual formation proceeds top-down: from more general to more specific concepts. Differing from the classical philosophical tradition and from Jackendoff, Mandler does not actually consider categories to be something other than concepts. She maintains rather that global categories are the first kinds of concepts children develop: they are the first kind of conceptual thought produced by infants and therefore also constitute a precondition for language learning (Mandler 1992). In addition, unlike Jackendoff she doesn't seem to think that categories must necessarily be *summa genera*. If fact, the categories she puts forward in her research are quite mixed as far as their generality level is concerned: in her writings she discusses mainly categories such as ANIMATE, INANIMATE, VEHICLES, PLANTS, FURNITURE (Mandler and McDonough 1998; Mandler 2000) CAUSE, PATH, CONTAINMENT, SUPPORT, AGENCY (see e.g. Mandler 1992), THING and MOTION (Mandler 2008).

Since according to Mandler, general categories do not match our adult superordinate concepts, her hypothesis suggests that the acquisition process starts with some broad and undifferentiated global categories which then allow for the development of the adult conceptual system, including specific and taxonomically organized adult superordinate concepts.

She argues that the early classifications carried out on the basis of categories (e.g. the classification of an object as *animate*) provides conceptual information about the object analogous to what Jackendoff calls 'conceptual constituents' (using Jackendoff's notation, these provide e.g. the information that the categorized object is [ANIMATE]). This information is a conceptual primitive which is essential for

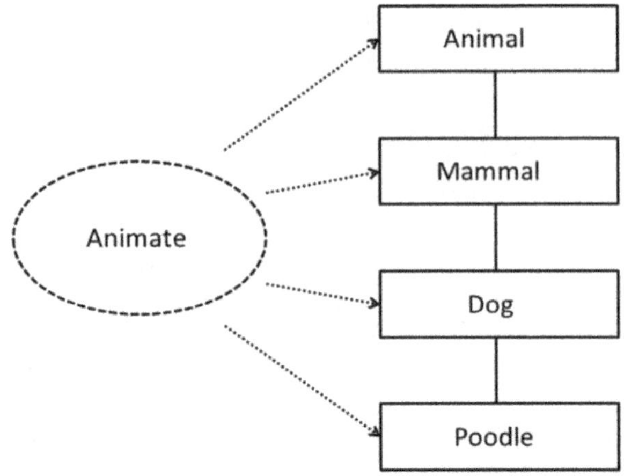

Fig. 3 Animate is a constituent at all levels of the hierarchy of these superordinate/subordinate concepts

carrying out more specific generalizations (like e.g. this [ANIMATE] object is a *dog*, a *mammal*, a *poodle* etc.) and they continue to be part of the concept they contribute to forming even after the complete development of the concept (Fig. 3).

This means that all our adult concepts of (e.g. dog, mammal and poodle will always also include the information [ANIMATE]).

This idea that conceptual organization starts from broad classifications is also suggested by George Lakoff's and Mark Johnson's studies which are, in fact, the main frame of reference for Mandler's investigations. Lakoff and Johnson deal with the issue of the origins of concept formation, but they don't make use of labels like 'categories' or 'categorical organization'. They rather speak of 'preconceptual organization of experience', suggesting that preconceptual organization of experience is a necessary condition for the formation of a conceptual system such as the one that characterizes human knowledge. Lakoff and Johnson maintain that, in order to develop a conceptual system, experience must first of all be organized according to some basic principles (in our terminology: categories). More specifically, it is organized according to very general prelinguistic schemas, which are in turn developed before linguistic and conceptual learning on the basis of our bodily experience and are then extended to other more complex experiential fields using metaphoric thought.

An example in this direction is that of physical containment:

> [...] let us consider briefly an ordinary instance of image schematic structure emerging from our experience of physical containment. Our encounter with containment and boundedness is one of the most pervasive features of our bodily experience. We are intimately aware of our bodies as three-dimensional containers into which we put certain things (food, water, air) and out of which other things emerge (food and water wastes, air, blood, etc.). From the beginning, we experience constant physical containment in our surroundings (those things that envelop us) (Johnson 1987: 21).

According to Johnson this experience of containment gives rise to a schematic image based on the idea of boundaries that leads people to distinguish between the inside and the outside of the body and this schematic image can be extended as a general criterion for the organization of experience. This schematic image allows us to organize every experience we have according to the principle inside/outside of some elected boundaries. The boundaries of the inside/outside can be chosen in different situations and depend on our position, perspective and on the experiences we need to organize. Because we develop this inside–outside image schema, we conceptualize the world as a set of entities that have an inside and an outside, that therefore have boundaries separating them from other things; and that are subject to restrictions and limitations: "physical in-out orientation involves separation, differentiation, and enclosure, which implies restriction and limitation" (Johnson 1987: 22). Also our spatial experience is organized in terms of external and internal spaces, while actions and events are organized in terms of movements from the inside of something to the outside of something or vice versa.

From this schema derive also more specific forms of organization of experience:

> There are thus at least five important entailments or consequences of these recurring experiential image-schematic structures for in-out orientation. (i) The experience of containment typically involves protection from, or resistance to, external forces. When eyeglasses are in a case, they are protected against forceful impacts. (ii) Containment also limits and restricts forces within the container. When I am in a room, or in a jacket, I am restrained in my forceful movements. (iii) Because of this restraint of forces, the contained object gets a relative fixity of location. For example, the fish gets located in the fishbowl. The cup is held in the hand. (iv) This relative fixing of location within the container means that the contained object becomes either accessible or inaccessible to the view of some observer. It is either held so that it can be observed, or else the container itself blocks or hides the object from view. (v) Finally, we experience transitivity of containment. If B is in A, then whatever is in B is also in A. If I am in bed, and my bed is in my room, then I am in my room. What I have just described are five of the most important consequences of the structure of in-out schemata. George Lakoff and I have called such consequences 'entailments' because they are implications of the internal structure of image schemata (Johnson 1987: 22; see also Lakoff and Johnson 1980: Chaps. 15–18).

The relevance of Lakoff's and Johnson's view with respect to our discussion on categorical organization is not easy to establish, because the examples they discuss are of very different kinds. In fact, they vary from very general and basic cases (like CONTAINMENT) which are analogous to if not identical with Mandler's global concepts, and therefore correspond to the idea of categories we are introducing, to very specific cases that cannot be considered categorical, but must rather be certainly considered conceptual. Take for example the discussion of how the concepts of ARGUMENT or LIFE or others are structured staring from perceptual experience through a metaphorical procedure. The common idea underlying these different kinds of examples is the thesis that understanding relies (in all its forms) on perceptual experience—driven not only by the external world, but also by the body of the experiencing subject—and on metaphorical procedures of extension or translation of the understanding of a certain experience to other analogous experiences. However, Lakoff and Johnson often discuss this general thesis not in relation to possible categories, but when considering (linguistic and culturally driven) abstract concepts. They suggest, for example, that *arguments are understood as wars* (as is shown e.g. by the use of expressions like: "Your claims are indefensible; He attacked every weak point in my argument. His criticisms were right on target. I demolished his argument. I've never won an argument with him": see Lakoff and Johnson 1980). Or they maintain that *life is understood as a journey*, in which "birth is arrival, life is being present here, and death is departure" (Lakoff and Turner 1989: 1) while "the person leading a life is a traveler. His purposes are destinations, the means for achieving purposes are routes [...]" etc. (Lakoff and Turner 1989: 3). Or further, *theories are understood as buildings* (they have foundations, need support, can collapse, have a framework, etc., see Lakoff and Johnson 1980: 46) *ideas are understood as resources* (ideas can run out, they can be useful or useless; thoughts can be wasted etc.; see Lakoff and Johnson 1980: 48) etc.

These are examples of ways in which more abstract concepts are understood on the basis of metaphors that analogize them to other more concrete experiences which we have a better understanding of. Every concept can be understood on the basis of different metaphors which lead to different characterizations of it; these depend largely on the cultural connotations of the concept: ideas can be understood e.g. also as *food* (e.g. half-baked ideas, raw facts, warmed-over theories, etc.) or as *people* (one can give birth to ideas, be the father of a theory; an idea can die off, etc.) or as *plants* (an idea comes to fruition; it can take years for an idea to come to full flower; an idea can be planted in the mind of someone, etc.) or as other things (see Lakoff and Johnson 1980: 46ff).

Since they discuss cases like these, in the literature Lakoff and Johnson are mentioned mainly for their attempt to perceptually ground abstract concepts. However, a relevant part of the investigations they carry out and the examples they discuss are rather about the much more basic image-schematic capacity of humans to order their experience according to some fundamental, broad preconceptual principles. The case of containment goes in this direction since it delineates something like a categorical partition of perceptual (i.e. 'proprioceptive': see Chap. 2, Sects. 3–6) origin, which precedes conceptualization and which leads us to organize experience according to very general principles such as, first and foremost, inside/outside and, as a consequence, separation: differentiation, enclosure, restriction, limitation, boundaries, etc.

Other examples that go in the same direction are the so called *orientational metaphors*, which, as the name suggests, are related to the organization of experience according to basic spatial principles which are driven by the way in which our body functions and are included in the external world like up–down, in–out, front–back, near–far, deep–shallow, central–peripheral (Lakoff and Johnson 1980: 14, 25). They suggest that these oppositions are constituted starting—among other things—from the directionality of the eyes (which give rise to a front and, in opposition, to a back), from the perception of kinesthesia, of gravitational force and again of the visual field (which gives rise to the opposition between up and down), from the perception people have of their body (as a basis for forming the opposition inside/outside, as already mentioned, but also as a starting point to constitute the opposition between deep and shallow), the viewer's point of view (as a starting point to constitute the opposition near-far as well as the opposition between center —which is where the observer is located—and periphery.

Lakoff and Johnson address these basic orientation schemas mainly because they have an influence on higher conceptual and linguistic thought and are applied metaphorically to express abstract states or conditions. For example, in English semantics the idea of 'up' is connected to positive things like health, having control, being conscious, etc., while the idea of 'down' is related to negative things like sickness, being subject to control, not being conscious (feeling up vs. being low; to be in top shape vs. to drop dead; having control over someone vs. being under control etc.; see Lakoff and Johnson 1980: 14ff). However, the idea that people develop such basic orientation principles starting from bodily information can also constitute evidence to support the hypothesis we are arguing for here: i.e. that the cognitive system produces a prelinguistic and preconceptual organization of experience—of objects, events, etc.—using organizational schemes which have been formed starting from information driven by bodily structure and bodily movements. This is the direction taken by Johnson in *The Body in the Mind* as well as by Mandler in her studies inspired by Lakoff's and Johnson's investigations (see e.g. Mandler 1998).

All the cognitive theories mentioned above which put forward the idea of a global (categorical), prelinguistic and preconceptual organization of experience (Jackendoff, Lakoff, Johnson, Mandler) introduce different possible categorical partitions. This shows that the issue of what categories our experience is organized into and how many categories there are is extremely difficult to address and cannot be answered on the basis of a single theoretical (i.e. philosophical) approach only. As e.g. Jackendoff also points out: "The existence of a particular ontological category is not a matter of physics or metaphysical speculation or formal parsimony, but an empirical psychological issue, to be determined on the basis of its value in explaining the experience and behaviour of humans and other organisms" (Jackendoff 1983: 52). However, even though philosophy alone cannot develop a complete list of categories, philosophical investigation can rely on empirical studies to put forward some hypotheses on the possible function of categories, on possible categorical partitions, on the way in which these are organized (i.e. how they relate to each other) and on their origin (suggesting e.g. whether they are innate, as e.g. Jackendoff maintains, or whether they are formed and how—this is the direction taken for example, by Johnson, Lakoff and Mandler).

In fact, the idea that our experience is organized first of all into general categories is compatible with multiple hypotheses on categories: among others, it is compatible with the idea that categories might in turn be related to each other in specific ways (e.g. some might be more general than others; or some might be formed sooner and constitute the starting point for the formation of others). In this chapter we will in fact develop this line of research and discuss (on the basis of some concrete examples of categorical partition) not only why we should assume that there is categorical organization which precedes the development of a conceptual system, but also how these categorical partitions relate to each other.

This last point raises another, equally important issue regarding categories, which philosophical research—with the help of empirical research—should address. Categorical organization, as we said, comes before and is a precondition for the development of the conceptual system. This opens the door for the question of whether categories are innate or are just formed before concepts on the basis of information and organizational structure which are available at a very early developmental stage. Jackendoff maintains e.g. that categories are innate; Mandler, Johnson and Lakoff suggest on the contrary that they might be formed starting from sensory experience using basic mechanisms for the analysis of perceptual information (we will come back to this issue later in this chapter). The analysis carried out in this chapter will also try to provide some clues to this issue. Specifically, we will argue for the thesis that there is no need to assume that categories are innately present in the human mind, but that their formation can be explained starting from basic mechanisms that work on bodily information. This hypothesis will not be discussed in general, i.e. with relation to a categorical system considered in its complexity. Unfortunately, so far we don't have a general, systematic hypothesis concerning the overall organization of categorical structure, or its formation and function with respect to the conceptual system; neither has something of this nature been proposed by other researchers yet. However, we will introduce and develop a particular hypothesis on a specific example of categorical partition, which has been discussed also by Mandler, related to folkbiological taxonomies. Specifically, we will discuss the example of the dichotomies between ANIMATE/INANIMATE, LIVING/NON-LIVING, BIOLOGICAL/NON-BIOLOGICAL. We will put forward the hypothesis that ANIMATE/INANIMATE is the first categorical dichotomy to be formed and that this constitutes the starting point for developing other more general as well as more specific classifications such as LIVING/NON-LIVING, BIOLOGICAL/NON-BIOLOGICAL on the one hand and basic concepts on the other. We will argue that the dichotomy LIVING/NON-LIVING is categorical only in part (in some aspects it is indeed conceptual), while the dichotomy BIOLOGICAL/NON-BIOLOGICAL is entirely conceptual.

3 The Example of Folk Biological Categories[6]

At first glance, the surrounding world seems to be characterized by a sharp distinction between two kinds of entities: living beings and objects. This distinction seems to belong more or less universally to people's naive intuitions about biology (Atran 1998: 567). This general intuition can be found in civilizations and cultures very different from our own, even though there are cultural differences in the criteria used to determine exactly which objects must be considered to be 'alive', i.e. which kinds of entities belong or

---

[6]Some parts of the following three sections of this chapter (Sects. 3–5) are taken from an article we already published under the title *Sensing aliveness* (Dellantonio et al. 2012). We thank the editor-in-chief of *Integrative Psychological & Behavioral Science*, Professor Jaan Valsiner, for giving us the permission to use this material.

don't belong to the set of 'living beings' (see e.g. Anggoro et al. 2005; Waxman and Lidz 2006; Waxman and Medin 2006). For instance, in the pre-Socratic tradition, old Greek culture identified all things that were able to reproduce themselves as 'living': they called 'psyche' the principle responsible for reproduction and developed a concept of 'life' inclusive of humans, animals and plants. They conceptualized living beings as 'empsychos' (endowed with '*psychè*') to differentiate them from mere objects ('*apsychà*'). Within the field of 'living beings', they further differentiated human life ('*bìos*') from non-human forms of life ('*zoé*'). An analogous situation can be found in traditional Arabic culture, where humans, animals and plants are included in the same concept of 'living things' since they are all considered to be provided with a 'vitalistic principle' in virtue of which they are able to reproduce themselves and to give contingent behavioral responses to environmental stimuli. Still, not all traditional classification systems share a concept of 'living things' that includes humans, animals and plants. In fact, plants are not always recognized as being 'alive'. For instance, in the Hebrew culture humans and animals were considered far more 'living' then plants, so a contrastive conceptualization has been formed that excludes plants from the set of 'living beings'. This is probably due to the influence of religion, because in Genesis, the first book of the Old Testament, it is stated that plants have been created to serve as nutrition for 'living beings', a category which is considered to include humans and animals only (Stavy and Wax 1989; Hatano and Inagaki 2000). Also in Japanese culture we find a contrastive conceptualization conveyed by the use of two different terms to express our verb 'to be': 'iru' is the verb used in reference to 'living beings' (i.e. humans and animals only), while 'aru' is used in reference to plants and nonliving things (Inagaki and Hatano 2002: 20). Nevertheless, a different situation can be found, for example, in Indonesian culture, where the kingdom of nature is designated by the words 'makhluk hidup' ('living beings'), and includes humans, animals and plants. Even if there is evidence that humans and animals are considered to be different from plants, no other concept or word has been developed by this culture to group humans and animals together in order to indicate that they are a different form of life from plants (Anggoro et al. 2008).

These examples show that—even though plants have an ambiguous taxonomic position and are sometimes included and at other times excluded from the class of 'living beings'—the distinction between 'living beings' versus 'non-living things' seems to be transculturally present and to have always been present in the course of human history. Its presence is so widespread that it seems legitimate to wonder whether this might reflect a basic fact concerning the deep structure of the human mind, i.e. concerning the way the mind develops knowledge of the world in a universal and necessary manner and more specifically concerning the way humans organize their knowledge according to specific prelinguistic categories. Although people's judgments about which entities should be considered 'alive' show some variance depending on cultural factors—primarily in terms of the variance in the taxonomic position of plants—the question concerning precisely what these universal elements of classification (i.e. these categories) consist of should be addressed.

Recent studies in developmental psychology support the idea that humans organize their knowledge according to specific categories related to the aliveness of objects, and they also help make sense of the ambiguous taxonomic role of plants.

Indeed, they provide evidence that already long before attaining linguistic mastery (i.e. before being subject to any sociocultural influence) infants exhibit the capacity to distinguish between things that are 'alive' and things that are not (see e.g.: Premack 1990; Woodward et al. 1993; Spelke et al. 1995). More specifically, infants seem to organize their perception of the world according to the differentiation between animate and inanimate objects, i.e. according to the dichotomy ANIMATE/INANIMATE. As for example Mandler maintains: the "distinction between animals and nonanimals (or things) is one of the earliest conceptual divisions that infants make" (Mandler 2002: 315).[7] On the contrary, the categorical dichotomy LIVING/NON-LIVING seems to be learned much later under the influence of language. Empirical research shows that young children up to 5 years of age do not understand the so called 'folkbiologic' (adult) classification according to which humans, animals and plants belong to the same category of living things. They group animals together showing that they consider all them very similar to each other, but they have difficulties associating animals with plants and understanding that they can also be grouped together by virtue of the fact that they are all living things (see e.g. Waxman and Medin 2006: 53). This result is also confirmed by a number of cross-cultural studies (Hatano et al. 1993).

This leads to the hypothesis that the categorical dichotomy ANIMATE/INANIMATE (i.e. the distinction between humans and animals on the one hand and plants and material objects on the other) is not only the first one to be learned, but possibly also the only one which is really universal and prelinguistic—i.e. immune from cultural influences and directly related to the way our mind spontaneously constitutes and classifies percepts. Other folk-biological categorical dichotomies such as LIVING/NON-LIVING (or BIOLOGICAL/NON-BIOLOGICAL) are instead developed at a later stage through language acquisition and through exposure to a belief system and constitute a further refinement and internal differentiation of this original categorical distinction. This further internal differentiation is carried out on the basis of the criteria each culture develops to determine exactly which instances should be considered 'living', 'alive' or the like. Indeed, the line of inquiry pursued by this study suggests that the ANIMATE/INANIMATE dichotomy is a basic and universal element of the human cognitive system that provides an essential precondition for further and more complex conceptualizations carried out under the influence of cultural factors.

Studies from developmental psychology offer important clues for the comprehension of the spontaneous processes of the human mind that lead to the organization of the perceptual universe according to the dichotomy ANIMATE/INANIMATE. Indeed, even though the conceptual system is formed before and independently

[7]While most authors take into account the categorical opposition ANIMATE/INANIMATE, Mandler speaks about the categorical opposition ANIMAL/NON-ANIMAL. However, if we look in detail at the cases the different studies deal with, it is easy to see that they consider ANIMAL and ANIMATE as synonymous since in both cases attention is paid to animal motility. In this regard, see for example Poulin-Dubois 1999. In this study, we prefer to use the terms 'animate' and 'inanimate' since the opposition ANIMAL/NON-ANIMALS sends to evoke a difference between humans and animals that leads erroneously to excluding human beings from the category ANIMAL.

from language, its structure is contaminated and modified by the acquisition of linguistic meanings which are highly dependent on cultural factors.

Significant evidence supporting the fact that infants are able to discriminate between animate and inanimate entities comes from research on neonatal mother–newborn interaction which specifically examine the imitative processes newborns carry out before having developed a figural representation of their own body. Indeed, already shortly after birth, infants exhibit specific imitative behaviours such as simple or selective (i.e. lateral) tongue protrusion. Furthermore, they try to imitate the general facial expression of adults, and to syntonize (tune) with them, responding to solicitations with vocalizations that become more and more accurate and modulated (see Meltzoff and Moore 1977, 1983, 1989; Tronik 1978; Stern 1985; Meltzoff 1988, 1995). Imitation and syntonization (tuning) are specific actions performed only in response to certain stimuli produced by other human beings; none of these activities can be elicited by showing the infant inanimate objects. In this sense, these experiments show that, already at a very early developmental stage, newborns have at least the capacity to differentiate human beings from inanimate things.

As for the problem of determining the origin of the capacity for distinguishing humans from inanimate instances, several studies on older infants carried out using more complex methods show that this capacity relies on movement indicators. In fact, these studies point out that animals are classified together with humans in the same category as animate instances because they share the same kinds of motion. In contrast, inanimate instances are considered to belong to a different class since they don't exhibit the same characteristics of movement. The specific characteristics of movement that seem to be considered in establishing whether an instance is animate or inanimate may be divided into three main kinds:

(i) self-propelled movement—only animate objects are capable of moving on their own, while physical objects move only when they are propelled by an external agent (see Premack 1990; Gelman et al. 1994; Leslie 1982; Leslie and Keeble 1987; Baillargeon 1995; Spelke 1991, 1994; Spelke et al. 1992, 1995; Poulin-Doubois et al. 1996);

(ii) non-mechanical movement—infants as young as 3-months-old are able to perceptually distinguish biological movements from similar more mechanical movements, which are biologically 'incorrect' (see Bertenthal 1993; Baron-Cohen 1995; Arterberry and Bornstein 2001);

(iii) goal-directed movement (intentionality)[8]—"animate motions have a quality of directedness or purpose, and biological control mechanisms make it possible for animates to respond (or adjust) to environments […] and adapt to unforeseeable changes in circumstances" (Subrahmanyam et al. 2002: 348; see also Meltzoff and Moore 1997; Woodward et al. 2001).

[8]Even if many authors refer to this aspect using the term 'intentionality', we should rather speak of voluntary and involuntary movements since the word 'intentional' in the field of the philosophy of mind assumes a different meaning connected with the concept of intentionality as defined by Brentano.

What these studies allow us to state is that—given that even very young children show the capacity to distinguish animate from inanimate objects—it is plausible that the categorical dichotomy ANIMATE/INANIMATE is a primary and original element of the way humans classify percepts (concerning this conclusion see also Subrahmanyam et al. 2002; Gallistel 1990; Gelman et al. 1995).

If this is the case, the question arises whether the capacity to apply this categorical distinction may be produced by the mind on the basis of information available to the conceptual system (empiricist view) or whether it arises from the inborn categorical organization of the cognitive system or from some core conceptual knowledge produced by some innate domain-specific learning mechanisms (nativist position). The point of the opposition between an empiricist and a nativist view is not whether this categorical distinction needs to be learned since not only empiricism but also nativism maintains the necessity of learning as an unavoidable means to move on from a potential to an actual capacity. Neither does it concern the problem of whether humans can learn this distinction because they are equipped with some mechanisms that allow them some form of organization of their experience.

The real difference between empiricism and nativism has to be traced back to the fact that according to nativist positions the differentiation between animate and inanimate objects is carried out on the basis of innate highly specialized, domain-specific mechanisms dedicated to the production of a particular core knowledge about animate and inanimate instances (see e.g. Carey and Spelke 1994; Carey 2000). On the contrary, empiricism states that it is possible to explain learning on the basis of domain general learning processes and mechanisms without assuming the existence of specialized inborn cognitive structures that allow for the appropriate organization of particular experiential data, since this same organization can be produced by the mind using the more general capacities it is endowed with.[9]

[9]The debate between nativism and empiricism is actually much more differentiated, and there is considerable disagreement about what exactly empiricism and nativism consist in and what their opposition is about. Some forms of nativism assume e.g. either that the early capacity to distinguish between animate and inanimate instances is due to the specific organization of conceptual knowledge in the brain produced through evolution (see Caramazza and Shelton 1998; Trevarthen 1979; Trevarthen and Hubley 1978) or that it can be ascribed to the availability in the conceptual system of primitive already formed conceptual 'atoms', which just need to be triggered by experience (see Fodor 1975, 1981, 1987; Spelke 1994). Recent studies in the field of developmental psychology and infant research favour instead a metaphysically and ontologically less binding form of the term nativism. In fact, they deny a 'final state nativism', according to which some concepts or categories are innate with a well-defined structure. Instead, they argue for a 'starting state nativism' maintaining that some functions for the organization of the information are available to the cognitive system prior to birth. According to this position, these innate elements allow, during development, the constitution of cognitive contents with a higher level of definition (see Gopnik and Meltzoff 1997; Meltzoff and Gopnik 1993; Meltzoff and Brooks 2001). The position upheld in this study is consistent with this hypothesis at least in the sense that some primary functions may be formed already in intrauterine life. About this aspect see, for example, Sheets-Johnstone (1999) and Haselager et al. (2012).

As far as the capacity of infants to distinguish between animate and inanimate instances is concerned, the data may be interpreted, in principle, according to either an empiricist or a nativist perspective.

(a) According to a nativist explanation, the capacity to distinguish between animate and inanimate instances could be built into the cognitive system, i.e. it could be due to some innate link between animate instances and certain movement characteristics. Children might 'know' (in some innate fashion) that *only* animate things exhibit specific kinds of movement like self-propelled, non-mechanical and goal-directed movement and they might single out animate instances on the basis of an analysis of the way they move: entities would be perceived as animate insofar as the way they move appears to be self-propelled, non-mechanical and goal-directed. According to the defenders of this position, it is only because the cognitive system already 'knows' that there are different kinds of instances, animate and inanimate, which exhibit different kinds of movements, that perceived objects may be organized and classified, depending on the kind of movement they exhibit, according to the dichotomy ANIMATE/INANIMATE.

(b) If we want to give an empiricist interpretation of the capacity to distinguish between animate and inanimate instances, we cannot assume that children are endowed with such knowledge, but we need to show that the distinctive movement properties of animate and inanimate objects can constitute the experiential basis for the development of the categories animate/inanimate by infants. The reliability of this explanation depends on the possibility of concretely demonstrating how the conceptual system can come to single out animate and inanimate instances on the basis of an appropriate organization of sensorimotor information. In fact, most of the empiricist views proposed by developmental psychology do not offer an exhaustive argument as to how sensorimotor information might be used in order to form the categories ANIMATE/INANIMATE (see e.g. Mandler 2002).

From an epistemological point of view, nativist explanations are based on the idea that the introduction of an inborn categorical distinction is the only possible way to account for the capacity exhibited by infants. However, if one could show that this capacity can be explained using other, non-innate cognitive resources, a nativist explanation would become unnecessary and would be superseded. Indeed, the hypothesis of the research we propose in this article goes in this direction and tries to give a genetic account of the categorical dichotomy animate/inanimate. The idea underlying this work is that—if the genetic hypothesis is supported–it will ipso facto (by a principle corollary to Ockham's razor) supplant the nativist position.

4 The Origin of the Categorical Dichotomy Animate/Inanimate

Our proposal concerning a possible explanation for the formation of the categorical dichotomy ANIMATE/INANIMATE in the infant mind is based on the hypothesis that the cognitive system and, more specifically, the conceptual system, does not work exclusively on information of exteroceptive origins: that is, its function is not limited to performing operations of composition, connection, comparison, etc. on information taken from 'external' objects. The cognitive system also works simultaneously on another kind of information that according to our thesis should be called *propriosensitive* or '*proprioceptive*' (for a detailed definition and discussion of this concept, see Chap. 2, Sects. 3–6).[10]

We have defined 'proprioception' in a way that does not entirely match the most common use of this word including only motoric and postural information; instead, we have embraced a Gibsonian notion of 'proprioception' (we always use this term in inverted commas to distinguish it from the more classical neurophysiological description of the notion: when we refer to the latter, we use *proprioception* without inverted commas) that includes all kinds of information detected through our internal, propriosensitive perceptual system. To briefly sum up those aspects of the definition of 'proprioception' introduced earlier in the book which are most important with respect of the theses in this chapter, we embrace the perspective that 'proprioception' describes both any kind of propriosensitive information or internal perception: its content consists of sensations with physiological origin related to the corporeity of the experiencing subject and in particular to all 'activities' connected to the body and to voluntary and especially involuntary movements of the body. For this reason, we use the word 'proprioception' in a wider sense, based on the synthesis of two different kinds of information produced by the body, which are generally referred to using the notions of proprioception (in a narrow sense) and 'interoception'. Indeed, the word proprioception classically defines those sensations originating from the muscular and motor system used by the mind to constitute perceptual space, to locate the position of the body in space and to identify objects external to the body. According to this definition, proprioception consists of sensory information generated by monitoring the muscular system, that is, a kind of perceptual 'sixth sense' which interacts and complements the sensations provided by the other five senses with information produced internally to the subject by the motor organs (see e.g. already Sherrington 1906).

According to the hypothesis considered in this research, 'proprioception' does not consist in explicit and objective knowledge of the body, but rather provides an early form of knowledge implicitly available to the cognitive system which enables the development of further perceptual abilities and more complex and specific forms

[10]About the importance of propriosensitive information for the constitution of perceived objects see also O'Shaughnessy (1995), Cole and Waterman (1991), Haselager et al. (2012) and Gallagher and Cole (1995).

of object-identification and discrimination. In fact, this kind of information is processed together with external information coming from the sensorium which carries data about the environment surrounding the subject. More specifically, according to this interpretation, propriosensitive information performs two functions: that of creating self-knowledge based on implicit control schemata which are available to the cognitive system from a very early stage in its development (probably even before birth[11]) (see Gallagher 1986; Gallagher and Cole 1995; Gallagher and Meltzoff 1996; Gallagher and Zahavi 2008; O'Shaughnessy 1995) and that of structuring internal working models (see Bretherton 1992, 1993), which integrate information originating in the body of the subject with information coming from the surrounding environment. These implicit control schemata and internal working models also produce implicit knowledge of a relational character concerning the relationship between the body and the environment.

In our perspective, 'proprioception' is not, however, the only means through which the cognitive system acquires tacit information about bodily states to be used for further perceptual and cognitive operations. Another kind of 'internal information' the cognitive system can rely on, usually defined as 'interoceptive', relates to changes of state in the qualitative dimensions of bodily experience (feelings, emotions, etc.). These qualitative changes of state are related to cold, warmth, joy, pain, stress, relaxation, excitement, etc. and manifest themselves through the modification of blood pressure, respiration, and the endocrine system (see e.g. already Gellhorn 1943: 15), which in turn have repercussions on muscle tone and general body posture. Since interoceptive information is related to spontaneous and involuntary bodily micro-movements, it can be considered similar and/or complementary to the information designated by the classical notion of proprioception which is also related to bodily movements, even though the latter concerns more macroscopic movements connected with the muscular and postural system. The use of the word 'proprioception' in the wider sense derives from the hypothesis that these two kinds of information interact with each other and complement one another, since they can be translated into each other and processed jointly by the cognitive system in terms of motor information. The working hypothesis we advance here suggests that these two kinds of information are a basic fund of the cognitive system and they can therefore be used in perceptual operations and in the constitution of the basic categories needed to classify percepts (see Phillips 1985; O'Shaughnessy 1995; Berthoz [1997] 2000; Berthoz and Petit [2006] 2008).

Defined thus 'proprioception'—we have argued—may be the key to explaining the formation of the categorical dichotomy ANIMATE/INANIMATE in the infant mind. An epistemological perspective that allows us to theoretically explain the genesis (i.e. the constitution) of the dichotomy ANIMATE/INANIMATE on the basis of this notion, is Edmund Husserl's phenomenological theory of knowledge and

[11]In fact, in the fetal condition it is possible to observe the development of muscular proprioception (9 weeks), as well as the presence of reflex movements and sound sensitivity (24–25 weeks): see Haselager et al. (2012).

particularly his "phenomenological egology", which concerns the relation between the body and the development of thought and knowledge.[12] According to Husserl, the elements constituting subjectivity cannot be investigated starting from 'the top', that is, with higher logical functions in their complex working, but needs to be investigated through a 'bottom up' approach that considers the experiencing of subjectivity in its development, drawing attention, in particular, to its somatic features. Husserl was critical of the nativist view maintained by some of the psychologists of his time. As we previously clarified (see this chapter, Sect. 1), he assumes the existence of basic categories ('regions' in his lexicon—see Husserl [1913] 1982), which are the fundamental elements on which the perceptual articulation of our world relies. Still, these categories are neither innate, nor explicitly or voluntarily constructed or deduced by the subject; but rather originally constituted in an implicit manner in sensory experience (Husserl [1913] 1980).

According to Husserl, the most basic experience consists of 'sensing' inside movement: before being subjects, we are always moving bodies; and we are moving bodies even before taking the voluntary decision to move the body. As Maxine Sheet-Johnstone clarifies, summarizing Husserl's position: "movement forms the I that moves before the I that moves forms movement. Spontaneous movement is the constitutive source of agency, of subjecthood, of selfhood, the dynamic core of our sense of ourselves as agents" (Sheets-Johnstone 1999: 135). The moving body can cause movement inside and outside itself, while sensing movements allows one to develop a first understanding both of the body and of what is perceived in the environment (Husserl [1913] 1989). Hence, movement information allows us to distinguish ourselves both from other subjects and from mere objects.

In fact, unlike humans, objects do not move autonomously, but can at best be moved or manipulated. Other subjects are on the contrary able to move like we do, but they are authors of movement that we can only observe from the outside, without experiencing them internally as happens with our own movements. According to Husserl our capacity to recognize in the world the presence of other subjects is due to a specific mental operation called analogizing apperception. For Husserl, analogizing apperception brings us to the recognition of the presence of other subjects on the basis of an "analogising transfer of an originally instituted (…) sense to a new case" triggered by a

> certain assimilative perception; but it by no means follows that there would by an inference from analogy. Apperception is not inference, not a thinking act. Every apperception in which we apprehend at a glance, and attentively grasp, objects given beforehand […] every apperception in which we understand their sense and its horizons forthwith, points back to a 'primal instituting', in which an object with a similar sense became constituted for the first time (Husserl [1929] 1977: 111).

[12]For a general discussion concerning the notion of phenomenological egology see Broekmann (1963). On the relation between phenomenological egology and psychology see Gallagher (2005); Berthoz and Petit [2006] 2008 and Gallagher and Zahavi (2008).

In the literature, this process has sometimes been interpreted as an operation of comparison between two mental images—that is between two image-like explicit and prospective representations: the representation of one's own body and the representation of another person's body (see Gallagher and Meltzoff 1996). Nevertheless, in our opinion this interpretation does not capture Husserl's idea since he explicitly rules out that this process consists of a form of direct inference.

Husserl suggests that the constitution of others as subjects starts from the constitution of others as objects, in the sense that the first step of the recognition of the alter ego is the constitution of the other person's body as a physical object. Only in a second step of object constitution is the other person's body recognized as a different and particular object compared to other types of inert physical objects and therefore constituted as "living body". Regarding the way this differentiation is carried out, Husserl explains that a physical body can show itself as a living body, as an animate organism, only through its behaviour (see Husserl [1929] 1977: 114), which triggers an "apperceptive transfer [of sense] from my animate organism", because "my animate organism is the only body that is or can be constituted originally as an animate organism (a functioning organism)" (Husserl [1929] 1977: 110). This operation "excludes an actually direct, and hence primordial, showing of the predicates belonging to an animate organism specifically, a showing of them in perception proper" (Husserl [1929] 1977: 111), even though "it is clear from the very beginning that only a similarity connecting [...] that body over there with my body there can serve as the motivational basis for the 'analogizing' apprehension of that body as another animate organism" (Husserl [1929] 1977: 111).

Since we ruled out the possibility of interpreting this process as a form of direct perception both of one's own body image and of the other person's body image together with an explicit comparison of their common traits, we propose to explain the working of analogizing apperception in terms of information processing carried out by the cognitive system at a sub-personal level, that is under the level of explicit awareness. More specifically, this process must be seen as an implicit recognition of some affinities between the information held in the percept and the information held in the perceiving system.

According to the view we propose, the human mind accesses knowledge first of all by constituting perceptual objects spatially. This means that both animate and inanimate objects are constituted first of all as mere material objects on the basis of their external, spatial properties. This hypothesis put forward by Husserl matches the idea presented in the previous chapter that concepts are made up (also) of perceptual information which describes the geometrical and topological properties of the objects (see Chap. 3, Sect. 5). Once they are spatially constituted, objects are differentiated from each other using further information coming from different sources, internal and external to the subject. It is at this stage that the process of analogizing apperception takes place picking out different kinds of properties of the concretely perceived objects: if the perceived object is animate, this process picks out its non-physical properties, that is the mental properties related to its status as a living creature.

The identification of these properties is made possible by the fact that our cognitive system can make use of propriosensitive (i.e. 'proprioceptive') information. This propriosensitive information originates from the way the cognitive system processes and apperceives the constitutive experience of the individual's own living body which enjoys an absolutely unique status since it is, on the one hand, a perceptible object in the world but, on the other hand, also an object that holds perceptual and qualitative information. The process of analogizing apperception 'couples' and 'analogizes' the propriosensitive information available in the first person with the information extracted from the perceived object, since the perceived object—that is the other subject—shows traits (reactions, movements etc.) which are very similar to the ones that the subject connects to his own perception of being alive.

The word 'apperception' designates the process through which the cognitive system tacitly constitutes the representation of the object. This process cannot be seen as a conscious and voluntary act of the cognitive system that explicitly and sequentially analyzes the properties exhibited by the object in order to categorize it. Apperception describes instead a mechanism for the recognition of information about a perceived external object in which this information is 'interpreted' by the system in relation to itself. In the act of apperception, the information about the object shows itself to the subject in the form of a representation in which the subject's implicit self-awareness is tacitly co-present: the subject's self-awareness co-constitutes the object (although without being co-objectivized). This whole process can be described as an operation of 'totalization' since the cognitive system completes the percept and makes it coherent using information that is not only driven by the object as it is perceived in its external appearance, but also by using information contributed by the perceiving subject.

The capacity to distinguish animate and inanimate objects may be seen as a by-product of analogizing apperception. This can be explained using propriosensitive information that the first person has about himself: propriosensitive information generated by this experience may indeed be 'projected' onto other 'instances' in the world which show kinetic features similar to the ones that the person associates with being alive. If 'animate' is used as the linguistic label to indicate all the properties that are extrapolated by a person's cognitive system from the experience of his/her own living body as a moving body, then 'animate' refers primarily to the information drawn from his own spontaneous and involuntary motility bound up with his vital functions, with eye movements, with respiration; it refers to sensations driven by the interaction and integration of body parts and the kinetic character of receptive activity, to the prospectivity of sensorial qualities and of emotional states. This information not only plays a part in forming the sense of self and in discriminating the self from the non-self (Bermudez 2000; Haselager et al. 2012), but is also the basis for the constitution of the category ANIMATE.

If we go back to the observations about the infant's capacity to distinguish between animate and inanimate entities, these last remarks allow us to state that the spontaneous and non-mechanical characteristics of movement (that is respiratory movement, movements of minimal deambulation, eye movements, facial expressions, reactions to stimuli, involuntary proxemics, etc.) of animate instances may be the information infants use to apply this distinction since this information is analogous to that which they experience constantly and tacitly in themselves. According to this account, humans are able to distinguish an animal from a non-animal exactly because only animals 'are analogized' as animate and are perceived as such on the basis of the way we implicitly feel in the act of living, that is in the act of moving. In this sense, the information we use to elaborate a criterion to discriminate between animate and inanimate instances, and consequently to form the categorical distinction ANIMATE/INANIMATE, is an implicit form of knowledge produced by the cognitive system about the movement of the subject's own body. The availability of this implicit knowledge allows us to discriminate perceived objects through the criterion 'similar in movement'/'dissimilar in movement'. For this reason, we share the position of Sheet-Johnstone, when she states that "aliveness is thus a concept [...] grounded in movement" (Sheets-Johnstone 1999: 135).

As mentioned earlier, the literature in the field of developmental psychology confirms that there is nothing new in explaining the capacity to differentiate animate and inanimate instances on the basis of motion and specifically on the peculiarity of the motion of animate instances. Still, the added value of the hypothesis we propose here consists in showing the importance of biological motion in identifying animate objects *without relying on innate knowledge or on domain specific mechanisms* that would allow us to associate specific characteristics of motion with animate and inanimate objects respectively.

The empirical data on the infant's capacity to distinguish animate and inanimate instances reported previously (see Sect. 2 points 1–3) may indeed be reconsidered as consequences of the view proposed here. (i) The child moves autonomously and therefore recognizes autonomous (self-propelled) movement as a peculiar characteristic of a creature that is similar to himself. (ii) Furthermore, it is the perception—or, more precisely the interoception integrated with proprioception—of the spontaneous character of his own body movements—from the macroscopic to the less evident ones—that makes it possible for the infant to distinguish on the basis of an analogical principle what is biologically correct in terms of mechanical movements. (iii) The capacity to identify goal-directed movements and to distinguish them from others may also be explained as applying an analogical principle that is primarily based on the perception of the individual's own movements as controlled and directed towards the achievement of specific goals and projecting the perceptual qualities of this movement to other kinds of movement observed in the external world which exhibit the same characteristics.

This explanation of the way humans comes to associate self-propelled, biological and goal directed motion with animacy and enabling them to distinguish on this basis animate from inanimate objects does not rely on any domain-specific mechanism, whose task is only to carry out this particular differentiation. What is needed is only some kind of analogizing mechanism to compare information coming from different sources.

5 Folk-Biological Categorical and Conceptual Hierarchies

The analysis carried out in the previous section suggests that the categorical dichotomy ANIMATE/INANIMATE is a basic element in human conceptual structure. Moreover, the explanation we offered for how the categories of ANIMATE and INANIMATE are developed helps to cast some light on the categorical organization of human knowledge as far as other folk-biological categories which sit on higher levels of generality like LIVING/NON-LIVING and BIOLOGICAL/NON-BIOLOGICAL are concerned.

In order to address these questions, we first need to consider which instances legitimately belong to which category and why. The categorical opposition ANIMATE/INANIMATE corresponds to the dichotomy ANIMAL/non-ANIMAL. The preference for the labels ANIMATE/INANIMATE is due to the fact that the word 'animal' tends erroneously to evoke a difference between humans and animals and leads us to think that humans belong to a different category from animals; on the contrary, the term 'animate' immediately suggests that both humans and animals should be considered as members. As far as the categorical opposition LIVING/NON-LIVING is concerned, the literature presents it as an extension of the ANIMATE/INANIMATE opposition in the sense that LIVING is supposed to include, not only animate instances, but also plants and sometimes even parts of plants such as fruits, vegetables and flowers. Still, more often the parts of plants are not considered as LIVING, but rather as belonging to the category BIOLOGICAL. So the category BIOLOGICAL includes not only properly living instances, but also fruits, vegetables, flowers, food objects. Biological objects are often contrasted not only with NON-BIOLOGICAL OBJECTS, but also with ARTIFACTS or TOOLS, even if strictly speaking artifacts and tools are only a subclass of NON-BIOLOGICAL OBJECTS.

The hypothesis we have proposed clarifies the relationships which exist among these categories and offers a genetic order of derivation starting from a basic categorical dichotomy and moving on to more complex ones. According to our view the most basic categorical opposition is ANIMATE/INANIMATE because (in the first place) humans and (in the second place) animals are the simplest and most natural subjects for infants to analogize to themselves by virtue of their characteristic motility. However, the categorical opposition ANIMATE/INANIMATE also provides a basis for the construction of other categories, first of all the category LIVING (which includes both animate objects and plants).

In fact, the inclusion of animals and plants in the same category and their opposition to non-living objects can be explained on the basis of some analogous characteristics in the movements animals and plants exhibit. The growth and the tropisms of plants can indeed also be interpreted as forms of self-propelled, non-mechanical and goal-directed motion. However, the growth of plants and their tropisms manifests differently from animate movements at least in the sense that they cannot be noticed in a single observational unit, but need more time in order to take place and be noted.

Therefore, it is plausible to assume that the opposition LIVING/NON-LIVING develops starting from the categorical dichotomy ANIMATE/INANIMATE through an extension of the idea of animacy to include in the same class all objects that move in a similar manner and will initially only include plant life—that is only plants that grow, that re-orient themselves towards the light—and not parts of plants like fruits, vegetables and flowers. Since the motility of plants and its analogy to the way humans move is more difficult to observe, it is plausible to assume that the opposition LIVING/NON-LIVING will be developed later in the cognitive development of the child. The categorical opposites ANIMATE/INANIMATE and LIVING/NON-LIVING can, furthermore, be considered to be the basis for the constitution of other categories like BIOLOGICAL or ARTIFACTS. BIOLOGICAL may be formed inferentially through the extension and the parceling of properties: i.e. all parts of living organisms might be included in the class of the biological objects. Analogously, ARTIFACTS may be formed inferentially intersecting the category of animacy (or the concept of human being which derives from an internal differentiation of the class of animate things) with the class of non-biological things, concluding that all non-biological, man-made things are ARTIFACTS.

This conclusion suggests that, cognitively speaking, the categorical dichotomy ANIMATE/INANIMATE is the most fundamental one and the only one which is entirely independent from any cultural and linguistic influence since it is developed entirely on the basis of specific movement characteristics of objects and on the analogy between these movement characteristics and the kinds of movements everyone experiences first hand according to (implicit and unconscious) associations: it moves like me—is animate; it doesn't move like me—is inanimate. Since the categorical dichotomy LIVING/non-LIVING is also developed staring from a similar, if weaker, analogy concerning the way in which living things move, we can consider it as *partially* independent from cultural and linguistic contamination too. However, because of the weakness of the analogy between the movement characteristics of animate things and the growth/tropisms of plants, the class of living things can undergo some cultural variation: different cultures can, for example, stress or attenuate this analogy giving rise to categorical dichotomies which are more or less separate from each other, or they can include in the class of LIVING other instances that show some kind of non-mechanical and self-propelled motion like e.g. water.

Going back to the definitions given previously, if we call 'categories' all and only those categorical partitions which are spontaneously produced by our cognitive system before language learning and independently from cultural factors in order to organize experience according to specific ontological regions, then ANIMATE/INANIMATE is univocally a categorical partition, while LIVING/NON-LIVING has an ambiguous status, being partially categorical and partially conceptual. Other folk-biological categories like BIOLOGICAL or ARTIFACTS are, on the contrary, not categorical at all. Indeed, according to this definition of categories, they are general concepts, in the sense that they are a further differentiation of the category of ANIMATE/INANIMATE and are formed inferentially under the influence of culture. The truly genetic problem concerns only the first step of the categorical formation: that is the formation of ANIMATE/INANIMATE.

The further *conceptual* oppositions that are built based on this core dichotomy, through its extension or through the application of inferential procedures, may well provide us with important elements in order to understand the organization of knowledge throughout its development, but they are irrelevant as regards the problem of understanding categorization at its more basic level. As far as these further conceptual oppositions are concerned, our hypothesis is compatible with the idea that they may be largely determined by cultural factors. In fact, it suggests that partitions like biological or artifacts are much more culturally and individually variable then ANIMATE/INANIMATE and LIVING/NON-LIVING which are contingently formed depending on whether a culture develops partitions like these and which instances they include.

As previously stated, folk-biology points out that all cultures share the same fundamental biological concepts, distinguishing between 'animate' and 'inanimate entities' and/or between 'living' and 'non-living entities' entities. Still, this idea does not exclude the possibility that the biological concepts developed by different cultures are based on more or less variable taxonomies: i.e. it is compatible with the idea that the concrete determination of the instances considered to be members of one or of the other concept may vary between cultures and may depend on cultural factors (like religion, traditions and so on). However, if the hypothesis we suggest is correct, the categorical dichotomy ANIMATE/INANIMATE is a universal element of the human mind and must therefore be transculturally shared. So, if we are right, all the different taxonomies underlying culturally determined biological concepts should share the same lowest common denominator: i.e. a distinction between animate entities that exhibit specific movement characteristics (they move as we do) on the one hand, and inanimate entities that do not exhibit the same movement features on the other. In fact, the very existence of disciplines like folkbiology and folkphysics that examine our key naïve physical and biological concepts in a cross-cultural framework could be traced back to the primary and basic distinction between respectively inert objects and moving biological objects. In fact, as already mentioned at the beginning of this discussion (see above Sect. 3), some cross-cultural studies that investigate the criteria used by people of different ages and cultural origins to distinguish between living and non-living things (i.e. humans–animals–plants vs. everything else) provide results that can be interpreted as a possible confirmation of our hypothesis. These studies stress that—even though it is fundamental from a folkbiological point of view—the concept of 'living thing' is difficult for everyone to acquire and young children (up to 5 years of age) in all cultures tend to spontaneously distort its structure, excluding plants and including instead non-living objects that appear to move on their own (see e.g. Carey 1985; Hatano et al. 1993; Keil 1983; Opfer and Siegler 2004; Stavy and Wax 1989; Waxman and Medin 2006).

A plausible interpretation of this distortion is that young children fail to 'align' the categorical dichotomy they developed before language acquisition with the application rules of the words 'living'/'non-living' in their languages, which are culturally determined and include all entities that are considered alive according to a specific culture (mostly plants, but possibly also other entities). This could be a confirmation of the fact that the opposition 'living'/'non-living' is not the basic one, while the primary and original categorical distinction children of all cultures develop consists of the dichotomy ANIMATE/INANIMATE; and this applies independently from the specific biological concepts they are going to learn thereafter from their culture.

Staring from this idea that the dichotomy ANIMATE/INANIMATE is a basic and truly categorical partition we can also make a further prediction: people will differentiate different degrees of 'vitality' according to the degree of similarity they recognize between the considered instance and themselves. This implies first of all that humans will not identify all animate things as equally 'animate': other human beings will be recognized as 'more animate' than animals; in turn mammals will be recognized as 'more animate' than inferior animals that are less similar to ourselves and move in a less similar manner: mammals will be considered e.g. as 'more animate' than insects; however, animals will be always considered as 'more animate' (i.e. as having a higher degree of vitality) than plants. This conclusion is also supported by a number of cross-cultural studies, which develop on this basis the hypothesis that "children's judgments about life status and the possession of the attributes of living things are [...] influenced by [...] the degree of similarity between a target entity and the best understood living thing, the human being" (Hatano et al. 1993: 51).

This categorical stratification hypothesis follows the line 'human–animal–plants', that is, a line that does not start from the simple in order to form the complex, but that starts from the complex (the human) to develop the idea of the simple (the non-human). Furthermore, in the view we propose, this categorical stratification must be conceived dichotomously, in the sense that the development of a category on the basis of specific properties (in this case, movements) implies automatically the genesis of its opposite through logical negation: the category of INANIMATE is formed in opposition to that of ANIMATE, since INANIMATE simply collects the instances that do not show the characteristics of movement peculiar to animate instances.

This hypothesis on the origin of folk-biological categories has important consequences as far as the putative hierarchal structure of our knowledge is concerned.

In fact, it challenges two main assumptions made by the mainstream research on concepts:

(1) the so called *basic level primacy* according to which the so called basic level concepts are cognitively primary with respect to more general and more specific concepts, and
(2) the idea that generalization and specialization proceeds in a linear manner respectively 'upward' and downward' from the basic level.

The general idea behind these assumption is that conceptual structure is "organized in a hierarchy from the most general to the most specific": "categories that are cognitively basic are 'in the middle' of a general to specific hierarchy" (Lakoff 1990: 13), while superordinate and subordinate concepts are developed through generalization and specialization processes starting from the basic level. This hierarchical structure of categories is considered universal, in the sense that all cultures appear to organize their knowledge according to this scheme (Berlin 1992; see also Murphy 2004: 233–235).

Words like e.g. 'bird', 'chair', 'flower' express so called basic categories. If we consider a classical structure like the one that proceeds 'upward' and 'downward' starting e.g. from the basic category 'bird', *animal*, *living thing* and *biological object* turn out to be superordinate categories at an increasing level of generalization (something like *jay* and *blue jay* would be subordinate categories at an increasing level of specialization - see Fig. 4 for a schematic representation).

However, the stratification hypothesis we previously put forward is not consistent with this idea, especially if the primacy of basic level is meant in a genetic sense. In fact, we suggest that animal—specifically the categorical dichotomy ANIMATE/INANIMATE—is formed *before* the (basic) concept of bird and is rather the necessary precondition for the development of this concept—and of basic concepts in general.

In fact, we make a straight distinction between categories (through which we organize our experience from an ontological point of view) and concepts (the mental representations that support our linguistic mastery). Categories are universal and prelinguistic, i.e. immune from cultural influences, and group things together in general macroclasses. Concepts on the contrary are developed through (parallel to) language acquisition and through exposure to a belief system. They consist in a further refinement and internal differentiation of this original categorical distinction which is carried out on the basis of the criteria each culture develops to determine exactly what its words refer to. Indeed, categories are a more general and fundamental element than concepts and are a necessary precondition for the development of more and more specific language-based concepts.

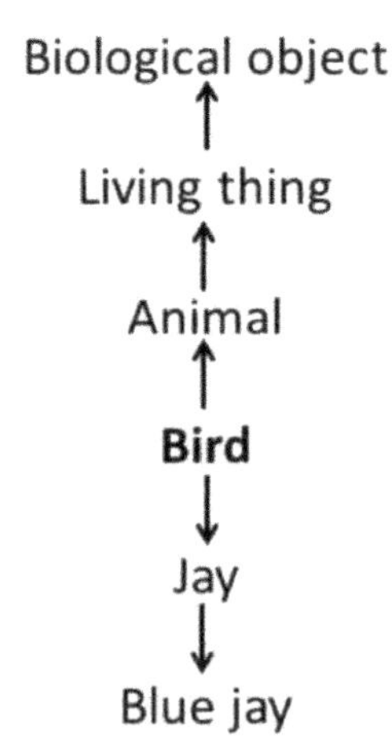

Fig. 4 Basic concepts as the starting point for the development of superordinate and subordinate concepts

The hypothesis we maintain in this research implies that the categorical dichotomy animate/inanimate is one of the basic categorical partitions the cognitive system in organized into and that it is essential—together with other categories and categorical dichotomies—for organizing our experience from an ontological point of view. *Thus, categories are cognitively primary with respect to all kinds of concepts*. In this sense, our study agrees with Mandler's thesis (already mentioned in Sect. 1) that the "conceptual system is categorical from its inception, with global categories being formed first, followed by subdivision of these broader categories" (Mandler 2002: 315). In the next section we will try to clarify this aspect of our hypothesis, specifying the relationship it entertains with the (almost unquestioned) thesis of the cognitive primacy of basic level concepts, and to explain its consequences as far as cognitive architecture is concerned.

6 The Cognitive Primacy of Categories

The so called *basic level primacy thesis* relies on many different factors connected with the so called basic level categories. First of all, it refers to the idea that basic level categories are more *simple* than the superordinate and subordinate ones with respect to their composition, that they are *preferred* to their subordinates and subordinates, and that they are also the most *advantageous* from a cognitive point of view.

Basic level categories like *cat*, *chair*, *car* etc. describe "the most inclusive level at which category members possess a significant number of common attributes" and "for which highly similar sequences of movements are made to category members" (Murphy 2004: 211).[13] Indeed cats are e.g. much more similar to each other both in their shape and in their movements than animals are. The same applies to *cars* with respect to *vehicles* or to *chair* with respect to *furniture*: *vehicles* exhibit shapes and movements that are less similar to each other's than the shapes and the movements of cars. Analogously, furniture can be very differentiated, while chairs bear a much closer resemblance to each other. Due to their similarity in shape, basic level categories can also be represented using a mental image (Rosch et al. 1976; Jolicoeur et al. 1984; Murphy and Brownell 1985). The same applies to subordinate categories, but not to superordinate ones. In addition, basic level categories are the ones people spontaneously use most frequently and easily.

People associates basic level categories with many concrete properties including —among other things—their parts. The same properties are ascribed also to subordinate categories; however, subordinate categories are harder to sort out, since they are very similar to each other and their distinctive properties consist only in minor perceptual or functional modifications of the properties they share with the corresponding basic level category. In turn, superordinates are linked with

[13]See also the original studies by Rosch (1978) and Rosch et al. (1976).

functional and abstract properties only.[14] Thus, basic level categories are advantageous with respect to both superordinate and subordinate categories, since they offer many cues to identify the instances that belong to them and at the same time they are well distinguished from other categories. Birds share e.g. many common properties that help to identify them, while animals are much more differentiated from each other and have only very few common attributes. In turn, different birds are highly similar to each other and very distinctive properties are needed to distinguish the various species, while only a few properties (like fly, have feathers, etc.) are sufficient to identify birds as a whole (Rosch et al. 1976). Basic categories respond therefore to a principle of cognitive economy, i.e. of parsimony, in the sense that they allow us to distinguish among objects insofar as they are highly differentiated from each other, without excessive specialization and without the need for creating/memorizing overly small categories (Rosch et al. 1976; Rosch 1978).

However, the primacy of basic categories must not only be linked to factors like simplicity, preference and cognitive advantage. Basic categories have been rather considered as cognitively primary also in two further and deeper senses: *they are seen as a universal way in which people belonging to different cultures organize their knowledge* and *they are also often considered as the first categories children form*. All the above mentioned aspects concerning the cognitive primacy of basic levels are considered nearly unquestionable by most of the scientific community. However, there are a few studies that dispute basic level primacy with respect to these last two (more compromising) variations, which implies a specific hypothesis on concept formation prior to linguistic learning.[15] The main research line challenging this thesis has been put forward by Mandler and colleagues in various writings and *it opposes the idea that basic level categories are universal and are the first ones children learns*. Our analysis on folk-biological categories comes to a similar conclusion.

As far as the supposed 'genetic primacy' of the basic level is concerned, this is usually traced back to the fact that it is at this level that instances belonging to the same category have a high within-category similarity and a low between category similarity, which makes it easy for children to learn them. On the contrary, because of their lack of internal homogeneity many superordinate categories appear to be particularly difficult for children to learn and they often appear to conceive of them just as sets containing multiple types of objects (see e.g. Horton and Markman 1980; Markman 1985; Rosch et al. 1976; Wisniewski et al. 1996). In line with this hypothesis some developmental studies suggest that basic level categories are the first ones children form and name (Anglin 1977; Horton and Markman 1980; Mervis and Crisafi 1982; Rosch et al. 1976).

---

[14]About the problem of determining what properties are associated with what kind of category see e.g. Rosch et al. (1976) and Tversky and Hemenway (1984).

[15]For specific criticisms, see e.g. Mandler (1998: 289, 2004: Chap. 6, 2008) and Mandler and McDonough (2000).

Mandler's aim is not to deny the validity of the results on the primacy of basic concepts *with respect to linguistic learning*—which are solid and well corroborated (for a brief review see e.g. Murphy 2004: 327–331). Neither does Mandler deny that the *words* referring to basic concepts are the first ones that children use, nor does she explicitly criticize the idea that basic level concepts are somehow privileged and cognitively more simple and economic than others. However, she does challenge the idea that an infant's experience begins to be conceptually organized through the learning of basic concepts. Indeed, according to Mandler, infant categorization seems to be top-down and start at a more global level:

> children appear to begin categorizing at a more global level; only gradually do these global categories become differentiated. We assume that such differentiation involves making conceptual distinctions that are initially absent, even when perceptual differentiation is present. The language adults use when labeling and talking about objects to children may provide some of the impetus toward the more detailed conceptualizations that basic-level concepts represent (Mandler et al. 1991: 294).

And further:

> The fact that concept formation is primarily top-down in nature, has a number of consequences for how the conceptual system develops [...]. I have already mentioned one: 'Basic-level' concepts inherit the properties of the global classes from which they are derived. [...] Because infants at first divide the world of objects into animals and nonanimals and because learning takes place within existing knowledge structures, every time a new category is learned within these global domains, it is necessarily learned as a subdivision of this larger division. The infant looking at dogs and cats does not at first conceptualize them as two different kinds, but as discussed below, most likely as two different-looking, self-moving interactors. When they do become conceptually distinct, it is as two self-moving interactors that vary in appearance and the sounds they make, that behave rather differently to petting them, and that have different names, but their animal membership is never in question (Mandler 2008: 210).

As Mandler reports, many developmental studies have found global categories developing in infancy and she suggests that this is not counterintuitive at all.[16] Mandler admits that at the beginning very young children's categorization processes might largely rely on perceptual similarity and that this apparently constitutes a reason to assume that global categories are learned later (since objects belonging to global categories are widely dissimilar from each other). However, she also suggests that infants have only little experience with specific types of object—e.g. with specific types of animals—while it will be easier for them to distinguish among larger classes:

> [...] if they have observed any animals at all, they will have experienced the characteristics that differentiate animals as a global class from nonanimals [...]. This information characterizing the difference between animate and inanimate objects is abstract, but

[16]More specifically, Mandler complains that basic-level effects in infancy have been found only by studies that rely on picture-looking tasks, while data collected using other techniques like object-handling tasks provide evidence in favor of the idea that global categories precede finer conceptual distinctions. See Mandler (1998) and Mandler and McDonough (1993).

> fundamental [...]. Even for normal adults, it is easier to conceptualize how an animal differs from a piece of furniture than to conceptualize how a lion differs from a tiger. And of course, insofar as my argument is correct that to have a concept of dog entails the larger concept of animal, there is a logical sequence involved in the order of early concept formation as well (Mandler 1998: 289).

According to Mandler, the only unequivocal evidence there is in favor of the primacy of basic-level categorization is that basic-level *terms* are the first ones that children learn. For Mandler this primacy can be interpreted only in a specifically linguistic sense and it depends on the fact that those terms are used more frequently than superordinates in daily speech and in speech to children. Mandler observes that:

> [...] children only learn the words that are spoken to them. Because the majority of object nouns addressed to children are at what is known as the 'basic level', the first such nouns that children learn tend to be of that type. However, the fact that children might learn the word 'dog' before the word 'animal', in itself that tells us little about their interpretation of the word's meaning (i.e. the concept they understand by it). This is shown clearly by the common phenomenon in early language of overextension, in which a word such as 'dog' is used to refer to animals in general (Mandler 2002: 316–317).[17]

Mandler argues that—at the beginning of linguistic learning—words are not understood and applied according to their specific social meaning: for a child the word 'dog' does not necessarily mean dog (it does not necessarily refer to the class of dogs as this has been identified by the English language). On the contrary, at the beginning of linguistic learning word understanding relies on forms of prelinguistic categorization developed during previous stages. According to Mandler, in an early, prelinguistic stage of their development infants learn to distinguish between animate and inanimate instances (i.e. between animals and non-animals); once they are exposed to the word 'dog' and need to understand/learn it, first of all they will rely on this dichotomy and associate the word 'dog' with animals—i.e. not only to dogs, but also to other animals. Only later will children learn that 'dog' has a more specific meaning and it denotes only a particular kind of animal, characterized by a certain form and by specific characteristics.

This explanation of the steps children undergo in order to learn a word like 'dog'—which assumes that at a first stage they have an 'overextended' understanding of it and only later readjust the meaning to pick out only the object the word correctly applies to—does not challenge the idea of the primacy of basic levels *tout court*: it does not deny e.g. that basic level concepts are the most frequently used ones and that cognitively they are the most advantageous ones. In fact, neither does the idea that global categories are learned first challenge the hypothesis that linguistic conceptual learning starts from the basic level. The hypothesis Mandler puts forward—which we share—suggests that conceptual formation undergoes two stages: in a first stage global categories are formed that

[17]Mandler's references for the phenomenon of 'overextension' are Clark (1983) and Rescorla (1980). For a detailed exposition of this aspect see Mandler (2004: 59–91).

organize the experience according to large ontological regions (for this notion see Sect. 1 of this chapter), while conceptual organization in a proper sense is carried out only later through linguistic learning.

This distinction between these two stages of conceptual formation confirms the need to clearly differentiate between the notion of 'category' and that of 'concept'. On the basis of its current use in the literature, at the beginning of this section we spoke of 'basic level categories', even though later we spoke of 'basic level concepts'. As a matter of fact, in the literature on concepts these two expressions are used interchangeably, since most authors do not differentiate between the notion of 'concept' and that of 'category'. However, the hypothesis we put forward here suggests that these notions do not describe one and the same thing. More specifically, the hypothesis we argue for suggests that in this case we should speak of 'basic level concepts' and that basic level concepts should be conceived as linguistic concepts which have been learned through language. These differentiation does not challenge any of the effects related to basic level concepts pointed out by the experimental literature on this topic. On the contrary, the analysis we carried out indicates that categories are not the same as concepts and cannot be simply seen as concepts with a higher degree of generality: i.e. as concepts hierarchically superordinate with respect e.g. to basic level concepts.

In our perspective, the classical hierarchical view according to which children learn basic concepts first and then develop more general and more specialized concepts is much too simplistic: we challenge the idea itself of a hierarchy which is developed linearly both downward and upward. In fact, we *do not* maintain that the classical hierarchy that goes from the basic level to superordinate concepts should be simply inverted in an opposite hierarchy that starts from very general concepts to basic level ones according e.g. to the schema (Fig. 5).

The genetic path we are suggesting is *double* and *not linear*. We suggested that experience is organized first of all categorically. Before learning any more specific distinctions (of a conceptual kind), children need to have the ability to situate what they perceive in specific ontological regions: they need to know that the experience they have is e.g. of an object, of an event, of a quality etc.; or they need to know that a certain experience originates in the outside world or in their body (according to the dichotomy internal/external), or they need to know whether a certain thing they are perceiving is animate or inanimate.

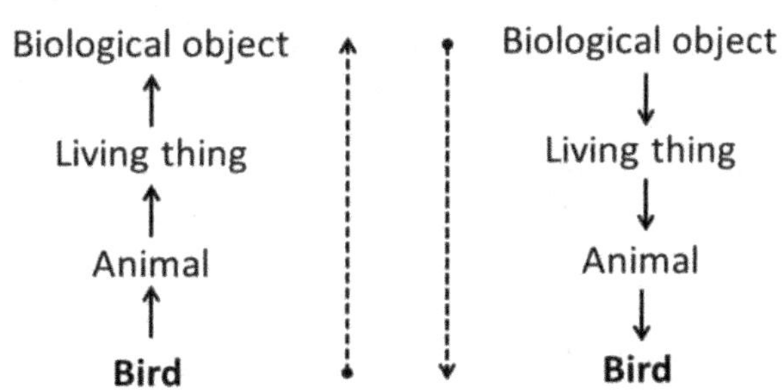

Fig. 5 Classical hierarchy (from the basic level to superordinate concepts) on the *left*; inverted hierarchy (from very general concepts to the basic level) on the *right*

This organization of the experience is a precondition in order for them to carry out other, more specific partitions through linguistic learning and through cultural influences mediated by language. Even though the conceptual (linguistic) system originates from the categorical one, it is also largely independent of it. Take for example the case of animate/inanimate: the hypothesis we put forward here suggests that the categorical dichotomy animate/inanimate is universal; however, this does not imply that all conceptual systems that have been developed by different cultures and languages all of which start from this dichotomy must include one (and only one) word to express exactly this opposition: e.g. the word animate as opposed to inanimate. On the contrary, it is possible that a language doesn't have any specific words to express the idea of animacy as opposed to the idea of inanimacy, but animacy and inanimacy are included in a larger conceptual opposition such as e.g. the one between living and non-living things (on this aspect see e.g. the examples at the beginning of Sect. 2 of this chapter). Or it is possible that a language does not have any single word to indicate all animals as a group (corresponding with the English word 'animal'), but only have various words to indicate specific kinds of animals.

This means that the existence of a categorical system which is distinct from and preliminary to the conceptual system does not offers a solution to the problem of establishing how a child comes to learn the specific conceptual system implicit to her/his language as well as the conceptual hierarchies which are implied by it. The specific conceptual system of a language is learned on the basis of a previously existent categorical system; however, the classifications children learn when they acquire their language might not be completely congruent with the categorical partitions. Depending on the language people acquire and on the culture the classifications of their language rely on, they will tend to consider animals as more or less similar to plants, or they will tend to consider different animals as more or less similar to each other, etc. However, since every conceptual system stems from the same categorical and prelingustic partitions, our hypothesis suggests that all conceptual systems will keep track of that initial categorical information. Since categories are used for the formation of concepts — the information they carry does not get lost in cognitive development: when subjects form complex concepts that allow them to distinguish e.g. between different animal instances—different persons, different animals etc.—all these concepts still carry the core information 'animate' which has been used to build them.

Furthermore, while conceptual organization can be thought of as a hierarchy of superordinate and subordinate concepts, our hypothesis suggests that categorical relationships might need to be conceived differently: as organized according to quasi-spatial relationships like inclusion, exclusion and relative size. Going back to the terms introduced in the first section of this chapter, if categories are ontological regions, the relationships among categories are not of a hierarchical kind, but rather concern the extension and overlapping of categories. From a merely geographical point of view, Italy is not subordinate to Europe, but it is rather a (smaller, internal) part of it. In the case of categories, the genetic line is not bound to any hierarchy and it follows neither an ascendant path (from the specific to the general) nor a descendant one (from the general to the specific). The genetic order of derivation follows completely different principles that—in our hypothesis—are not of logical, but of a bodily and egological kind. In the case of animate/inanimate it starts from what is more similar to us and ends with what is more different from us.

Once again, this 'spatial' idea of categorical organization should not be understood as a critique of the thesis that *conceptual* system is hierarchically organized. This hypothesis suggests rather the need to differentiate as clearly as possible between categories and concepts. Categories can be seen as ontological regions or macro-classes that are not hierarchically organized, but are related to each other according to relationships of inclusion/exclusion and can be compared from the point of view of their extension. In contrast, the conceptual system is hierarchically organized into superordinate and subordinate concepts that tend to converge with the hierarchical organization of the lexical system of the language the system is related to.[18] On the other hand, superordinate concepts are not categories, as e.g. the Aristotelian tradition would also suggest. Other than categories, in our hypothesis superordinate concepts should be considered as culturally determined generalizations of more basic concepts (which probably start with basic concepts since these are—as the literature shows—particularly easy to form on the basis of the perceptual similarity of the instances that fall under them and on the basis of the fact that they can rely on a perceptual scheme understood as a proxytype—on this see Chap. 3, Sects. 5 and 6).

Like any other kind of concept, superordinate concepts are also formed including categorical information. However, the organization of experience in categories does not (necessarily) correspond to the classifications described by superordinate concepts since the categorical partitions are universal, while the superordinate concepts are culturally specific and are learned through a language.

In our hypothesis, categories have outlines which are much vaguer than concepts. One can e.g. be able to locate an entity in the region of the animate instances without knowing that this entity is an animal and therefore has all the properties of animals (eats, breathes, is born, dies, etc.). *The thesis that the capacity to locate an instance in a category precedes all other conceptual capacities implies that already at a very early stage of their development infants distinguish different ontological regions and differentiate (in an entirely implicit manner) between animate and inanimate objects, in the sense that they interact differently with them. This thesis does not imply however that they have any quasi-propositional or quasi-explicit knowledge of the macro-classes their experience is organized into; in fact, only concepts are mastered in this way.*

This hypothesis might also find an empirical validation in the neuropsychological literature. According to our view categorical capacities like the one which distinguishes 'animate' from 'inanimate' objects are developed at a very early stage

[18]We speak of convergence instead of coincidence because—as we made clear previously—people never learn and master their language in a perfect way. The concepts people develop which support their linguistic understanding never entirely correspond to the sociolinguistic definitions of words. About this see Chap. 3, Sects. 1–3.

and therefore they constitute the precondition for language learning and for the acquisition of more specific classifications. For this reason, they can be considered as *fundamental elements of human knowledge organization.* Because categorical capacities are fundamental, it might be plausible to assume that they could be the last to be lost due to degenerative diseases or due to extended and disabling neural damage. So if the prediction we make is correct, a patient may completely lose his capacity to identify an instance in its individuality (e.g. to identify a dog as a dog, a monkey as a monkey, etc.), but s/he will still be able to recognize the 'animality' of animals (of the dog, of the monkey etc.) he perceives. This means that people might lose their conceptual and linguistic capacities, but still be able to make non-verbal categorical distinctions and to recognize e.g. the characteristic animacy of animals in comparison to non-animate instances, since this capacity has very deep roots and should be particularly resistant to damage within the cognitive system. Since this residual capacity is supposed to be non-linguistic, only non-linguistic tasks would be appropriate for testing whether these capacities are indeed preserved.[19]

According to this hypothesis, the deconstruction of our categorical and conceptual capacities proceeds in an inverted manner with respect to their construction: the most basic capacities (which are the precondition for the development of more complex capacities) should be the first ones to appear and the last ones that be lost in the case of degenerative diseases. A position like this is also proposed by Mandler: "Because of the nature of the learning process, in which each new distinction is learned as a subdivision of a superordinate category, the superordinate distinctions should be the most firmly established and therefore one would expect a pattern 'first in, last out' […]. The available literature suggests exactly that pattern" (Mandler 2004: 211). On this topic see also Patterson and Hodges 1995; Warrington 1975; Mandler 1998, 2002: 335, 2009: 200–201).

Mandler maintains that neuropsychological data on the breakdown of the conceptual system under brain damage in adulthood confirms this pattern: "The literature on semantic dementia suggests that breakdown of conceptual knowledge about objects goes from the loss of specific details to more general categories in quite an orderly fashion" (Mandler 2004: 208). However, the finding that often broad superordinate information is preserved, while finer grained semantic knowledge is lost can be explained in different ways. Indeed, even though it might reflect the way in which our knowledge is organized (see e.g. Collins and Quillian 1969; Shallice 1988), it might also be the consequence of a less revealing phenomenon: when semantic information degrades, specific distinctions are lost first, while more general information is preserved for a longer time (see e.g. McClelland and Rumelhard 1985; Rapp and Caramazza 1989, 1993).

[19]The fact that linguistic and non-linguistic knowledge cannot be investigated using the same kind of evidence has been pointed out also by Sartori et al. (2002) and by Farah and Wallace (1992). For this reason, all the studies on category-specific deficits that use linguistic tasks (like asking the patients to name specific objects or to verbally provide some knowledge regarding them) are irrelevant with respect to our hypothesis.

Nevertheless, starting from the genetic hypothesis we put forward in this chapter it is possible to sketch a much more precise prediction of the pattern of impairment one should find to support the thesis that non-linguistic categories are more basic than the conceptual system—and that they are therefore also less vulnerable to brain damage. In fact, to confirm the pattern 'first in, last out', during the progressive decline of semantic memory we should be able to observe a stage in which the patient's linguistic capacities are already clearly impaired and the patient performs very badly in all kind of tasks involving words (naming, word–picture matching etc.), even though s/he is still capable to discern animate and inanimate objects in a non-linguistic fashion and e.g. to group together pictures of animals distinguishing them from pictures of inanimate things. The capacity to distinguish animate from inanimate things should be particularly resistant. Specifically, it should be more resistant than the capacity to distinguish between living and non-living things and between biological and non-biological objects. At this regard it should be noticed once more that—if categories are interpreted as a pre-linguistic means of classification—then evidence about the progressive decline of the capacity to use/understand *words* referring to specific classes of things shouldn't be considered relevant with respect to the problem of determining whether the ability to organize experience according to general categories is indeed the last one humans lose.

References

Anggoro, F. K., Waxman, S. R., & Medin, D. L. (2005). The effects of naming practices on childern's understanding of living things. In *Proceedings of 27th annual meeting of the cognitive science society* (pp. 139–144). Hillsdale, NJ: Erlbaum.

Anggoro, F. K., Waxman, S. R., & Medin, D. L. (2008). Naming practices and the acquisition of key biological concepts: Evidence from English and Indonesian. *Psychological Science, 19*(4), 314–319.

Anglin, J. M. (1977). *Word, object and conceptual development*. New York: W.W. Norton.

Aquila, R. (1983). *The representational mind. A study of Kant's theory of knowledge*. Bloomington: Indiana University Press.

Aristotle. (1963). Categories. In J. L. Ackrill (Ed.), *Aristotle, categories and de interpretatione* (pp. 3–42). Oxford: Oxford University Press.

Arterberry, M. E., & Bornstein, M. H. (2001). Three-month-old infants' categorization of animals and vehicles based on static and dynamic attributes. *Journal of Experimental Child Psychology, 80*(4), 333–346.

Atran, S. (1998). Folkbiology and the anthropology of science: Cognitive universals and cultural particulars. *Behavioral and Brain Sciences, 21*(4), 547–609.

Baillargeon, R. (1995). Physical reasoning in infancy. In M. S. Gazzaniga (Ed.), *The cognitive neurosciences* (pp. 181–204). Cambridge, MA: MIT Press.

Baron-Cohen, S. (1995). *Mindblindness: An essay on autism and theory of mind*. Cambridge, MA: MIT Press.

Berlin, B. (1992). *Ethnobiological classification: Principles of categorization of plants and animals in traditional societies*. Princeton, NJ: Princeton University Press.

Bermudez, J. L. (2000). *The paradox of self-consciousness*. Cambridge, MA: MIT Press.

Bertenthal, B. (1993). Infants perception of biomechanical motions: Intrinsic image and knowledge-based constraints. In C. Granrud (Ed.), *Visual perception and cognition in infancy* (pp. 175–214). Hillsdale, NJ: Erlbaum.

Berthoz, A. ([1997] 2000). *The brain's sense of movement* (G. Weiss. Trans.). Cambridge, MA: Harvard University Press.

Berthoz, A. & Petit, J.-L. ([2006] 2008). *The physiology and phenomenology of action* (C. Macann Trans.). Oxford: Oxford University Press.

Bretherton, I. (1992). The origins of attachment theory: John Bowlby and Mary Ainsworth. *Developmental Psychology, 28*(5), 759–775.

Bretherton, I. (1993). From dialog to internal working models: The co-construction of self in relationships. In C. A. Nelson (Ed.), *Memory and affect in development. The Minnesota symposia on child psychology* (pp. 237–264). Hillsdale, NJ: Erlbaum.

Broekmann, J. M. (1963). *Phänomenologie und Egologie. Faktisches und transzendentales Ego bei Husserl*. The Hague: M. Nijhoff.

Caramazza, A., & Shelton, J. R. (1998). Domain-specific knowledge systems in the brain: The animate-inanimate distinction. *Journal of Cognitive Neuroscience, 10*(1), 1–34.

Carey, S. (1985). *Conceptual change in childhood*. Cambridge, MA: MIT Press.

Carey, S. (2000). The origins of concepts. *Journal of Cognition and Development, 1*(1), 37–41.

Carey, S., & Spelke, E. S. (1994). Domain-specific knowledge and conceptual change. In L. A. Hirschfeld & S. A. Gelman (Eds.), *Mapping the mind: Domain specificity in cognition and culture* (pp. 169–200). Cambridge: Cambridge University Press.

Clark, E. V. (1983). Meanings and concepts. In J. H. Flavell & E. M. Markman (Eds.), *Handbook of child psychology. Volume 3: Cognitive development* (pp. 787–840). New York: Wiley.

Cole, J., & Waterman, I. (1991). *Pride and a daily marathon*. Cambridge, MA: MIT Press.

Collins, A. M., & Quillian, M. R. (1969). Retrival time from semantic memory. *Journal of Verbal Learning and Verbal Behaviour, 8*(2), 240–247.

Dellantonio, S., Innamorati, M., & Pastore, L. (2012). Sensing aliveness: An hypothesis on the constitution of the categories 'animate' and 'inanimate'. *Integrative Psychological and Behavioral Science, 46*(2), 172–195.

Drummond, J. J. (2008). *Historical dictionary of Husserl's philosophy*. Lanham, Toronto, Plymouth: Scarecrow Press.

Farah, M. J., & Wallace, M. A. (1992). Semantically bounded anomia: Implications for the neural implementation of naming. *Neuropsychologia, 30*(7), 609–621.

Fodor, J. A. (1975). *The language of thought*. New York: Crowell.

Fodor, J. A. (1981). The present status of the innateness controversy. In J. A. Fodor (Ed.), *Representations philosophical essays on the foundations of cognitive science* (pp. 257–316). Brighton: The Harvester Press.

Fodor, J. A. (1987). *Psychosemantics. The problem of meaning in the philosophy of mind*. Cambridge, MA: MIT Press.

Gallagher, S. (1986). Body image and body schema: A conceptual clarification. *The Journal of Mind and Behavior, 7*(4), 541–554.

Gallagher, S. (2005). *How the body shapes the mind*. Oxford: Oxford University Press.

Gallagher, S., & Cole, J. (1995). Body image and body schema in a deafferented subject. *The Journal of Mind and Behavior, 16*(4), 369–390.

Gallagher, S., & Meltzoff, A. N. (1996). The earliest sense of self and others: Merleau-Ponty and the recent developmental studies. *Philosophical Psychology, 9*(2), 213–236.

Gallagher, S., & Zahavi, D. (2008). *The phenomenological mind*. London, New York: Routledge.

Gallistel, C. R. (1990). *The organization of learning*. Cambridge, MA: MIT Press.

Gardner, H. E. (1985). *The new science of mind*. New York: Basic Books.

Gellhorn, E. (1943). *Anatomic regulations: Their significance for physiology, psychology, and neuropsychiatry*. New York: Interscience Publications.

Gelman, S. A., Coley, J. D., & Gottfried, G. M. (1994). Essentialist beliefs in children. In L. A. Hirschfeld & S. A. Gelman (Eds.), *Mapping the mind. Domain specificity in cognition and culture* (pp. 341–365). Cambridge, MA: Cambridge University Press.

Gelman, R., Durgin, F., & Kaufman, L. (1995). Distinguishing between animates and inanimates: Not by motion alone. In D. Sperber, D. Premack, & A. J. Premack (Eds.), *Causal cognition: A multidisciplinary debate* (pp. 150–184). Oxford: Clarendon Press.

Goldstone, R. L., & Kersten, A. (2003). Concepts and categorization. In A. F. Healy & R. W. Proctor (Eds.), *Comprehensive handbook of psychology* (pp. 599–621). New York: Wiley.

Gopnik, A., & Meltzoff, A. (1997). *Words, thoughts and theories*. Cambridge, MA: MIT Press.

Haselager, W. F. G., Broens, M., & Quilici Gonzalez, M. E. (2012). The importance of sensing one's movements in the world for the sense of personal identity. *Rivista internazionale di Filosofia e Psicologia, 3*(1), 1–11.

Hatano, G., & Inagaki, K. (2000). Domain-specific constraints of conceptual development. *International Journal of Behavioral Development, 24*(3), 267–275.

Hatano, G., Siegler, R. S., Richards, D. D., Inagaki, K., Stavy, R., & Wax, N. (1993). The development of biological knowledge: A multi-national study. *Cognitive Development, 8*(1), 47–62.

Horton, M. S., & Markman, E. M. (1980). Developmental differences in the acquisition of basic and superordinate categories. *Child Development, 51*(3), 708–719.

Husserl, E. ([1913] 1980). *Ideas pertaining to a pure phenomenology and to a phenomenological philosophy. Third book: Phenomenology and the foundation of sciences* (T. E. Klein & W .E. Pohl Trans.). Den Haag: M. Nijhoff.

Husserl, E. ([1913] 1982). *Ideas pertaining to a pure Phenomenology and to a phenomenological philosophy. First book: General introduction to a pure phenomenology.* (F. Kersten Trans.). Den Haag: M. Nijhoff.

Husserl, E. ([1913] 1989). *Ideas pertaining to a pure phenomenology and to a phenomenological philosophy. Second book: Studies in phenomenology of constitution* (R. Rojcewicz & A. Schumer Trans.). Den Haag: M. Nijhoff.

Husserl, E. ([1929] 1977). *Cartesian meditations. An introduction to phenomenology* (D. Cairns Trans.). Den Haag: M. Nijhoff.

Inagaki, K., & Hatano, G. (2002). *Young children's naive thinking about the biological world.* New York: Psychology Press.

Jackendoff, R. (1983). *Semantics and cognition.* Cambridge, MA: MIT Press.

Jackendoff, R. ([1987] 1994). *Consciousness and the computational mind* (4th ed.). Cambridge, MA: MIT Press.

Jackendoff, R. (1995). *Semantic structures.* Cambridge, MA: MIT Press.

Jackendoff, R. (2002). *Foundations of language. Brain, meaning, grammar, evolution.* Oxford: Oxford University Press.

Johnson, M. (1987). *The body in the mind: The bodily basis of meaning, imagination, and reason.* Chicago: Chicago University Press.

Jolicoeur, P., Gluck, M., & Kosslyn, S. M. (1984). Pictures and names: Making the connection. *Cognitive Psychology, 16*(2), 243–275.

Kant, I. ([1787] 2003). *Critique of pure reason.* Translated by N. Kemp Smith. New York: Palgrave MacMillan.

Keil, F. C. (1983). On the emergence of semantic and conceptual distinctions. *Journal of Experimental Psychology: General, 112*(3), 357–385.

Lakoff, G. (1990). *Women, fire, and dangerous things. What categories reveal about the mind.* Chicago: Chicago University Press.

Lakoff, G., & Johnson, M. (1980). *Metaphors we live by.* Chicago: Chicago University Press.

Lakoff, G., & Turner, M. (1989). *More than a cool reason. A field guide to poetic metaphor.* Chicago: Chicago University Press.

Leslie, A. M. (1982). The perception of causality in infants. *Perception, 11*(2), 173–186.

Leslie, A. M., & Keeble, S. (1987). Do six month old infants perceive causality? *Cognition, 25*(3), 265–288.

Mandler, J. M. (1992). How to build a baby II: Conceptual primitives. *Psychological Review, 99* (4), 587–604.

Mandler, J. M. (1998). Representation. In D. Kuhn & R.S. Siegler (Eds.), *Handbook of child psychology. Volume 2: Cognition, perception and language* (5th ed., pp. 255–308). New York: Wiley.

Mandler, J. M. (2000). Perceptual and conceptual processes in infancy. *Journal of Cognition and Development, 1*(1), 3–36.

Mandler, J. M. (2002). On the foundations of the semantic system. In E. M. Forde & G. W. Humphreys (Eds.), *Category specificity in brain and mind* (pp. 315–340). New York: Psychology Press.

Mandler, J. M. (2004). *The foundation of mind. Origins of conceptual thought*. Oxford: Oxford University Press.

Mandler, J. M. (2008). On the birth and growth of concepts. *Philosophical Psychology, 21*(2), 207–230.

Mandler, J. M. (2009). The conceptual foundations of animals and artifacts. In E. Margolis & S. Laurence (Eds.), *Creations of the mind. Theories of artifacts and their representation* (pp. 191–211). Oxford: Oxford University Press.

Mandler, J. M., Bauer, P. J., & McDonough, L. (1991). Separating the sheep from the goats: Differentiating global categories. *Cognitive Psychology, 23*(2), 263–298.

Mandler, J. M., & McDonough, L. (1993). Concept formation in infancy. *Cognitive Development, 8*(3), 291–318.

Mandler, J. M., & McDonough, L. (1998). On developing a knowledge base in infancy. *Developmental Psychology, 34*(6), 1274–1288.

Mandler, J. M., & McDonough, L. (2000). Advancing downward to the basic level. *Journal of Cognition and Development, 1*(4), 379–403.

Markman, E. M. (1985). Why superordinate category terms can be mass nouns. *Cognition, 19*(1), 31–53.

McClelland, J. L., & Rumelhard, D. E. (1985). Distributed memory and the representation of general and specific information. *Journal of Experimental Psychology: General, 114*(2), 159–188.

Medin, D. L., & Rips, L. J. (2005). Concepts and categories: Memory, meaning, and metaphysics. In K. J. Holyoak & R. G. Morrispon (Eds.), *The Cambridge handbook of thinking and reasoning* (pp. 37–72). Cambridge: Cambridge University Press.

Meltzoff, A. N. (1988). Infant imitation and memory: Nine-month-old in immediate and deferred tests. *Child Development, 59*(1), 217–225.

Meltzoff, A. N. (1995). Understanding the intentions of others: Re-enactment of intended acts by 18-month-old children. *Developmental Psychology, 31*(5), 838–850.

Meltzoff, A. N., & Brooks, R. (2001). "Like me" as a building block for understanding other minds: Bodily acts, attention, and intention. In B. F. Malle, L. J. Moses, & D. A. Baldwin (Eds.), *Intentions and intentionality. Foundations of social cognition* (pp. 171–191). Cambridge, MA: MIT Press.

Meltzoff, A. N., & Gopnik, A. (1993). The role of imitation in understanding persons and developing a theory of mind. In S. Baron-Cohen, H. Tager-Flussberg, & D. Cohen (Eds.), *Understanding other minds: Perspectives from autism* (pp. 335–366). New York, Oxford: Oxford University Press.

Meltzoff, A. N., & Moore, M. K. (1977). Imitation of facial and manual gesture by human neonates. *Science, 198*(4312), 75–78.

Meltzoff, A. N., & Moore, M. K. (1983). Newborn infants imitate adult facial gestures. *Child Development, 54*(3), 702–709.

Meltzoff, A. N., & Moore, M. K. (1989). Imitation in newborn infants: Exploring the range of gesture imitated and the underlying mechanisms. *Developmental Psychology, 25*(6), 954–962.

Meltzoff, A. N., & Moore, M. K. (1997). Explaining facial imitation: A theoretical model. *Early Development and Parenting, 6*(3–4), 179–192.

Mervis, C. B., & Crisafi, M. A. (1982). Order of acquisition of subordinate, basic, and superordinate level categories. *Child Development, 53*(1), 258–266.

Murphy, G. L. (2004). *The big book of concepts*. Cambridge, MA: MIT Press.

Murphy, G. L., & Brownell, H. H. (1985). Category differentiation in object recognition: Typicality constraint on the basic category advantage. *Journal of Experimental Psychology A: Human Experimental Psychology, 53A,* 962–982.

O'Shaughnessy, B. (1995). Proprioception and body image. In J. Bermudez, A. Marcel, & N. Eilan (Eds.), *The body and the self* (pp. 175–203). Cambridge, MA: MIT Press.

Opfer, J. E., & Siegler, R. S. (2004). Revisiting preschoolers' living things concept: A microgenetic study of conceptual change in basic biology. *Cognitive Psychology, 49*(4), 301–332.

Paton, H. J. (1936). *Kant's metaphysics of experience*. London: George Allen & Unwin.

Patterson, K., & Hodges, J. R. (1995). Disorders of semantic memory. In A. D. Baddeley, B. A. Wilson, & F. N. Watts (Eds.), *Handbook of memory disorders* (pp. 167–186). London: Wiley.

Phillips, C. (1985). *Movements of the hand.* Liverpool: Humanity Press.

Poulin-Dubois, D. (1999). Infants' distinction between animate and inanimate objects: The origins of naive psychology. In P. Rochat (Ed.), *Early social cognition. Understanding others in the first months of life* (pp. 257–280). Hillsdale, NJ: Erlbaum.

Poulin-Doubois, D., Lepage, A., & Ferland, D. (1996). Infants' concept of animacy. *Cognitive Development, 11*(1), 19–36.

Premack, D. (1990). The infant's theory of self-propelled objects. *Cognition, 36*(1), 1–16.

Rapp, B., & Caramazza, A. (1989). General to specific access to word meaning: A claim reexamined. *Cognitive Neuropsychology, 6*(2), 251–272.

Rapp, B., & Caramazza, A. (1993). On the distinction between deficits to access and deficits of storage: A question of theory. *Cognitive Neuropsychology, 10*(2), 113–142.

Rescorla, L. A. (1980). Overextension in early language development. *Journal of Child Language, 7*(2), 321–335.

Rosch, E. (1978). Principles of categorization. In E. H. Rosch & B. B. Lloyd (Eds.), *Cognition and categorization* (pp. 27–48). Hillsdale, NJ: Erlbaum.

Rosch, E., Mervis, C. B., Gray, W., Johnson, D., & Boyes-Braem, P. (1976). Basic objects in natural categories. *Cognitive Psychology, 8*(3), 382–439.

Sartori, G., Job, R., & Zago, S. (2002). A case domain-specific semantic deficit. In E. M. Forde & G. W. Humphreys (Eds.), *Category specificity in brain and mind* (pp. 25–49). New York: Psychology Press.

Shallice, T. (1988). *From neuropsychology to mental structure*. Cambridge: Cambridge University Press.

Sheets-Johnstone, M. (1999). *The primacy of movement.* Amsterdam: John Benjamins.

Sherrington, C. S. (1906). *The integrative action of nervous system.* New Haven: Yale University Press.

Smith, D. W. (1995). Mind and body. In B. Smith & D. W. Smith (Eds.), *The Cambridge companion to Husserl* (pp. 323–393). Cambridge: Cambridge University Press.

Smith, D. W. (2004). *Mind world. Essays in phenomenological ontology*. Cambridge: Cambridge University Press.

Smith, R. (2012). Aristotle's logic. In E. N. Zalta (Ed.), *The stanford encyclopedia of philosophy.* http://plato.stanford.edu/archives/spr2012/entries/aristotle-logic/.

Spelke, E. S. (1991). Physical knowledge in infancy: Reflections on Piaget's theory. In S. Carey & R. Gelman (Eds.), *The epigenesis of mind: Essays on biology and cognition* (pp. 37–61). Hillsdale, NJ: Erlbaum.

Spelke, E. S. (1994). Initial knowledge: Six suggestions. *Cognition, 50*(1–3), 431–445.

Spelke, E. S., Breinlinger, K., Macomber, J., & Jacobson, K. (1992). Origins of knowledge. *Psychological Review, 99*(4), 605–632.

Spelke, E. S., Phillips, A., & Woodward, A. L. (1995). Infant's knowledge of object motion and human action. In D. Sperber, D. Premack, & A. J. Premack (Eds.), *Causal cognition: A multidisciplinary debate* (pp. 44–78). Oxford: Clarendon Press.

Stavy, R., & Wax, N. (1989). Children's conceptions of plants as living things. *Human Development, 32*(2), 88–94.

Stern, D. (1985). *The interpersonal world of the infant.* New York: Basic Books.

Subrahmanyam, K., Gelman, R., & Lafosse, A. (2002). Animates and other separably moveable objects. In E. M. Forde & G. W. Humphreys (Eds.), *Category specificity in brain and mind* (pp. 341–373). New York: Psychology Press.

Trevarthen, C. (1979). Communication and cooperation in early Infancy: A description of primary intersubjectivity. In M. Bullowa (Ed.), *Before speech. The beginning of interpersonal communication* (pp. 321–347). Cambridge, MA: Cambridge University Press.

Trevarthen, C., & Hubley, P. (1978). Secondary intersubjectivity: Confidence, confiding, and acts of meaning in the first year. In A. Lock (Ed.), *Action, gesture and symbol: The emergence of language* (pp. 183–229). London: Academic Press.

Tronik, E. Z. (1978). The infant response to entrapment between contradictory messages in face-to-face interaction. *American Academy of Child Psychiatry, 17*(1), 1–13.

Tversky, B., & Hemenway, K. (1984). Objects, parts, and categories. *Journal of Experimental Psychology: General, 113*(2), 169–193.

Warrington, E. K. (1975). The selective impairment of semantic memory. *Quarterly Journal of Experimental Psychology, 27*(4), 635–657.

Waxman, W. (1991). *Kant's model of the mind. A new interpretation of transcendental idealism.* Oxford: Oxford University Press.

Waxman, S. R., & Lidz, J. (2006). Early word learning. In D. Kuhn & R. Siegler (Eds.), *Handbook of child psychology. Volume 2: Cognition, perception and language* (6th ed., pp. 299–335). New York: Wiley.

Waxman, S. R. & Medin, D. L. (2006). Core knowledge, naming and the acquisition of the fundamental (folk)biologic concept 'alive'. In *Proceedings of the 28th annual conference of cognitive science society* (pp. 53–55). Mahwah, NJ: CSS Press.

Wisniewski, E. J., Imai, M., & Casey, L. (1996). On the equivalence of superordinate concepts. *Cognition, 60*(3), 269–298.

Woodward, A. L., Phillips, A., & Spelke, E. S. (1993). Infant's expectations about the motion of animate versus inanimate objects. In *Proceedings of the 15th annual conference of cognitive science society* (pp. 1087–1091). Hillsdale, NJ: Erlbaum.

Woodward, A. L., Sommerville, J. A., & Guajardo, J. J. (2001). How infants make sense of intentional action. In B. F. Malle, L. J. Moses, & D. A. Baldwin (Eds.), *Intentions and intentionality. Foundations of social cognition* (pp. 149–169). Cambridge, MA: MIT Press.

Chapter 5
Internal States: From Headache to Anger. Conceptualization and Semantic Mastery

Abstract Here we ask if we can also apply the distinction between referential and inferential competence we introduced in Chap. 3 to words that do *not* refer to things that are perceived using the external senses, especially to words/concepts that denote bodily experiences (such as pain, thirst, hunger, etc.) or emotions. We introduce and discuss the hypothesis that—even though such words/concepts do not refer to intersubjectively identifiable entities in the external world—they do have a kind of referent that can be accessed via direct perception, more specifically 'proprioception', as we have defined it in terms of all propriosensitive information we can consciously access. In the first part of the chapter, we specifically consider terms denoting bodily experiences such as 'pain' or 'hunger' and argue that their referents are identified and classified from a first-personal point of view on the basis of four main characteristics: their specific intensity, their localization in the body, their co-occurrence with other signals and above all their specific qualitative sensations. Emotions are addressed in the second part of the chapter. We suggest that there is a continuity between bodily experiences and emotions. In particular, we argue for a perceptual theory of emotions in line with that proposed by James and Lange at the end of the 19th century and developed more recently by authors such as Damasio (see also Chap. 2, Sects. 5 and 6). The hypothesis we put forward is that the referential information that supports the categorization of emotions and therefore also our mastery of terms referring to emotions consists in the perception (i.e. the 'proprioception') of those bodily states and changes in bodily states which constitute our emotional experience. In the context of this discussion we examine some objections to this line of reasoning that arise from a cognitivist perspective and following authors such as Oatley, Johnson-Laird and Frijda, we distinguish between basic emotions that can be identified and classified solely on the basis on how they feel and complex emotions whose identification and classification additionally depends on cognitive factors. To describe how emotions are identified and classified on the basis of how they feel, we rely on Marcel and Lambie's distinction between an 'emotion state' and an 'emotion experience'. Both notions indicate kinds of feelings that we consciously experience. However, they describe first-order and second order emotion awareness respectively. The emotion state is the feeling we have of the bodily states and changes that occur when we are experiencing an

S. Dellantonio and L. Pastore, *Internal Perception*, Studies in Applied Philosophy, Epistemology and Rational Ethics 40, DOI 10.1007/978-3-662-55763-1_5

emotion, while the emotion experience is the fully developed and integrated emotion we both experience and are, with reflection, aware of experiencing. On the basis of this differentiation, we also show that the same characteristics that aid in the identification and classification of bodily experiences (specific qualitative sensations; somatic localization; specific intensity; presence/absence of specific concomitant sensations) can also be used for the identification and classification of emotions—at least basic emotions. In the last two sections of the chapter we present some clinical evidence on the semantic competence of people who suffer from Alexithymia and Autism Spectrum Disorder which supports the conclusions of our preceding analyses.

1 Referential Competence and Internal States

In his book *Lexical competence* (1997: Chap. 3; see also Marconi 1987, 1995) Diego Marconi argues that our capacity to understand and use the words in a language consists of (at least) two different abilities, which he calls *inferential* and *referential competence*. As we have already seen (Chap. 3, Sect. 3), inferential competence is "the ability to manage a network of connections among words, underlying such performances as semantic inference, paraphrase, definition, retrieval of a word from its definition, finding a synonym, and so forth" (Marconi 1997: 59), while referential competence "is the ability to map lexical items onto the world." (Marconi 1997: 60) These two competences are partially independent from each other, in the sense that they make use of different kinds of knowledge and information. The capacity to manage a network of connections among words is a competence of a propositional kind that requires an explicit and verbalized knowledge of instances, their properties and their reciprocal relationships. The capacity to link a class of observable objects with a word is instead related to perception and—according to Marconi (1991: 28, 74)—especially to vision, since he shares the widespread view that for sighted people vision is the most important perceptual channel for the recognition of referents in the external world.

In the case of terms that have an external observable referent (such as e.g. 'cat'), Marconi suggests that a person will fully master the corresponding word if and only if s/he exhibits both referential and inferential competence. A person will be, for example, fully competent with respect to the word 'cat' if s/he is able to report that cats are pets, four-legged, mammals, that they have hair, sharp ears, etc. and if s/he can, at the same time, also identify and perceptually classify cats in the real world, distinguishing them from other kinds of things. However, in the case of words that do not have an external referent that is directly accessible to perception, it is not clear whether it is possible to identify a specifically referential component of competence. In other words, it is not clear whether mastery of the word meaning in such cases still requires knowledge of a referential kind *independent of inferential knowledge* or whether it may consist only in inferential competence as mediated by language that links certain words to other words.

Marconi addresses this problem in relation to theoretical terms. He develops his position in analogy with that of Gilbert Harman and maintains that, at least in the case of theoretical terms, our semantic competence is only of an inferential kind: "it is possible that there are certain sorts of theoretical terms, like 'quark', that play no role in perception at all, so that the content of the concepts they express is determined entirely by inferential role" (Harman 1987: 61; see also Harman 1990). Thus, according to Harman as well as to Marconi our referential competence with respect to words such as 'quark' is entirely based on our capacity to make inferences. Consequently, our concept of a quark is only composed of propositional information, more specifically, of information included in the definition that physical science has developed for the corresponding word. In fact, since language has a compositional structure, it is plausible to assume that we learn and understand some terms by relying on their observable referents, while we acquire others through the mediation of their definitions, i.e. by means of other words that we already know which directly or indirectly rely on linguistic elements that have been learned on the basis of perception. So, the meaning of words that do not have an external observable referent *might be an inferential construct* and one could hypothesize that the mastery of such words might not require a specifically referential competence, *separate and independent from inferential competence*. According to this position, it is only in the case of words that refer to observable entities in the external world that we can speak of *specific* referential competence, i.e. a capacity of a *perceptual* kind that is *relatively independent* of language and the propositional (i.e. inferential) knowledge people have of word meaning (for a more detailed explanation of this difference see Chap. 3, Sect. 3).

This distinction—which is implicit to Marconi's view—between words that do have observable referents in the external world and words that do not, also appears to be sustained by Prinz's notion of proxytype (driven by Jackendoff's view on the function of 3D models) according to which conceptual formation starts from the development of a perceptual model or schema of external objects—the proxytype—that describes their common perceptual and especially visual characteristics: shape, parts, proportions of the parts, spatial relations, movements, etc. (see Chap. 3, Sect. 5). If our capacity to perceptually identify and classify instances relies uniquely on perceptual schemata of the external objects such as those described by proxytypes, then we can have specifically referential competence only in the case of words denoting observable objects in the external world that have spatial features.[1]

[1]In the literature, there are several different views suggesting that we can directly access non-relational properties of our perceptual experience and that these play an essential role with respect to our ability to identify phenomena or objects that fall into a specific conceptual class or that are denoted by a specific word. Those views which maintain the existence of non-conceptual contents and fine-grained components of our experience are particularly akin to the perspective we put forward here and provide further evidence to support the analysis we carry out here. The notion of 'non-conceptual content' basically describes a content that can be acquired, identified and recognized by a subject without recourse to any explicitly inferential or relational strategy. The fine-grained components of our experience consist in those properties of subjective experience that are perceptually salient and allow for purely perceptual discrimination. Both of these lines of

This view leads to a sharp separation between two kinds of words (as well as between two kinds of concepts that support our semantic mastery of these words). On the one hand, there are those terms that refer to observable entities in the external world: an individual can only be considered to have fully mastered the semantics of these words if s/he exhibits both referential and inferential competence. On the other hand, there are words that do not denote observable entities in the external world, for which understanding depends entirely on linguistic information; when it comes to these words, inferential competence is all that is needed for full semantic mastery.

Indeed, this is a plausible conclusion in relation to theoretical terms such as 'quark' (for a more detailed discussion of cases like this see Chap. 6). However, there are other terminological classes that appear to be much more ambiguous and problematic, because—even though, like theoretical terms, they do not have an external referent which is observable through the external senses (visually or through the auditory, gustatory, olfactory or tactile channels)—they do seem to have a referential counterpart. At first glance, a particularly problematic classes of words in this sense are those which individuate *internal states* and more specifically *bodily experiences such as pain, thirst, hunger, itch* or *emotions such as fear, anger, disgust, happiness* or *love, friendship, envy*. Terms like these do not refer to identifiable entities in the external world that can be perceived through the senses and observed in an intersubjective manner. However, they also do not seem to be similar to theoretical terms that can be characterized in a purely inferential (propositional) way and that are not directly linked to perceptual information. In fact, these terms seem to have some kind of referent that can be accessed through direct perception and—at least to some extent—classified on the basis of this referent without recourse to propositional information.

Even though Marconi does not specifically discuss the case of word classes denoting internal states or emotions, there is a relevant reason why he would most probably reject the idea that their mastery requires purely referential competence. The referential component of these words would be internal to the subject and accessible only to the first person, but Marconi embraces Ludwig Wittgenstein's externalized view of language mastery according to which our knowledge of words referring to internal states cannot rely on internal, subjective information. In fact, according to Wittgenstein we master a word referring to an internal state only when we know the inferential rules that determine its appropriate use in the language; i.e. when we know under what external circumstances we can apply these rules (Wittgenstein [1953] 1986: 89ff). As Wittgenstein maintains in his *Philosophical Investigations*: "The essential thing about private experience is really not that each

(Footnote 1 continued)

research intertwine with the issue of referential competence as well as with the grounding problem as we address it here (see especially Chaps. 2 and 3). For a general overview see Gunther (2003). For a more specific discussion of the issue starting from different perspectives see Peacocke (2001), Schoemaker (2002), Tye (2002, 2005, 2006), Raftopoulos and Müller (2006), Bermudez (2008), Raftopoulos (2009) and Schmidt (2015).

person possesses his own exemplar, but that nobody knows whether other people also have this or something else" (Wittgenstein [1953] 1986: 95). Wittgenstein's view is behavioristic, while Marconi embraces a functionalist position. Even though behaviorism and functionalism are expressions of different models of the mind, as we discussed previously (see Chap. 1), they both maintain that internal states need to be explained on the basis of their link to external factors. Thus, Marconi also thinks that the referents of (any kind of) words should be characterized by a certain objectivity and should have their normative grounding in social criteria that determine how the word must be used.

According to functionalism, internal states such as e.g. 'pain' (which is a very particular internal state for reasons we will clarify and discuss below) cannot be individuated on the basis of their subjective, qualitative components (for the notion of 'qualitative', see Chap. 1, Sect. 3). The individuation of internal states is carried out through inferences that describe them propositionally and explicitly in their causal relations with other elements such as the situation/condition that caused them (e.g. in the case of 'pain': I injured myself with a knife), other internal states that interact with them (such as the belief that in this situation I can freely express my pain and go to the hospital) and the behavior it brought about (screaming and going to the hospital). Thus, this view supports the (in many respects unlikely) hypothesis that the properties that allow us to identify what 'pain' is, are exclusively extrinsic instead of intrinsic and that we know our pain because we know what causal role it plays with respect to our behavior. Like the behaviorist approach, the functionalist approach also rules out the idea that subjective, qualitative experience could play a role with respect to the identification of internal states (see e.g. Gopnik 1993; Gopnik and Meltzoff 1998), leading to an attitude that might be described as *qualoskepticism* or more radically as *qualophobia* (Goldman 1993b: 377). In this view, people do not directly discriminate their bodily sensations and emotions, but rather recognize them by assuming a third person perspective on themselves which allows them to individuate the causal role of their states.

This view has important consequences for the way in which semantic competence with respect to words referring to internal states is explained. In the functionalistic approach, the meaning of terms referring to internal states is not hypothesized to rely on a referent that has specific intrinsic properties, directly and privately accessible to the first person only which are then conveyed by a concept/word. In contrast, the quality of the subjective experience corresponding to internal states is not considered to influence the conceptual classification of this state and, consequently, the information included in this feeling is not considered to play any role with respect to the mastery of the corresponding words. In fact, the possibility of using 'intrinsic criteria' to fix the meaning of these terms is ruled out, as their semantics is defined in a relational manner, starting from the position that each word enjoys in the linguistic network that includes it and that regulates its use.

Marconi adopts this approach because he thinks that the referents of words should be 'objective' and publicly (intersubjectively) identifiable. For this reason, extrinsic properties must always come first, while intrinsic properties must be disregarded. In his approach, the semantic competence exhibited by human beings with respect to words referring to internal states cannot be based on a purely referential component deriving from the private sensations of the first person (understood as the informational core of a corresponding conceptual representation that supports word mastery—see Chap. 3). Consequently, referential competence with respect to these words cannot be considered autonomous or independent from inferential competence; instead, the referential aspect of the competence is considered to depend on the inferential one. In this view, I do not identify e.g. ‘anger’ on the basis of the sensations I experience when I am in this state (since these sensations are private and therefore can only be verified indirectly), rather what I identify as ‘anger’ depends on how this word is defined and on the inferential relationships it entertains with other words that determine the appropriate circumstances and behaviors for its use.

This idea that the reference of terms cannot be something ‘private’ that individuals experience for themselves leads to a ‘disembodied’ and ‘a-subjective’ semantics that explains our understanding of words denoting emotional and bodily experiences in a highly counterintuitive manner. Indeed, we think that we know what ‘pain’ means because we have experienced what pain feels like. However, in the framework of an ‘a-subjective’ theory, the fact that people have privileged (direct) access to their internal states and to their subjective, qualitative experiences (to their bodily sensations and emotions) does not imply that they have specific referential competence with respect to the words describing these states. Their subjective experience cannot be considered to be a form of information analogous to that derived from the perception of external objects.

However, in our everyday use of language as well as in the clinical literature there is evidence indicating that the direct experience people have of their internal states does provide a special knowledge of themselves which is essential for developing a full mastery of the words denoting these states. In this chapter, we will use Marconi’s differentiation between inferential and referential competence in order to challenge the idea that the reference of such words can only be external and based exclusively on extrinsic properties. We will introduce and discuss the hypothesis that terms that refer to the internal states of subjects are analogous to words that have an observable referent in the external world: their full mastery requires not only the potential for inferential competence, but also the capacity for specifically referential competence which is relatively independent of inferential competence. We will suggest that—as in the case of concepts referring to observable entities—concepts that refer to internal states are also characterized by a *core* made up of a particular form of *perceptual information*, i.e. the *subjective* and *qualitative* experience people have of these states. In the view we propose, this form of perceptual information is also organized in a perceptual schema (see Chap. 3, Sect. 6) that itself plays a central role with respect to the classification (identification and recognition) of internal states and therefore to the mastery of the corresponding words.

In this chapter, we will deal specifically with two word classes denoting internal states, i.e. with words referring to bodily experiences such as itch, pain, thirst, hunger etc. and with words denoting emotions. The argument we propose is twofold. On the one hand, we try to address some thorny problems related to the idea that reference might also be internal and that we can distinguish among and therefore classify various types of subjective, qualitative experience, each of which has distinctive intrinsic properties. On the other hand, we will refer to the clinical literature to corroborate the intuition that subjective qualitative experience is indeed essential in developing a full semantic mastery of words denoting internal states.

2 What are the Criteria We Use to Classify/to Conceptualize Our Internal States?

What is the *explicit* and *implicit* information people use to form conceptual representations of their internal states, i.e. their bodily experiences as well as their emotions? Is it inferential information of a propositional nature only or should we assume that our capacity to master words denoting internal states is characterized also by referential competence? If the latter is true, what kind of information might this referential competence rely on? In order to address these issues, we must first investigate what criteria humans use to classify their internal states and to apply the corresponding words in their language. From this point of view, the question that needs to be answered is: when people use words denoting internal states to express e.g. something like 'I am thirsty', 'I have a headache', 'I am angry', 'I am in love' what is the strategy they use to identify the state they are in (i.e. to identify that they are experiencing a state with specific characteristics) and to recognize and classify it (i.e. to recognize that a state with these specific characteristics belongs to a certain type) and thus to classify it as belonging to this type (is it thirst, a headache, anger, love)? Is this a *direct* (*immediate, intrinsic*) or an *indirect* (*mediated, relational*) procedure?

If this procedure is indirect, then a person's identification, recognition and classification of a specific state is of inferential kind, i.e. it is a form of "knowledge by description".[2] In this view, people individuate their states using a strategy similar to that we considered previously in relation to behaviorism and functionalism (see Chap. 1). The most important aspect of this procedure is that the individuation of internal states (e.g. anger) does *not* rely on a characteristic trait of the state (such as the feeling one experiences when s/he is in that state: e.g. the *feeling of anger*) that is used as a stable referent to recognize that state and express it in one or more words (e.g. 'anger', 'fury', 'indignation', 'contempt', etc.). Instead, a state must be identified using the intersection of the properties related to the external conditions that usually cause such a state (e.g., the state of being angry) and the reactions that

[2]Bertrand Russell distinguished two different forms of knowledge: a "knowledge by description" based on the properties of the objects—on how it is—and a "knowledge by acquaintance" which we will consider more closely below that relies on a direct awareness of the object of our knowledge. See Russell (1910–1911, 1912). A similar point of view is found in Lewis (1929), where the issue is indirectly addressed in a very interesting discussion on the notion of "qualia".

follow from it. Thus, people recognize that they are angry because the external conditions and the reactions they experience correspond to those typical for anger, i.e. they correspond to the characteristics people usually rely on to describe anger. This is an indirect strategy of identification because in order to know that they are in a certain state, people must reason about the properties that individuate a specific state. This is the strategy suggested e.g. by Edward Tolman—a second-generation behaviorist—in a nicely philosophical fashion that echoes Wittgenstein *ante* Wittgenstein.

> Suppose we turn to the young child learning to talk. He hears the words 'anger' or 'angry' in certain situations. I behave angrily and he is told that I am 'angry'. The child cannot get inside me nor directly experience my anger. The one thing therefore which can ultimately identify the word for him and give it its meaning, must be something about my behavior or the situation which produces it, or both. Or, again, he exhibits anger in his behavior and is told that he is 'angry'. Again the word must get its identification and meaning from something about the behavior-situation and from that only. But, perhaps you will say that since in this second case the child does directly experience his own conscious emotion, as such, this time the meaning of the word will get directly attached by him to the conscious state as such. Very good. But, be it noted, this conscious state when you tell him that he is angry no more gets directly into you than did your conscious state, when you were angry, get directly into him. It was only because he was behaving in such and such a way that you knew that he was angry and told him so. You could not see his anger any more than he could see yours. In both cases it was only what may perhaps be called the behavior-analogue of the emotion which was the common term mediating between you. Hence, ultimately it seems to be this behavior-analogue which defines, if not your meanings for, at any rate your uses of, the word. When an introspecting organism reports that he is angry, what you thereby know and identify is not his conscious state as such (which never gets into you) but rather that he is in a condition which makes him likely to behave in a certain way (Tolman 1923: 218–219).

By contrast, using a *direct* strategy of recognition, a person knows s/he is in a certain state because s/he has "knowledge by acquaintance" of that state in the sense that s/he is "directly aware, without the intermediary of any process of inference" (Russell 1912: 78). As Russell elsewhere specifies: "I say that I am acquainted with an object when I have a direct cognitive relation to that object, i.e., when I am directly aware of the object itself. When I speak of a cognitive relation here, I do not mean the sort of relation which constitutes judgment, but the sort which constitutes presentation" (Russell 1910–1911: 108). A direct form of knowledge like this presupposes the existence of properties directly accessible to the subject that allow him/her to identify and recognize the state s/he is experiencing. An approach based on direct knowledge, means that a state is identified on the basis of its *intrinsic* instead of its *extrinsic* (contextual, relational) properties. We call *intrinsic* those properties that can be used as something like a rigid designator in the sense that they directly link the information about the existence of a certain state with its identification and recognition. *The hypothesis we develop in this chapter is that the identification and recognition of internal states is intrinsic instead of extrinsic: intrinsic properties are something that can be the object of an internal ostension which can be used as a referent for the words denoting internal states. This assures the semantic stability of words denoting internal states as well as the speakers' ability to have both referential and inferential mastery of them.*

A view that quite accurately expresses the hypothesis we have in mind and that we will use therefore as a starting point to develop our position has been proposed by Alvin Goldman. This view specifically concerns the way in which people come to master the so called *mentalistic vocabulary*, i.e. the set of words we use to refer to our internal subjective states.[3] The central issue Goldman addresses is: "how an ordinary person understands the various mental words that he or she uses" (Goldman 1993a: 82). Or, to ask the question differently, "How do people detect and classify their own (current) mental states?" (Goldman 2006: 223). From a cognitive point of view, this classification must be made on the basis of some sort of information about the state at issue: "If a person classifies himself as currently having a headache, or an itch, his cognitive system must have received some sort of information that matches his concept of *headache* or his concept of *itch*" (Goldman 1993a: 84). However, the information people use for this aim might be of different kinds. In particular, as we have already mentioned, we could classify our mental states using either:

- an indirect strategy, i.e. extrinsic, *inferential information* about the state—i.e. explicit verbal information about headaches (e.g., regarding the other internal states 'headache' relates to such as e.g. the desire to get rid of the headache etc.) or about the situational elements that usually cause or contribute to causing this state to arise (e.g., alcohol, high blood pressure, stress etc.) and about the effects brought about by the state (the typical behavioral reactions to headaches, e.g. the disposition to take a painkiller);
- or a direct strategy, i.e. intrinsic, *referential information* about what the state is —such as e.g. the sensory experience corresponding to the state: what we 'feel' when we are in that state.

Goldman is persuaded that the strategy people use to classify their states is not inferential, but direct and indeed based on sensory experience. To support this view (presented also as *intrinsicalism*, see Goldman 1993b: 367), he offers a significant example:

> Suppose I wake up in the morning with a headache and immediately classify my state as having a headache. How do I make this classification? Surely I don't do it by identifying an actual cause of the headache, since I do not have (nor does my cognitive system have) any relevant information about the headache's cause. There hasn't been time to obtain such information since awakening. Similarly, the classification cannot be based on information about any effect of the headache, because the classification may precede any effect that

[3]For a definition of mentalism see Chap. 1, Sect. 1. The theoretical framework of Alvin Goldman's analysis is very different from ours. As for Goldman's approach, the discussion of the mentalist vocabulary is a corollary of a larger argument which aims to epistemologically legitimize his approach to mindreading, i.e. his particular version of so-called simulation theory. In this context, Goldman develops the hypothesis that subjective, first person experience is essential for understanding at least some specific kinds of words denoting internal states.

> could be detected. Thus, there is no way my cognitive system could arrive at the belief that a current state of mine fits the specification F(headache), if this requires attention to actual causes or effects of the current (token) state (Goldman 1993a: 84–85).

The case of the morning headache one recognizes one has immediately after s/he wakes up univocally shows that a behavioral, dispositional or functional strategy is inadequate to explain how the first person identifies his/her own internal states. The identification of the headache cannot be of a functional kind, since a person might know s/he has a headache without having much explicit and verbal information "on the causes and effects of headache. In fact, immediately after awakening typically people do not recall information about anything." As Goldman notes: "Is it plausible that one always possesses this kind of functional information whenever one self-ascribes headache (whenever one either says 'I have a headache' or is prepared to say this)? Not at all, I suggest" (Goldman 1993a: 84).

Self-ascription or self-attribution of a state does not always require—this is the point of Goldman's argument—an indirect identification procedure based on behavioral, dispositional or functional information. More precisely, according to Goldman this relational and subject-independent information is neither necessary nor sufficient to identify such a state: it is not necessary because we can classify the state even when this information is not available. I do not need to immediately recall that maybe yesterday I drank too much or I slept too little to know that I have a headache. Moreover, this extrinsic information is also not sufficient because, if we only had inferential information about the state, we would not know that we have a headache, since we would not feel the headache. In cases such as that of morning headache the identification occurs directly; thus, the classification, i.e. the conceptualization of the state, is carried out solely on the basis of its *intrinsic* properties, i.e. of the feeling—of the sensory information—corresponding to the state. Further, the first person has special authority with respect to any judgment concerning the fact that s/he is experiencing a state as well as the identification of the state s/he is experiencing.

Thus, using the terminology we have adopted in this work, even though people usually also have some inferential information about headaches and therefore also some inferential competence with respect to the word 'headache', they also have referential information about headache experiences and therefore also some referential competence regarding the word 'headache'. In this sense, the mastery of words denoting internal states such as 'headache' is analogous to that of terms denoting observable things, insofar as it includes a referential component which is directly perceived by the subject and is relatively independent of the inferential knowledge the subject has of the term. If words referring to observable objects in the external world are mastered in the same way as terms denoting internal states that can be directly perceived by the subject, then the conceptual representations that support the mastery of both kinds of words are plausibly analogous from a structural point of view. Like concrete words (see Chap. 3), words that denote internal states may include both a core concept consisting of perceptual information that allows the referential identification of the instances belonging to the concept and inferential information about the 'context' of the state (its possible causes, effects, etc.).

The issue we need to deal with is what kind of information does this referential core concept exactly consist in—or, to put it differently—what kind of information does our referential competence with respect to these kinds of words rely on. Goldman's suggests that: "The best candidates, it would seem, are so-called qualitative properties of mental states—their phenomenological or subjective feelings (often called 'qualia'). Certainly it is highly plausible that one classifies such sensations as headaches or itches on the basis of their qualitative feel." (Goldman 1993a: 87; for a more detailed description of the notion of 'qualitative feel'/'qualitative information' see Chap. 1, Sects. 5 and 6) Goldman's position challenges the classical behavioristic and functionalistic position that qualia are not essential for carrying out classifications. As e.g. Tolman points out discussing the issue of qualia with relation to color experience:

> If what looks green to you, looked the complementary red to me, I doubt if there would be any way in which that fact could ever be discovered. But I do not necessarily wish to stress this point. The only two things I do wish to insist upon are, first, that there is no way of proving that your qualia and my qualia are the same, and, second, that whether they are the same or not in no way affects the results of our science (Tolman 1922b: 143).

When we learn the names of colors from other people, we cannot determine what they see. But this does not matter, because to learn the name of certain feelings we need only to understand what relationships they entertain with respect to each other:

> When I present the spectrum [of colors] to a fellow man and seek to discover his color experiences, […] I ask him to indicate which colors look alike to him and which different. That is, I ask him to arrange and sort the parts of the spectrum so as to indicate their relative degrees of similarity and difference. […] In so far, to be sure, as he names and describes them in the same way that I should, it indicates also that his qualia and my qualia show the same similarities and differences and that we have been taught the same language. But this is all that it indicates. His behavior […] can never indicate to me just what his immediate qualia, as such, may be. All that it shows is the nature of the term-characters of those qualia and whether or not those term-characters are the same as my term characters. […] That is, the world of stimuli presents a particular system of similarities and differences for each organism. In the case of a fellow human this system may be the same as it is for me. In the case of a visitor from Mars it may or it may not be the same. But in either case it is only this system of similarities and differences which I learn about and never the immediately experienced qualia as such (Tolman 1922b: 143–144).

According to the classical behavioristic and functionalistic view, qualitative experiences do not play any role in determining how we learn words and how we conceptualize things, but in Goldman's view they are essential at least in order to understand and to categorize our internal states. In his perspective, the basic information we use to form our conceptual representations of internal states is the qualitative feeling which is available to the first person only and which corresponds to the *phenomenal* or *qualitative* component of the state—*the feeling one experiences when s/he is in that state*, i.e. *the 'what it is like' of the state*. According to Goldman, when it comes to words and concepts referring to sensations and bodily states, we have to admit that their "qualitative character is an essential part of [their] content" (Goldman 1993c: 21).

In fact, in Goldman's view, many kinds of internal states are somehow known by acquaintance. Goldman maintains that his position applies to all kinds of internal states, i.e. not only to bodily experiences such as pain and itch or to emotions, but also to mental states in general such as doxastic states (i.e. propositional attitudes: beliefs, desires etc.) and thoughts, because they are all known directly (introspectively). However, as e.g. Vincent Picciuto and Peter Carruthers note, equating all these forms of direct knowledge of internal states is problematic and possibly unjustified (Picciuto and Carruthers 2015). Indeed, we should distinguish at least between the internal perception we have of certain bodily signals in proprioception and interception (see Chap. 2) and the direct (privileged and authoritative) knowledge we have of higher order mental states (beliefs, desires, thoughts, etc.). While proprioception and interception are forms of internal perception (i.e. detection and processing of sensory signals) which we consider to be due to a propriosensitive perceptual system (see Chap. 2, Sects. 4 and 5), the capacity to grasp our internal, propositional mental states might be due to a higher order mechanism which does not have much in common with this sensory system. Thus, even though it is very plausible that higher order mental states (e.g. doxastic states) also have a qualitative component that is essential for their identification (the qualitative component might be essential e.g. for distinguishing that we believe something rather than desire it), it is appropriate to differentiate between a propriosensitive sensory system that detects and processes internal sensory information (proprioception and interoception) and some internal 'higher order inner sense'—to adopt Picciuto and Carruthers' terminology—whose function is to 'introspect', i.e. detect and represent internal propositional states. In this work, we only focus on the propriosensitive perceptual system (proprioception and interoception) while we do not consider the issue of a higher order inner sense. This is the reason why here we focus uniquely on bodily experiences such as pain, hunger, itch etc. and on emotions (which, on the basis of a perceptual theory, we conceptualize in continuity with bodily experiences).

In the view we suggest here, concepts/words referring to internal states such as bodily experiences and emotions are analogous to concepts/words denoting observable objects in the external world insofar as both kinds of concepts have an experiential core concept consisting of perceptual information and both kinds of words, to be fully mastered, also require specifically referential competence. Thus, the problem of understanding how we come to classify our internal states should be explained in a way that is largely analogous to that of explaining how we come to classify the things in the world we experience through our external senses (see Chap. 3, Sects. 4–6). This analogy with concepts denoting observable objects suggests that two steps in the categorization process need to be explained. In fact, even though the state we are feeling right now is absolutely unique, because we will never experience it in exactly the same way again (as we will never perceive e.g. a particular dog again in exactly the same way that we perceive it now), we are able to both identify this state through a *token conceptual representation* and to classify it through *a type conceptual representation* (see Chap. 3, Sect. 4).

In order to clarify how this identification and classification might be carried out, we need to specify precisely:

- what kind of qualitative/phenomenal information we make use of to form a *token representation* of our internal states that allows us e.g. to identify the specific throbbing, localized pain we are feeling in our head this morning, distinguishing it from other kinds of headache we have experienced at other times;

  and

- what kind of qualitative/phenomenal information we make use of to form a *type representation* of our internal states that allows us e.g. to identify that what we are experiencing is (a kind of) headache and to distinguish headaches from other different internal states such as tooth pain, itch or hunger.

In a later discussion of this topic, Goldman more specifically addresses the issue of qualitative information and the intrinsic properties of this information that people might rely on to identify and classify their internal states. Goldman suggests that qualitative information is a particular kind of perceptual information and must be interpreted by analogy with perceptual information: we directly perceive our internal states exactly as we directly perceive things in the external world by the senses. In line with the hypothesis on the existence of a propriosensitive ('proprioceptive') system we previously investigated in this work (see Chap. 2, Sects. 3–6), Goldman maintains that humans are equipped with an internal perceptual system—analogous to the external perceptual system—that continuously detects and processes information produced by our body. It is this system that provides us with the qualitative information that serves as a referential component for the conceptual representations people use to identify and classify their internal states and to master the words referring to these states. Goldman describes the activity of this system as a particular kind of 'introspection': "One answer is what I'll call the special method view […] One conception of the special method, a conception I'll defend, features the thesis that the special method is that of 'introspection', 'inner sense' or self-monitoring" (Goldman 2006: 224).

The fact that Goldman speaks of introspection should not lead one to think that he is suggesting we restore some sort of old-fashioned introspective method, as this was codified in classical psychology or in Cartesian philosophy.[4] Goldman is not interested in the classical notion of introspection as an intuitive source of *objectively true knowledge*. He argues for a "not very orthodox" (Goldman 2006: 224) view of

[4]For a general description of the notion of introspection in its 'orthodox' sense, see e.g. Dellantonio and Pastore (2010). For a recent overview of the notion of introspection that includes several definitions of it (without however including the sense we are considering here) see e.g. Prinz (2004b).

introspection characterizing it as a *quasi-perceptual modality* that allows us to 'feel', to 'perceive', the information produced by our internal states in the form of a qualitative sensation (of a phenomenal experience). According to this interpretation, introspection is the channel through which we gain access to that particular kind of information we previously defined as qualitative. More specifically, introspection is a system designed to gain information that helps the cognitive system build representations and acquire knowledge about the internal environment.

Further, introspection might not be the ideal word to define this capacity. The point is not so much the overlap with the old introspection-based psychology and its method, but rather the confusion between the perceptual and non-verbal introspective awareness of our internal feelings (which is what is actually intended by Goldman) and introspective access to our verbal thoughts in the sense of interior monologue. For this reason, we will favor the notion of 'proprioception' or of 'propriosensitive information'.[5] In spite of this terminological difference, the concept of introspection called into play by Goldman is extremely close to the idea of 'proprioception' we have suggested. Moreover, Goldman's considerations on introspection so conceived allow us to note some analogies and differences between 'proprioception' and perception.

Even though 'proprioception' might be compared with perception (i.e. exteroception), since both describe the awareness we have of our internal and external environment respectively, the difference lies in the kind of 'objects' perception and 'proprioception' inform us about. While perception presupposes the presence of an external, subject-independent object, intersubjectively observable and identifiable, which supplies the sense organs with information, in the case of 'proprioception' the relevant information comes from the inside of the body and produces representations whose referent is entirely subject-dependent. In other words, the reference of the conceptual representation developed starting from 'proprioception' is entirely egocentric and its reference cannot be identified on the basis of any allocentric criterion. This is the reason why 'proprioception' is only a *quasi*-perceptual procedure.

However, Goldman suggests that the general working procedure of the quasi-perceptual modality that we call 'proprioception' is quite analogous to perception. Like perceptual information, proprioceptive (propriosensitive) information also consists of a selection of all the information internally available to the subject. This information is processed by the proprioceptive (propriosensitive) system and enters together with other kinds of information in the constitution of the token and type representations through which internal states are identified and classified (see e.g. Goldman 2006: 246). In the same way as perception allows us to develop representations that are appropriate for identifying, recognizing and classifying

[5]Here 'proprioceptive' and 'propriosensitive' are mostly used as synonyms. When we use proprioception in this sense, we put the word into brackets to indicate that we do not understand it in accordance with the definition given by the current neuropsychological literature, but congruently with the description of this notion given by Gibson to refer to all kinds of body information we can become aware of. On this topic see Chap. 2, Sect. 3.

instances in the external world, 'proprioception' allows us to develop representations that are appropriate for identifying, recognizing and classifying internal states of the subject. As Goldman specifies:

> The process of identifying one's pains, emotions, or beliefs is modelled on perceptual processes of seeing or hearing objects or sounds. [...] it classifies a mental state token by recognizing in it features or criteria associated with the selected mental type. For example, one classifies a target state as 'thirst' or 'anger' by recognizing in it features criterial to one's concept (or conception) of thirst or of anger (Goldman 2006: 225).

Goldman's proposal could be seen as a research project whose starting point is the idea that the concepts we have to denote internal states are formed on the basis of internal information derived from the internal monitoring of our bodily changes through 'proprioception'. To develop this research project in full, a number of issues should be addressed concerning not only the way 'proprioception' might work (we discussed this issue above—see Chap. 2, Sects. 3–6), but also how this internal information can be used as *reference* to build concepts and learn the corresponding words. Going back to the differentiation made in the previous chapter: we need to determine what our referential competence with respect to these concepts consists in: i.e. *what kind of information the 'proprioceptive' system might use to build conceptual representations that allow people to identify their internal states.* This chapter will focus mainly on this last issue. Furthermore, even though Goldman's 'research project' concerns mental vocabulary in general (he would like to determine "how an ordinary person understands the various mental words that he or she uses", Goldman 1993a: 82), as already pointed out previously in this section, we will limit our discussion to two specific kinds of mental states, i.e. to bodily experiences such as 'pain', 'itch', 'hunger', 'thirst' on the one hand and to emotions on the other. Other mental states such as e.g. doxastic states (I believe, I think, I know) will not be considered here, although we will include a partial discussion in the next chapter.

3 The Classification of Internal States and the Analogy Between Perception and 'Proprioception'

The concepts we use in categorization processes and which support our semantic mastery do not only consist in information from the external world obtained by the sense organs but—and this is the main thesis of this work—they also consist of propriosensitive ('proprioceptive') information from inside the body that we become aware of. In particular, in the previous section we suggested that referential information of a propriosensitive origin is indispensable in order to develop at least those concepts that allow us to classify internal states such as bodily experiences (such as e.g. hunger, thirst, headache etc.) and emotional states (such as e.g. anger, joy, sadness etc.) and to semantically master the corresponding words. This internal information constitutes the perceptual core of the concepts denoting internal states, i.e. the "perceptual schema" that allows us to identify and classify the instances (i.e. the states) that fall under the concepts and to master the corresponding words from a referential point of view (on the notion of perceptual schema see Chap. 3, Sects. 5–7). Without these perceptual schemata we wouldn't be able to identify and to classify (thus also to label) what we feel.

The fundamental hypothesis we began discussing on the basis of Goldman's analysis is that internal states are categorized analogously to observable objects in the external world; the only difference is that—unlike external observable objects—internal states are not categorized on the basis of sensory information, but rather classified relying on propriosensitive information. The main difficulty one encounters, when trying to detail this hypothesis and make explicit those aspects of propriosensitive information that the cognitive system uses to classify internal states into token and type concepts, is that the elements constituting 'proprioception' are internal and therefore—unlike those that characterize perception—difficult to identify. In fact, unlike conceptual representations of external objects, the representations that result from the processing of propriosensitive information do not 'represent' anything in the external world. They do not have any allocentric reference and the origin of the stimulation cannot be ascribed to an external, intersubjectively identifiable and subject-independent source; thus, they must be understood as structures whose focal point is entirely centered on the subject. Because of the lack of an allocentric reference, it is particularly difficult to explain what aspects of propriosensitive information are used to form the type and token concepts of our internal states.

More specifically, in the case of external objects we hypothesized that their referential component categorization processes rely on perceptual schemas developed on the basis of specific observable features of objects such as their geometrical, topological and movement properties (e.g. cats are identified/classified together on the basis of their overall shape and movement characteristics as well as on the basis of other exterior properties such as having four legs, a tail, pointed ears, etc.—see Chap. 3, Sect. 5). To sustain the idea that the categorization of internal states is also characterized by a referential component, we need to specify what its 'perceptual schema' might consist of—i.e. what are the specific features of the state detected through 'proprioception' that might allow us to identify and to classify it from a referential point of view? In the next part of this section we will try to develop such a hypothesis beginning with the description of 'proprioception' we put forward previously (see Chap. 2, Sects. 5–7) building on Gibson's view.

First, 'proprioception' is understood in analogy with perception and it describes both the ability to detect signals originating in our body and the awareness of body information as a result of the processing of these signals by means of a propriosensitive perceptual system. Like perception, 'proprioception' also provides us with a continuous flow of information. In fact, using all the sensory modalities (vision, hearing, touch, but also kinesthesia and information related to the motoric system) perception produces a constant, dynamical and analogical mapping of the external environment. What we become aware of is the final product of this mapping which is a synthesis of all the information collected by the various sensory modalities. The kind of perception we have (i.e. what we actually perceive at any instant in time) depends on the contributions of these various modalities to the final mapping. In the same way, 'proprioception' includes all the information provided by the receptors internal to the body and convey signals that map bodily experiences and their changes. To be more precise, the idea we develop in Chap. 2 is that exteroception and 'proprioception' result from the joint work of the exteroceptive and the propriosensitive perceptual systems and are both part of the same overall perceptual process. The distinction between external and internal perception is purely functional: while perception offers a constant flow of information that maps the external environment, 'proprioception' provides a constant flow of information that maps our 'internal bodily environment', i.e. the complexity of our bodily experiences and bodily changes.

In the case of both perception and 'proprioception', we have conscious access only to the final product of the mapping which we experience as unitary and as simple. Both kinds of mapping are usually carried out without the subject being (completely or permanently) conscious of it and only a small part of the uninterrupted information flow produced by the mapping is actually consciously experienced by the subject. In fact, even though the internal monitoring system is permanently active and is always present to the cognitive system as a (psychic or mental) *basso continuo*, not all the bodily experiences and modifications reach awareness and are explicitly detected by the subject. Only when the monitored bodily signals become salient—plausibly because they exceed a certain 'threshold'[6]—they become conscious and attract the attention of the subject. As specified in Chap. 2, attentive mechanisms are extremely important for determining what are we aware of: consciousness—in the sense of the feelings we consciously experience—is always influenced by attention (on this see e.g. also Prinz 2002: Chaps. 2 and 3; Prinz 2010).

As already discussed previously (see Chap. 2, Sects. 3–6), attentional focus could also modify 'proprioceptive' monitoring, bringing to consciousness e.g. signals under the activation threshold or acting selectively and foregrounding certain signals in a way that is partially independent from their activation threshold. If I focus e.g. on the sensations of my digestive apparatus, I might notice I have a certain hunger I hadn't noticed until now. However, if the hunger increases, the fact that the activation exceeds a specific threshold is in itself sufficient for the perceived state to become conscious independently of my attentional focus. Even though I might try not to pay attention to my hunger, if it does exceed a certain threshold, I can hardly ignore it.

These observations indicate in the first place that the mapping of the 'proprioceptive' signal distinguishes between specific intensities of feeling. More

[6]Francis Crick and Christof Koch developed a neuroscientific hypothesis to explain how some aspects of subjective experience become conscious. According to this hypothesis, a component of our experience must be represented in short term memory in order to reach consciousness and this is possible only when the neural activity that implements this representation is above a specific activation threshold. See Crick and Koch (1990, 1991, 2003).

specifically: the modulation of such feelings can be detected as being at different (higher and lower) *intensities*: a headache can be e.g. of different degrees of intensity; this means that the very same feeling of headache (the same quality of pain) can occur at different intensities, just as the same sound might be heard more or less loudly. *Thus, intensity appears to be one of the features of internal feelings that is perceived by human beings and that might be used to classify them.*

Internal mapping can be seen as an incessantly working monitoring system that detects events related to bodily experiences and their modifications and that combines and integrates information coming from different parts of the body (see, e.g., Damasio 2010; Craig 2010; see also Chap. 2). In some respects, this mapping is analogous to that producing our so-called *body schema*—i.e. to the "system of processes that constantly regulate posture and movement: *sensory-motor processes* that function without reflective awareness or the necessity of perceptual monitoring" (Gallagher 2007: 273).[7] Like the monitoring that produces our body schema, 'proprioceptive' monitoring also operates at a preconscious and subpersonal level without the need for any conscious control and tacitly and constantly detects and integrates information coming from the various parts of the body. Furthermore, as in the case of the body schema, the outcome of 'proprioceptive' mapping can also be used by the cognitive system to carry out other functions or to control specific intentional acts. Not only is this 'proprioceptive' mapping similar to the body schema in the way it works, but it also uses the body schema as part of its own mapping in order to *localize the origin of the 'proprioceptive' signal.* As e.g. Michael Tye specifies:

> One cannot feel a pain without feeling the pain in a particular bodily region. The region may not be precisely delimited, but nonetheless, each pain is felt essentially as localized in a part of the body of the owner. Where there is no general bodily feeling in a given body region, it is not possible to feel pain there. What is true here for pains is true for itches, tickles, and other such bodily experiences. It is thus a phenomenological requirement for feeling a pain that one feel the bodily region in which the pain seems to be (Tye 2003: 63).

When a person has e.g. a headache, the stimulation s/he feels is not just a bodily signal, it is a signal localized in a specific part of the body—the head—and even in a specific part of the head (e.g. on the sides). For this reason, the mapping of 'proprioceptive' states provides information not only about the bodily experiences (e.g. on the ache), but also about the specific location of this state in the body (e.g. in the temples). *In fact, localization appears to be another central feature that helps classify a state since it tells us where the bodily changes detected by the propriosensitive perceptual system are occurring.* The fact that 'proprioception' also provides information about the position in the body of the detected signals does not mean that internal states must always have a single position/origin in the body. In

[7]The body schema is different from the body image which is often conscious and consists of a "system of perceptions, attitudes and beliefs pertaining to one's own body" (see Gallager 2007: 273). Indeed, some authors link 'proprioception' with the body image suggesting more or less the same idea we are arguing for here (see, e.g., Tye 2003: Chap. 2). However, we think that 'proprioception' makes use of and is connected with the body schema rather than with the body image.

fact, more often the propriosensitive perceptual system detects a number of signals from various bodily parts at the same time (see e.g. Tye 2003: Chap. 2) and the co-occurrence of these signals is essential in knowing what I am feeling or knowing why I have a certain feeling. For example, the feeling of being tired is not specifically localized in a specific position of the body schema but rather results from *the co-occurrence of a number of signals* spread across various bodily parts. Moreover, sometimes a persisting headache is accompanied by other widespread signals of tiredness and this conjunction helps me to understand that the headache is due to tiredness. *Thus, one important feature we use to classify our internal states is the co-occurrence of various signals.* If one of the signals is more salient than others, then these others might serve to specify it (e.g. to specify its cause), as in the case of the headache due to tiredness. In other cases, it is possible that all the co-occurring states (e.g. headache and tiredness) have the same salience. If so, it is the co-occurrence of them that determines what kind of feeling we perceive.

Finally, and most importantly, 'proprioceptive' feelings such as e.g. tiredness or headache also have another feature which is essential to their identification, i.e. they are perceptual experiences with a specific phenomenal character; i.e. characterized by a specific quality or by a specific what-it-is-like. The qualitative character of experience is not unique to internal experience, rather any kind of perception, be it internal or external, is associated with a distinctive qualitative feeling (see Chap. 1, Sect. 5; see also Buck 1993). In fact, the processing of sensory signals by every sensory modality produces corresponding qualitative information that people experience as such: "each perceptual experience that belongs to a modality has a distinctive modality-specific phenomenal character" (O'Callaghan 2015: 141), i.e. it gives rise to specific, corresponding qualia. Thus, when we e.g. see something, it might be the specific qualia produced by sight that inform us about the fact that we are seeing this something instead e.g. of hearing it or of touching it. Qualia might be the clue we use to identify through which channel we are perceiving the object of our perception.[8]

This view is not contradictory to the idea that perception is always multimodal (O'Callaghan 2008). Indeed, as e.g. Casey O'Callaghan in his study on sensory

[8]Qualia are one of the four criteria identified by Paul Grice in his classical paper Some remarks about the senses (Grice 1962) for individuating the senses. On the basis of qualia, we should be able to answer questions such as 'How many senses are there?' or 'What makes the senses different?'. These questions are maybe now more important than ever for the debate on perceptual experience since cognitive science is providing new evidence for the deep interaction among the senses and the necessity of considering 'proprioception' to be a sensory modality. For an overview of both the classical and more recent views on this issue see e.g. Macpherson (2011). However, the point we make here goes in a different direction: the idea we put forward is that the qualitative character of experience is what allows us to identify from a first person point of view the main source of any specific perceptual information we are experiencing. Thus, the issue is not how many senses there are from an objective point of view, but rather how a person can identify the perceptual experience s/he is having as e.g. visual, olfactory or auditory. On the function of qualia see, e.g., Ross (2001), Fürst (2004), Bigelow and Pargetter (1990), Muller (2008), Aranyosi (2003) and Musacchio (2005a, b).

modalities and the phenomenal character of experience points out: "Most times while you are awake, your perceptual experience is associated with more than one modality" (O'Callaghan 2015: 132). Any simple act we do is essentially multi-sensorial. Think for example of cooking pasta: we see the boiling water in the pot, we hear the boiling water, we smell the pasta in the water, we feel the spoon we use to stir it and we taste the flavour of the pasta when we test whether is it salty enough or cooked to our liking. Furthermore, we also experience the warmth of the vapour on our skin and our appetite is likely to increase as we smell the aroma, and we might feel a bit tired after standing at the stove for some time. Thus, cooking pasta involves a number of multisensory phenomenal experiences connected with the various forms of perception that go along with this act.

Not only is every experience we have characterized by the co-occurrence of multiple sensory experiences, but sometimes this co-occurrence is also essential to processing the sum of perceptual information. This is the case for instance with our capacity for locating objects in space: in order to localize the objects that we perceive in space, we need information on how our bodies, our heads and our eyes are positioned with respect to this object. As e.g. Tye points out: "In experiencing things visually, we experience their orientations and directions, and these are set in part by our own bodies. We use our own bodies to assign orientations and directions to things, as manifested in the visual phenomenology" (Tye 2003: 70). In this sense, visual information is always obtained jointly with kinesthetic and motoric information. As a consequence, the qualitative experience that characterizes vision cannot be understood to be a product of vision as a single sensory modality, but must rather be considered to be the result of the intersection of multiple sensory modalities. This does not apply to vision only; the same kind of reasoning can be applied to any kind of experience that involves more than one perceptual modality.

A different example that points in the same direction is taste.[9] As Stephen Biggs, Mohan Matthen and Dustin Stokes clarify:

> To the taster, it seems that flavour is a single composite quality of food, detected primarily by the tongue. The truth is more complex […]. Work by cognitive scientists has established that there are multiple sources of flavour perception, including not only taste receptors on the tongue (which account for the simple taste components—salt, sweet, sour, bitter and perhaps a couple more), but also olfactory receptors in the nasal tract (which react very differently when odorants emanating from the mouth flow over them in the retronasal, or opposite-to-sniffing direction), and the trigeminal nerve, which is responsible for sensation in the face and for the motor control of biting and chewing (It is the stimulation of the trigeminal nerve in the face that causes mint to have its characteristic coolness and chillies their characteristic heat—yet, the coolness and the heat are sensed *in the foods*, not in the face). As well, of course, there is the complex experiencing of eating or drinking, of which flavour is only one part, with texture and mouth-feel as distinct but well-integrated components. Flavour, though experienced as if it were delivered in a single perceptual act of sampling, is actually a composite of outputs from all these sources (Biggs et al. 2015: 2).

[9]For an overview of taste—i.e. of taste perception as an intrinsically complex and multi-modal perception—see, e.g., also Smith (2015).

The conclusion that can be drawn from both the examples of flavor and of spatial orientation is that, at least in some cases, the perceptual experiences we have are intrinsically multisensory. Thus, if each modality is associated with a distinctive phenomenal character, multisensory experience must be characterized by multiple qualities. However, since qualitative experiences are always perceived as unitary by subjects, *they must be the composite effect of a multiplicity of qualities which derive from the various sensory modalities involved in the perception.*

In fact, some people suggest an even more radical pluralization of the qualities that are blended together in the phenomenal experience we have at any instant of time. Indeed, it is possible that every sensory modality produces more than one qualitative experience that is distinctive for that modality. To experience redness or brightness could give rise e.g. to two different qualitative experiences of a visual kind. However, when we see something which is both red and bright, we have a unitary phenomenal experience resulting from the conjunction of the two composing qualities which occur co-consciously. In this case, different visual experiences could also be qualitatively different from each other because they consist of different blends of visual qualities (O'Callaghan 2015). This idea that each modality could be associated with a number of different but distinctive phenomenal experiences is consistent with the hypothesis proposed previously that we might rely on the qualitative aspects of our experience to identify the modality through which we receive any given information. Indeed, even if vision gives rise to different qualities, if they were nevertheless all distinctive for this modality, we would still be able to know that we are seeing something (instead of e.g. hearing it or smelling it). Thus, both alternative views (i.e. each modality produces one single distinctive quality versus each modality produces a plurality of distinctive qualities) allows us to link qualities to modalities.

A further aspect that needs to be explained concerns the fact that—even though our experience is always multimodal and complex—what we consciously experience is an organic unity. Indeed, many hypotheses have been proposed to explain how this unity might be realized: it could be conceived as a mere conjunction of the composing qualities which become unitarily co-conscious or as a novel quality originating from this conjunction or even as a fast circulation among the modalities which changes every instant in the stream of our consciousness so that at any given brief interval of time we are aware of only one single quale (for an overview of the literature on this issue see O'Callaghan (2015) who argues in favor of the novel quality as well as Spence and Bayne (2015) who argue in favor of continual switching). Even though the evidence we have at present might not be sufficient to determine which of these hypotheses is most plausible (the evidence adduced e.g. in O'Callaghan 2015 and in Spence and Bayne 2015 points in opposite directions), the idea that the conjunction of different qualities may give rise to a novel quality seems particularly plausible. In support of this view, O'Callaghan discusses the example of flavor.

> To perceptually experience flavor requires stimulating both taste buds and smell receptors. That is why food seems bland when you're stuffed up with a cold. The tongue and taste

> system suffice to perceptually experience the basic tastes of salty, sweet, bitter, sour, and umami, but retronasal olfaction is required to perceptually experience flavors, such as that of butter, fried chicken, pineapple, or cardamom. Odors sensed after traveling up through passages at the back of your mouth (retropharynx) are "referred" to the mouth and contribute to apparent flavor. […] Flavors are complex features. Tastes are one aspect of flavor. […] Olfactory attributes also are aspects of flavor. Somatosensory attributes also are a crucial aspect of flavors. Capsaicin gives chilis their pungent flavor by activating nociceptors; nicotine is bitter at low concentrations but burns at increasing concentrations; tannins give pomegranates and pecans their astringent flavor; even salt is an irritant at high concentrations. Flavor involves the coinstantiation of such gustatory, olfactory, and somatosensory features. To perceptually experience a flavor therefore may require a multimodal perceptual experience. Thus, no mere perceptual experience of any modality itself is a flavor experience. Flavor is a novel feature accessible only through multimodal perceptual experiences (O'Callaghan 2015: 158).

An analogy might be useful in understanding the sense and the possible implications of this description. Think of zabaione cream: zabaione is made of egg yolks, sugar and sweet wine (usually Marsala); the egg yolks are whipped with sugar; then they are mixed with the wine to obtain a cream. The apparent flavor of the zabaione is not identical to the conjunction of the flavors of the single ingredients which individually taste quite different than they do in the zabaione mixture. Thus, there is more to the zabaione than what is found in the single ingredients: zabaione gives rise to a novel quality—a new unitary and organic flavor—that is more than the mere co-instantiation of the flavor of sugar, egg yolks and Marsala.

Up until this point the zabaione example merely elaborates the view suggested by O'Callaghan. However, it might allow us to go a step further and suggest a slightly different position than that presented by O'Callaghan. Indeed, even though the quality of zabaione is not the mere co-instantiation of the qualities of its ingredients, starting from the flavor of the zabaione it is still possible to trace the ingredients it is made from. Indeed, one can even taste whether one of the ingredients is predominant or excessive: the zabaione may be too sweet or the flavor of the eggs or of the wine might be excessive. What this further development of the analogy suggests is that the novel quality generated by the blending of different individual qualities might still retain traces of them. Thus, if we draw our attention to the 'aspects' of the novel quality we might still decompose it into its components, at least insofar as their 'intensity' (i.e. their weight in the mixture) exceeds a certain threshold. Thus, even in this case, the intensity of the various qualities is an element of crucial relevance: indeed, it is indispensable to determine both the quality of the final blend (which depends on the intensity of each component quality) and the possibility of identifying the original individual qualities of the ingredients in the blend (if they have sufficient 'intensity'). These observations could possibly explain why—when we pay attention to our experiences—we seem to only be aware of one perceptual modality at a time (Spence and Bayne 2015). Indeed, it is possible that, as in the case of zabaione, when we pay attention to the aspects of the compound quality we are experiencing, we can instead focus on the quality of one ingredient at a time.

On this basis, we might speculate about a possible further analogy between perception and 'proprioception'. We defined 'proprioception' as a complex propriosensitive monitoring system that processes various kinds of bodily information concerning the respiratory system, the circulatory system, the digestive system, the musculoskeletal system, the endocrine system etc. The complexity of this propriosensitive monitoring suggests that—just as we have many perceptual modalities for the detection and processing of external information—'proprioception' might also be characterized by different modalities; in fact, proprioception understood in the strict classical sense as the perception of body position and body movements, interoception and nociception might consist in three different internal sensory modalities. Not unlike what occurs in the case of perception, we can hypothesize that each internal sensory modality produces specific corresponding qualia. Even if internal perception would be better understood as a single modality, it is highly plausible that it gives rise to a multiplicity of different perceptual experiences (qualia), all of which are distinctively internal.

In this sense, the considerations related to external perception also apply to 'proprioception' (now understood in the broader sense defined here, i.e. as the process and result of propriosensitive monitoring). Like perceptual signals, 'proprioceptive' signals are also experienced at any instant in time as an organic unity, however this unitary experience could originate from the co-occurrence of multimodal qualities or at least from a plurality of 'proprioceptive' qualia. If different modalities/component qualities had different weights/intensities in the production of the novel qualia compound, its specific *what-it-is-like* (qualitative feeling) would be determined mainly by the predominant modality (if you put too much sugar in the zabaione, sweetness will be the predominant final taste). Think e.g. of the case of tiredness. The sensation of tiredness is a unitary feeling resulting from the co-occurrence of various signals detected in a number of bodily parts from various channels, each of which contribute to the final sensation with a specific *quale*. When none of these signals is dominant, we feel a global tiredness which is perceive in a rather a-specific manner. By contrast, if the signal coming e.g. from the musculoskeletal system is dominant, the state we experience feels different: it feels like a physical tiredness whose specific quality might depend primarily on the musculoskeletal system as its primary source. This quality informs us, among other things, that the tiredness we are perceiving is possibly due to physical activity or to physical weakness (instead of e.g. a lack of sleep).

In general, these considerations suggest that qualia play a role in the identification of the signals we perceive through both perception and 'proprioception' (Buck 1993; Musacchio 2005b). However, this qualitative aspect could not be equally relevant in the cases of perception and 'proprioception'. The difference between them lies in the fact that, while in the case of perception what it is ultimately detected are objects, in the case of 'proprioception' we detect the *qualitative feelings themselves*. Seeing a red ball gives rise to a specific qualitative experience; however, my perception is not (or at least not primarily) about this qualitative experience because the object of my perception is the red ball. On the other hand, experiencing a headache also brings about a specific qualitative feeling;

but, in this case, the feeling is also the object of my internal perception in the sense that I am experiencing nothing but the specific quality of this headache. For this reason, other than in the case of the perception of an external object, in the case of 'proprioception' the qualitative feeling is not merely one aspect—possibly even a secondary aspect—of the perceptual act but it is rather the primary object of the perception. *Thus, the qualitative aspect of the internal feelings appears to be one of the more important—if not even the most important—feature we use in order to identify what we are experiencing.*

In conclusion, starting from this discussion of the analogies between perception and 'proprioception' we defined a few relevant features that might serve as criteria to identify and to classify our internal experience. Inverting the order of the arguments, (i) we argued that *qualitative feeling is one of the most important properties we can rely on to identify and categorize our internal states* (for an overview see Russell 2003; Bermond 2007). Moreover, we observed that the mapping of the 'proprioceptive' states also provides information about (ii) *the specific intensities of the feelings* and (iii) *the specific location of* states in the body and we suggested that this information might also be used to classify our internal states. (iv) Finally, we speculated that 'proprioception' often detects a number of concomitant states and that this co-occurrence of states might also support their identification and recognition. Our hypothesis is indeed that these four features could be the ones used by our cognitive system to process internal states and to identify/classify them. To sum up:

(i) First of all, internal states—or more specifically the 'proprioception' we have of them—are qualitatively characterized, this means that they correspond to specific qualitative feelings (headache or hunger correspond e.g. to very specific sensations and the two are qualitatively very different from each other). We call this feature *the specific quality of the sensation.*

(ii) Secondly, 'proprioception' does not only detect the specific quality of a sensation, but also its intensity. A pain can e.g. be more or less intense. We refer to this feature as *the specific intensity of the sensation.*

(iii) Thirdly, propriosensitive mapping keeps track of the specific localization of the detected bodily changes when these exceed a certain activation threshold and are sufficiently relevant to become conscious. In other words, 'proprioception' allows us to identify the origin of the internal sensations and to know e.g. that a certain pain is localized in the temples, at the back of the neck or in the knee. We label this property *the bodily localization of the sensation.*

(iv) Fourthly and finally, propriosensitive mapping detects a number of different bodily changes that occur at the same time; this information about the co-occurrence of various changes gives rise to a more complex qualitative sensation that provides additional information about the state (its cause and its global nature). We refer to this property as *the presence/absence of concomitant sensations.*

Our suggestion is that the identification and the classification of our internal states from a referential point of view might be carried out through the intersection of these four kinds of features; this information could constitute the perceptual schema included in concepts denoting internal states which enables referential competence with respect to these states. As in the case of external perception, also in that of 'proprioception' the subject is not aware of the features his/her cognitive system uses for categorization. In fact, we are not aware that we are using these features to identify and classify our internal states, but we resort to them in an unconscious and implicit manner.

To describe more closely how the identification and classification of internal states might be carried out on the basis of these features, it might be useful to consider a specific example. In the following section we discuss the case of thirst.

(i) First of all, thirst is characterized by a specific qualitative sensation connected e.g. with burning and dryness of the mouth and throat. We experience a qualitative sensation that qualifies thirst in general, and if we do not pay particular attention to it, we just become aware of being thirsty. However, if we focus on our thirst, we are able to identify this sensation in a more specific manner and to differentiate various kinds of thirst, especially when it is considered jointly with other concomitant sensations (see point iv).

(ii) Secondly, thirst has a specific bodily localization. We do not feel thirst in the knees; we feel thirst only as localized in the mouth and throat. Moreover, depending on the specific kind of thirst we experience in a certain moment, it may be localized more in the mouth than in the throat or the other way around.

(iii) Thirdly, thirst is also always characterized by a specific intensity of the sensation that allows us among other things to evaluate the urgency of the situation and to prepare for a more or less immediate action. In the specific case of thirst, the intensity might be useful mainly for distinguishing between various kinds (i.e. tokens) of thirst. However, in other cases this feature can also play a primary function in distinguishing between different types of internal states: it might serve e.g. to differentiate between 'hunger' and 'starvation'; or between 'discomfort' and 'pain'.

(iv) Fourthly and finally, thirst is also characterized by the presence/absence of specific concomitant sensations. The thirst we experience after a long run is characterized e.g. by the presence of specific concomitant sensations related to the muscles or the skin and by the absence of other concomitant sensations (e.g. in the stomach) that occur in the case of other kinds of thirst (e.g. such as the one we experience after having a hearty and too salty meal). In this case, the presence/absence of concomitant sensations helps identify the cause of thirst at a certain moment and is the basis for distinguishing between different tokens of thirst. However, it is plausible that in some cases these features may also be used to classify different types of states. In the case e.g. of 'hunger' versus 'starvation' some concomitant sensation of global extreme weakness might be a crucial in determining that the state we are experiencing is 'starvation' and not just hunger.

Among these four characteristics, the first one is surely the most important, since in a certain way it subsumes all the others. In fact, considering again the case of 'thirst', concomitant sensations that contribute to forming a specific kind of thirst give rise to a complex but unitarily perceived global quality that keeps track of all the specific qualities of the sensations located in the various body parts. Moreover, this quality also results from the intensity and from the localization of the sensations that compose this global sense of thirst. For this reason, the qualitative dimension reflects somehow all the other features, even though these remain independent in the sense that they are identifiable separately from the global quality of the state. Indeed, if one pays attention to his/her sensations of thirst, s/he can identify at least vaguely some features other than the mere qualitative sensations related to localization, specific intensity and the presence of concomitant sensations. While the specific quality of thirst remains a primitive element which cannot be reduced to others or decomposed, other components of the sensation of thirst concerning the localization, the intensity and the simultaneity with other akin but not identical feelings can be separated from the qualitative element through a process of selective attention.

It is because of the complexity of the information detected by the self-monitoring system that we can perceptively (i.e. 'propriosensitively') identify a certain internal state as e.g. a token/type of thirst by cross-checking the various characteristics of the available information. The fact that—on the basis of a voluntary process of selective attention—we can distinguish the main different characteristics of the qualitative information, does not imply that the identification and classification of internal states is based on conscious and voluntary processes. Once again, this complex and structured information constitutes merely the rough material used by the cognitive system to categorize internal states in an implicit, automatic and unconscious manner.

To conclude, getting back to the parallelism between 'proprioception' and perception, we could say that categorization of internal states on the basis of propriosensitive information is quite similar to the categorization of external states on the basis of perceptive information. The identification and the classification of an external object such as e.g. a dog is carried out mainly by relying on some spatial properties of the objects that have been detected primarily through vision (see Chap. 3). Similarly, the identification and the classification of an internal state such as e.g. thirst is carried out on the basis of propriosensitive information detected through internal mapping mainly on the basis of the four features we discussed above: the specific quality of the sensation, its specific intensity, the bodily localization of the sensation and the presence/absence of concomitant sensations. Thus, on the basis of the analogy between perception and 'proprioception', we can suppose that our referential competence for internal states relies on a perceptual schema that codifies these four kinds of features.

4 Perceptual Theories of Emotion: Can Emotions Be Identified on the Basis of How They Feel?

The previous description of how we might categorize internal states and develop a semantic competence with respect to words referring to them relies on examples that uniquely concern a particular type of states, i.e. bodily experiences such as thirst, hunger, pain, itch, etc. Till now we have not mentioned more complex kinds of states that originate in the body and especially emotions. In fact, emotions will be the subject of our attention in the following sections where we will investigate how we categorize and how we develop semantic competence with respect to words referring to emotions. Specifically, we will examine how we develop our emotional concepts and what function internal information plays in this case. We will consider whether we can rely on specific characteristics of internal feelings to classify emotions and to develop a specifically referential competence for emotional terms.

Since the literature on emotions is as immense as the number of theories on emotions that have been put forward, it is useful first of all to limit the area of interest of our discussion. Considering the aim of this study, emotions are relevant because they are salient examples of internal states analogous to—even though more complex than—those discussed in the previous sections (bodily experiences such as pain, thirst, hunger etc.). The literature we discussed in Chap. 2 considers emotions to be a kind of subjective experience originating in the internal environment of an individual as a consequence of changes in various bodily states and suggests that the knowledge we have of our emotions is due—at least in part—to the bodily feelings we experience when we undergo an emotional episode. Thus, the issue we will take into consideration is how these bodily feelings contribute to the conceptualization of our emotions. We will address problems such as: how do people come to identify and classify emotions? Do they categorize emotions primarily 'from the inside', i.e. in the first person starting from internal information? Is there a specifically referential competence for words referring to emotions? What is the relationship between referential and inferential competence with respect to words referring to emotions?[10]

The identification and the labeling of emotions is an age-old problem that was first addressed in philosophical inquiries (for an overview see, e.g., Solomon 1976;

[10]The issue we raise in this part concerns the classification and categorization of emotions or, stated differently, emotion concepts. This issue is not new in literature, even though it is only a secondary and mostly overlooked aspect of the research on emotions. Usually, however, the debate on categorization of emotions focuses on topics that only partially overlap with those we discuss here since it mainly concerns the issue of whether emotional concepts are culturally determined or whether at least some of them are universal and innate, as e.g. the basic emotions theory maintains. In addition, most frequently this debate does not consider the problem of emotional stimuli from the perspective of the perceptual experience emotions rely on, but instead starts with other questions like what elicits emotional responses: for a review see e.g. Brosch et al. (2010) and Scherer (2009). A study addressing this topic which offers a similar perspective can be found in Niedenthal (2008).

Landweer and Renz 2008), and later taken up in psychological research using empirical methods (Galati 2002; Dixon 2003). To be more precise, the hypotheses that had initially been put forward by philosophers were empirically assessed based on the outcomes of physiological research carried out in the field of medicine (this happened for example in the case of William James and Carl Georg Lange who based their views above all on Mosso's (1881, [1884] 1886) and Mantegazza's (1885) research). Later, when psychology gained the status of an autonomous and independent science, these hypotheses were also tested on the basis of experimental research carried out in the field of psychology using methods and techniques specific to this discipline (Wundt 1890).

One of the most problematic traits of emotions according to the classical philosophical tradition is that they are connected with bodily sensations, even though these sensations appear to be very different from others. In fact, sensations connected with emotions influence our thinking processes in a much deeper manner than other kinds of (bodily) sensations: they have the power to intrude into our thoughts and to pervasively condition our choices and behaviors, as happens in the case of knowledge. This is the reason why emotions have been classified in so many ways. In some cases they were considered to be forms of *beliefs* or *opinions* (e.g. in the stoic tradition, see Graver 2007: 31ff; Buddensiek 2008: 72ff); in other cases they were identified with *confused thoughts* which include aspects whose origin is opaque that are not under the control of the will (e.g. according to the Leibniz-Wolff, as well as to the Spinozian and Hegelian tradition of thought, see Galati 2002: 34, 54ff); in still other cases they were thought to be *forms of perception*, i.e. something of lower standing than intellectual knowledge, but still capable of carrying out an auxiliary function with respect to it (this is the case e.g. in the Platonic and Aristotelian tradition).[11]

[11]The main interest of classical philosophical investigations into emotions lay primarily in determining their nature and function. It is on the basis of the answers provided to these questions that we can distinguish the views put forward by the various philosophical traditions. As for the nature of emotions, the issue concerned what kind of phenomenon emotion represents: are they some kind of thought or rather some form of sensation? Since emotions occur rapidly and involve specific changes to an individual's bodily state, they were often considered to be a particular class of perceptions which revealed mainly whether a certain situation was pleasant or unpleasant (i.e. dangerous or damaging). Such a position can be traced back already to Plato's Philebus (1983) and to Aristotle's Rhetoric (2010). This view opens the way for the idea that emotions are per se meaningful, i.e. that they convey important information for the individual and that they have therefore a specific epistemological function and value. On the other hand, however, the fact that emotions are related to somatic sensations gives rise to an opposite perspective, in which they are considered as disruptive factors with respect to the orderly flow of thoughts. An example of this view can be seen in the stoic theories of emotions (e.g. in the perspective put forward by Cicero in the third and fourth sections of his Tusculanae Meditationes according to which emotions are perturbationes animi, i.e. perturbations of the soul): emotions condition individual behavior and above all they condition decision processes (and therefore the will) having a detrimental impact on them (see Cicero 2002). Since only intellectual factors can influence our intellectual faculty, then emotions must be intellectual entities. However, because they are deeply related to somatic states, they cannot be considered as entities that convey true knowledge, but only as unstable and fleeting

The fact that the emotional experience is characterized by specific somatic sensations and is basically passive and reactive raises the issue of what brings it about. Is it the perception of an environmental situation or is it the cognitive appraisal of the situation that triggers an emotion? Moreover, is it a mental state that causes bodily changes that correspond to emotions or do the bodily changes occur prior to and independently of the mental states? And above all, what are the internal or external cues that allow a person to identify and label his/her own emotional states?

Traditionally, the answer to these questions gave rise to two very different families of theories called "perceptual" and "cognitive" which described emotions respectively as *perceptions* (i.e. as subjective and qualitative experiences) and as *cognitions* (as judgments and as cognitive appraisals of situations). In fact, these two families of theories are not merely alternatives to each other; they rather occupy different positions on a continuum that goes from purely perceptual to purely cognitive theories. According to the first, emotions are uniquely perceptions free from any cognitive contamination, while according to the second they are purely cognitive phenomena that do not involve any bodily components. Nearly no theory of emotion takes either pure form, since as a matter of fact emotional experience is characterized by an inextricable interweaving of sensations and cognitive elements. However, the various theories assign different weights to the perceptual and cognitive factors, assuming positions that are more or less close to one pole or to the other. Because of this, they also give different answers to questions like what emotions are and how they are identified. To understand what the positions on this continuum are it is essential to address the issue of how emotions are conceptualized. We will begin this section by illustrating the perceptual theories and in the next go on to describe the cognitive views and then we will highlight the major differences between the two.

The first attempt to give a systematic, empirically and theoretically sound description of emotions as perceptions can be traced back to the research of William James and Carl Georg Lange. Even though the development of their theories follows different and autonomous paths, both consider the physiological component of emotions to be the crucial component in emotional experience, i.e. the cause of this experience as well as the element that people use to differentiate between various emotional experiences. Their view marks the first departure from the old philosophical debate on what causes emotional experiences: the *cause* of an emotional experience cannot be traced back to any intellectual activity or to processes such as the cognitive appraisal of an event or a situation. What determines both the occurrence and the recognition of an emotion are the bodily changes corresponding to that emotion and the awareness we have of them.

(Footnote 11 continued)

sensations. For this reason, they are useless if not even detrimental opinions. These two opposite perspectives on emotions (they are useful vs. detrimental) has persisted across the entire history of philosophy up until late modernity. For a good overview see, e.g., Landweer and Renz (2008) and Dixon (2003).

As for James, his theory relies on a view of the body as an *organism* which is globally capable of perceiving the environment. He conceives of the body as a "*sounding board*" which can change itself in specific ways in relation to the various modifications occurring in the external environment. The emotions are nothing but the perceptions that we have of these changes: "the bodily changes follow directly the perception of the exciting fact, and […] our feeling of the same changes as they occur is the emotion" (James 1884: 189–190). And further: "every one of the bodily changes, whatsoever it be, is felt, acutely or obscurely, the moment it occurs" (James 1884: 192). In this view, the emotion consists in the perception of a bodily change determined by a co-occurring stimulation/perception of a specific object or event. Since the emotion is (i.e. consists in or is identical with) our perception of the somatic change, then this is also the element on the basis of which we identify an emotion as such. "If we fancy some strong emotion, and then try to abstract from our consciousness of it all the feelings of its bodily symptoms, we find we have nothing left behind, no 'mind-stuff' out of which the emotion can be constituted and that a cold and neutral state of intellectual perception is all that remains." (James 1884: 193). Thus, in James' view, what allows a person to know that s/he has an emotional experience is that s/he perceives the somatic modifications occurring to him/her, i.e. that s/he is aware of these modifications and of what they feel like—of their global *phenomenic quality* (Slaby 2008: 551).

It would be impossible for us to know that we are feeling an emotion and thus identify what emotion we are experiencing to label it, if we were not aware of the somatic feelings that characterize it. James exemplified this point by using the case of the emotion called 'fear': "what kind of an emotion of fear would be left, if the feelings neither of quickened heart-beats nor of shallow breathing, neither of trembling lips nor of weakened limbs, neither of goose-flesh nor visceral stirrings, were present, it is quite impossible to think" (James 1884: 193–194). Thus, fear manifests itself through a complex of perceptual and somatic factors that we become aware of because of the subjective qualities they give rise to. The emotional experience is therefore a specific perceptual experience that involves the whole body and specifically those component of the bodily experience that later Charles Sherrington would more precisely qualify as *proprioception* and *interoception* (see Chap. 2, Sect. 3).

An analogous perspective that nevertheless includes some important differences, was proposed by Carl Georg Lange. Lange also considers that emotional experience is related to an awareness of the physiological changes that take place in the body. The example he uses is similar to James' example. It concerns the description of the experience of 'terror': "If from one terrified the accompanying bodily symptoms are removed, the pulse permitted to beat quietly, the glance to become firm, the color natural, the movements rapid and secure, the speech strong, the thoughts clear,—what is there left of his terror?" (Lange [1885] 1912: 675). Lange, however, focuses his attention on the possibility of providing a 'componential' taxonomy of emotional experience. Indeed, in his view emotional experience results from a combination of specific factors: the activation of the voluntary and involuntary muscles, vascular contraction and dilation and more or less motoric coordination. For example, a state such as disappointment is described in terms of "weakness of voluntary innervation" associated with "vascular contraction", while a state such as joy is described in terms of an "increase of voluntary innervation" associated with "vascular dilatation" (Lange [1885] 1922: 57ff). Lange's idea is that the somatic feelings we experience when we have an emotion can be described in a purely physiological manner by identifying the individual, specific somatic changes that gave rise to it. Consider for example 'disappointment': the weakness of voluntary innervation associated with the vascular contraction that occurs in our body when we are disappointed explains why this emotion gives rise to the specific qualitative character that correspond to it, i.e. to a feeling of falling and of cold. On the other hand, the increase of voluntary innervation associated with vascular dilatation that occurs when we experience joy gives rise to the typical qualitative feeling that corresponds to this emotion: increased body temperature and the propensity to move.

James and Lange place emotions in the class of perceptual experiences: emotions are experienced as bodily modifications and are identified on the basis of the feelings these give rise to. Thus, the James–Lange theory views emotions as perceptions, even though they are certainly different from perceptions of the usual kind. In fact, in the case of emotions what is perceived is not an object located externally to the subject that stimulates a specific sensory channel. In James' view, the reason why emotions are different from other kinds of perceptions and particularly difficult to investigate from an empirical point of view, is that, unlike perceptions of the usual kind (acquired through the senses), there isn't any *specific* or distinct sensory channel devoted to the detection of emotional 'stimuli' that can be made the subject of a specific and separate investigation. Emotions consist in a complex of feelings related to sensations and stimulations that we perceive and that we are aware of James (1890: 291ff). What is perceived is located internally to the body and for this reasons emotions are quite similar to bodily experiences (hunger, pain, tension etc.). Consequently, the identification, recognition and labeling of emotions also has much in common with the identification, recognition and labeling of bodily experiences.

What causally determines our emotional experience and constitutes the condition for its identification and recognition are bodily changes and the related sensations—what somewhat later, as already mentioned above, Charles Sherrington called *proprioceptions* and *interoceptions*.[12] Indeed, as James points out, an emotion that does not have any somatic correlate, would be a non-entity, i.e. it would be something purely verbal, a mere product of thought (*ens rationis*) that would not have any content, referent or empirical meaning (James 1890: 299ff). Therefore, somatic states are the necessary condition for an emotion to be known, identifiable and thinkable (Galati 2002: 172). In this respect, James is very explicit and he even suggests some sort of mental experiment: if we were able to control by will any of

[12]For a recovery in contemporary terms of James and Lange's theory of emotion see especially the theories of emotion put forward by Prinz (2004a) and Damasio (1994).

the muscles and movements related to an emotion we were experiencing, it would de facto rapidly fade away and disappear (James 1884: 197ff).

The idea that emotions wouldn't exist without their somatic counterparts is shared by Lange, even though there are some subtle but relevant differences between him and James that can be made explicit using Sherrington's later distinction between proprioception (the perception of the bodily movements due to changes in the tension of the musculoskeletal system) and interoception (the perception of sensations related to changes in the internal organs, especially the viscera). While James views emotions as largely related to bodily movements (i.e. to proprioceptive information), Lange focuses his attention primarily on the consequences of vasoconstriction and vasodilatation on the functioning of the organism and therefore on interoceptive factors as well as the hedonic tones related to them (i.e. general feelings of pleasantness and unpleasantness generated by changes in the physiology of circulation).

The comparison with Sherrington is not only useful in highlighting some of the differences between James and Lange, but also for understanding some intrinsic limitations to both of their perspectives. Sherrington's investigations on the functioning of the nervous system and in particular on the reflex arc indicated that the central nervous system unifies and integrates the peripheral stimuli conveyed by the autonomous nervous system. This has important implications for the investigation of emotional experience since it gives rise to a braincentric (or centralist) view of the genesis of the sensations related to the body and its states. Both James and Lange's research was carried out shortly before the first systematic publication of Sherrington's results. Even though at first glance Sherrington's perspective appears to confirm James and Lange's intuitions on emotional experience, it actually calls them into question. In fact, even though Sherrington confirms that proprioceptive and interoceptive information are an essential component in determining what emotion we are experiencing, his perspective challenges James and Lange's view that the autonomous nervous system and peripheral stimulation play a central role with respect to the genesis of emotional experience.

A brief overview of how James conceptualizes the origin of emotional episodes might help to clarify this point (a similar account is also offered by Lange): the presence of an object (which is per se neutral from an emotional point of view) generates sensory stimulation which directly reaches the cerebral cortex giving rise to the perception of the object; in turn, this perception directly causes visceral and muscular changes in the subject. What reaches awareness in the case of a perceived emotion is nothing more than the reafferent set of nervous signals coming from the muscles and the viscera to the cerebral cortex. This is the reason why James' (and Lange's) view can be described as perceptual and above all as peripheral: the sensations that we perceive to be emotional originate at a peripheral level, and are directly communicated to the cortex of the autonomous nervous system travelling along the spinal cord.

Charles Sherrington's research on the nervous system (Sherrington 1906) challenges exactly this aspect of the views put forward by James and Lange: it undermines the idea that there is a direct link between the viscera, the peripheral organs and the brain instead identifying a location in the encephalon as the area where emotional experience originates. Sherrington's results rely on animal testing: he observes that even though emotional reactions such as anger and fear can still be distinguished in animals with a surgically resected spinal cord, they do not give rise to the appropriate organized behavioral reactions such as assault or escape. In Sherrington's view, once the spinal cord has been resected, the reactions of the animals are only pseudo-emotive: some aspects of the emotional reaction still occur, but they are not sufficient to constitute an emotion as such. From these observations Sherrington comes to the conclusion that the sensations generated by the peripheral nervous system are necessary conditions for having an emotion, but they are not sufficient to determine an actual emotional experience. His studies lead to the hypothesis that the stimuli and sensations of peripheral origin assume a truly emotional valence and give rise to distinctive behavioral expression in the brain.[13]

Still, apart from these limitations of James and Lange's views due to their incomplete understanding of the physiology of emotions, they researchers deserve credit for bringing attention to the fundamental role of somatic sensations and the perceptual qualities of emotional experience for the identification, recognition and labelling of emotions from a first personal point of view. In spite of its historical relevance, this perspective has never been corroborated by empirical and experimental research. The reasons for this are twofold. (i) The first one is methodological and concerns the method that a scientific psychology should adopt; (ii) the second is theoretical and is related to criticism of the idea that the peripheral nervous system has a predominant role with respect to emotions.

(i) The methodological issue is historically linked to the attempt made by behaviorism to build a scientific psychology that is part of the natural sciences and therefore to describe psychological objects as the natural sciences describe objects, i.e. avoiding the attribution of any qualitative and subjective properties (see Chap. 1). Emotions are no exception to this; they cannot be described on the basis of subjective qualities (feelings), but they must be identified on the basis of some objective—intersubjectively controllable—element. In line with this, so-called first-generation behaviorism, whose manifesto is Watson's article of (1913) *Psychology as the behaviorist views it*, argues that each emotion can and must be identified with/described in terms of specific observable behavioral patterns elicited by similarly specific environmental stimuli.

However, this approach soon revealed a patent difficulty: indeed, it is difficult to link any emotion to a specific and unique behavioral pattern because the same emotion can give rise to very different behaviors, especially if it occurs at different degrees of intensity. Think for example of fear: (various degrees of) fear might trigger very different behavioral responses such as e.g. fighting, escaping or freezing. On the other hand, the same behavioral pattern may be triggered by

[13]Later, on the basis of further research, Sherrington comes to locate the center of emotions in that internal part of the brain called the mesencephalon, thus supporting an even more centralist hypothesis on the origins of emotional experience (Woodward and Sherrington 1904).

different emotions: a fight can be e.g. a reaction to fear as well as a reaction to anger. This kind of difficulty—which concerns both the description and the identification of emotions—led the behavioristic psychologists to integrate behavioral elements with other additional quantitative and controllable parameters, i.e. physiological responses to emotions (Watson and Morgan 1917; Watson and Rayner 1920; Watson [1924] 1931; Tolman 1922a). According to this approach various kinds of emotion can be identified and differentiated by virtue of the fact that each relies on a unique pattern of physiological responses (i.e. measures) (Duffy 1941).

This view only apparently harks back to the perspective on emotions offered by James and Lange. Even though bodily changes (physiological parameters) are again considered to be a crucial component of emotions, the focus of the behavioral theory is no longer on the feelings they give rise to—i.e. on emotion perception; it instead concentrates uniquely on the physiological parameters per se. In fact, in the view put forward by James and Lange the individuation and classification of emotions is carried out from a first-person point of view on the basis of the specific feelings that correspond to each emotion. The bodily changes that characterize our emotional episodes (visceral and muscular contraction, change of perceived temperature, of the heartbeat etc.) are merely the physical causes of these feelings, i.e. what brings them about. On the contrary, the behavioral theory hypothesizes that it is the bodily changes themselves—or more precisely their physiological measures—that determine what emotion we are experiencing at any moment since there is a one-to-one correspondence between specific physiological parameters and specific types of emotions.

Even though this approach is more accurate and complex than that of first-generation behaviorism and provides a strong incentive for the development of physiological research, its limitations also soon became quite clear. In fact, like the view that emotions are based on the behavioral patterns, the view that emotions are based on physiological responses turns out to be insufficient for uniquely identifying different kinds of emotions. Again, this is because analogous physiological reactions (visceral or muscular contractions, changes in blood pressure, body temperature, etc.) are shared by a number of emotional experiences. Further, even the intersection of behavioral and physiological patterns is insufficient to uniquely identify specific kinds of emotion.

The failure of the behavioristic view to account for emotional experience shows that it can be explained only by beginning from a more integrated account that includes not only behavioral and physiological but also 'mental' factors. The question remains, however, what mental factors should be included: whether—as James and Lange suggest—emotional feelings are decisive in terms of how we identify and recognize our emotions or whether this identification and recognition is instead based on cognitive elements of some kind. As we will see in the next section, later studies preferred this second option which is not only more compatible with an experimental (quantitative) approach to emotions, but is also consistent with the general perspective of the cognitive psychology movement that followed behaviorism.

(ii) The second reason why the James–Lange theory did not really take root in experimental research in psychology is that it gave excessive importance to the peripheral stimulation which was considered to be the autonomous and direct cause of the genesis of the emotional experience. As we already mentioned, this aspect was challenged by Sherrington's research which led to a *brain-centric* or *centralist* perspective on emotions. One of the authors who gave this perspective a further decisive boost was Cannon (1927, 1929). Cannon like Sherrington before him, considered that emotional experience was not primarily caused by visceral changes: not only because the stimulation these generate is not directly transmitted to the cerebral cortex, as the peripheral theory suggested, but also because these changes do not seem to be necessarily related to emotional experience. Visceral changes can also occur in the case of non-emotional experiences (e.g. typically for some feeling of unease); visceral changes might be induced artificially without generating an emotional episode; and even though visceral changes can be automatic components of some emotional experiences, they start later than the actual emotional experience. More generally, if we isolated the somatic components of each emotional episode, we would obtain a set of a-specific physiological activations that are per se not sufficient for identifying an emotion or for differentiating it from others. Somatic sensations are a-specific in the sense that they are shared by many emotions as well as by many somatic states: increased heart and respiratory rates together with muscular contraction are typical of several emotions, but they are also typical somatic reactions e.g. to physical effort and to stress. These are the reasons why Cannon denies that peripheral and visceral somatic changes can be considered per se to be the causes of the emotions.[14]

The main point of disagreement between Cannon and the so-called James–Lange theory concerns where emotions originate. Cannon cites Shannon's research on animals and suggests that the thalamus is the seat of the emotions, an even more central locus that that suggested by Sherrington. The thalamus plays a double causal role: it activates the cortex on the one hand and the viscera and the muscles on the other. The activation of the cortex allows for emotional experience to be conscious, while the activation of muscles and viscera gives rise to the somatic sensations that accompany these emotions (and which might occur after the emotional experience per se has reached awareness). In Cannon's view, the thalamus is the locus where the emotion as an automatic response of the organism is organized. The idea that this has a double causal role definitively separates the emotional experience from the somatic sensations which are at most co-occurring phenomena with respect to the genesis of emotion awareness. Cannon's line of research was very fertile and opened the door to many other studies, which mainly focused on the issue of the real physical locus where emotions originated in the belief that an answer to this

---

[14]The idea that somatic changes do not have per se an emotional character or that any independent somatic change does not on its own indicate the occurrence of an emotion is certainly compatible with James' and Lange's position, in which emotions are not detected by a specific sensory channel, but correspond to a pattern of signals and sensations. James' and Lange's theories incompatible with Cannon's insofar as Cannon rejects peripheralism.

question would also automatically lead to a solution to the problem of how various emotions can be differentiated by the first person.

We cannot enter here into the details of the long and complex history of the physiological research that aimed to identify the origins of emotional experience. However, it is important to point out that after Cannon, the issue of where the central seat of emotional experience might be became progressively more complex. While according to Cannon emotions originate in a single locus—the thalamus—later research suggested that the place where emotional experience is organized might actually be the hypothalamus which collects signals that are not only transmitted by the body through the nerves but also transmitted through the blood by hormones. This opens the way to a *systemic* view of emotional experience: emotion is conceived as a complex phenomenon which results from a stimulation that spreads across the whole organism traveling through various channels collected, integrated and organized at a central level by the brain (Galati 2002: 177–207). Still, a characteristic trait shared by these various lines of research is that they consider peripheral bodily changes to be relatively irrelevant. Independently of the kind of signal they generate and of the channel that brings these signals to be integrated at a central location, their function in the identification and recognition of emotions remains in the background. They are conceived almost as epiphenomenal with respect to emotions: they are a-specific sensations that co-occur with emotional experience without, however, having any direct or indirect causal function with respect to their genesis. In consequence, the function that qualitative information of a perceptual kind related to these somatic sensations plays in the identification and recognition of emotions is also not acknowledged.

More generally, the predominant trait of the lines of research on emotions that resulted from Sherrington's criticism of the James–Lange view is the idea that somatic manifestations (i.e. the physiological activations) related to emotions are a-specific. From a bodily point of view, what we observe during an emotional episode is only a generic state of arousal which does not play any causal role with respect to it and which does not have any relevant properties useful for its identification or recognition. The arousal only indicates that some changes are occurring in the body, but does not allow the subject to know that s/he is experiencing an emotion (instead of e.g. a bodily state of some kind) and even less to identify and recognize what emotion s/he is experiencing (on arousal in relation to the issue of how we identify emotions, see also Prinz 2004a: 160ff).[15]

[15]The target of Prinz's criticism here is actually a later and slightly more complex conception of emotions called a circumplex model (see e.g. Russell 1980; Watson and Tellegen 1985) which suffers from the same difficulty as Lindsley's theory. According to this model, the classification of emotions cannot occur on the basis of arousal only, but there is also a second measure that seems to be essential in defining emotions and that is therefore needed for their classification, i.e. valence. Indeed, we experience emotions to be necessarily (more or less) positive or negative. For this reason, for example, the circumplex model suggests that the parameters for the classifications of emotions need to be three-dimensional and consider both arousal and positive/negative valence. However, as Prinz also points out, this additional measure does not solve the problem we introduced before. Two emotions may be similar in arousal and valence and still be of different kinds

The generic character of arousal raises an issue: what allows us to know that we are experiencing an emotion and to identify/recognize what emotion we are experiencing? The rapid alternation between neurophysiological hypotheses on where and how emotions are processed in the brain has prevented research from investigating this issue. In fact, knowledge of the brain processes underlying emotions does not reveal anything about how the first person can identify that s/he is experiencing an emotion and recognize which emotion s/he is having. If the bodily changes that people experience in relation to emotions cannot be used to identify and recognize emotions, what can? This problem was investigated in the field of so-called cognitive psychology. This line of research supported a variation of the centralistic hypothesis, in which the cause of the emotions we experience as well as what allows us to know what emotions we are experiencing is not the physiological activations corresponding to somatic changes, but *the knowledge and thoughts that accompany them*. This position will be considered in more detail in the next section of this chapter.

However, the criticism against James and Lange's peripheralism has also lead to the development of a different line of research with respect to that described above led by Donald Lindsley. Lindsley's psychophysiological studies (1950, 1951) show that it is extremely difficult to discriminate between activations corresponding to emotions and activations corresponding to mere bodily experiences such as hunger and thirst. From a physiological point of view, these studies largely confirm the hypothesis of the a-specificity of the arousal (Lindsley 1970), although they also indicate that emotions can occur at different *intensities*. Emotions correspond to a generic state of arousal (i.e. to a physiological activation that has no specific quality that links it to individual kinds of emotion) that can, however, be located along a continuum that goes from zero (no emotion) to very high (intense emotion). In fact, the same electroencephalogram might correspond e.g. to various kinds of emotions; what we can differentiate using these physiological measures is not what kind of emotion we are experiencing, but rather the level of arousal of the emotion we experience. This is the reason why Lindsley's view is usually known as the *activation theory of emotions*.

The new element that characterizes this theory does not concern the idea that arousal cannot be used for identifying and recognizing an emotion. Two states might be similar in arousal but differ in their 'nature': jealousy and anger might be analogously high in arousal, but remain different emotions. On the other hand, emotions may differ in arousal and still be the same emotion: our jealousy might be more or less intense without becoming a different emotion since its degree does not affect its 'nature'—its object, the situation that gives rise to it, the reaction we have, etc. The significant change underlined by this view is rather the way in which it addresses the issue of *how people identify and recognize their emotions from a first*

(Footnote 15 continued)

(the example of jealousy and anger also applies in this case since they can be considered similar in terms of arousal (both are equally high) and valence (both are very unpleasant).

personal point of view provided that humans clearly have this capacity and that they cannot rely on arousal (or arousal intensity) to carry out this identification and recognition. Indeed, in Lindsley's perspective the capacity that humans exhibit to identify and to recognize their own emotions is not cognitive, but *perceptual*. Each emotion is characterized by an overall *sensory quality* that is sufficiently specific to allow us to identify and recognize what emotion we are experiencing. This quality derives plausibly from the synthesis and from the integration of various bodily changes that occur in correspondence with the emotion. Taken individually, no one of the several qualitative sensations generated by the bodily changes that occur during an emotion can qualify as an emotional episode. This means that an emotion is not characterized by the mere summation of a number of different bodily sensations; and yet, the synthesis and integration of all these sensations brings about a novel unitary quality, on the basis of which the emotional episode can be identified as such (and distinguished e.g. from bodily experiences such as hunger or thirst), classified and thus also labelled. While the somatic changes that correspond to an emotion are generic as are the sensations they generate, the overall quality that derives from all of them blended together is not. In this respect, Lindsley's studies introduce an element of great importance for research on the nature of emotional experience since they shift the focus of attention towards the qualitative sensations corresponding to emotions and lead us to reconsider introspective factors (internal qualitative experiences) as elements people rely on for identifying and classifying their emotions.

5 The Cognitive Theory of Emotion: Why Cognition is Needed to Identify Emotions

According to an influential part of the philosophical tradition of the 20th century, emotions are particular kinds of perceptions. Emotions occur *spontaneously* (we can neither decide to have an emotion at a given time nor can we prevent it from occurring), *reactively* and *passively* (something must cause the emotion to occur) (see, e.g., Gordon 1987). Their function is to drive the subject's attention to specific elements or features of a situation that are particularly relevant for him/her (De Sousa 1987; Rorty 1980). This information is not processed—at least initially—in a propositional format (Thalberg 1977), but consists of a physiological response that reaches awareness in the form of somatic feelings. These give rise to qualitative experiences which represent—directly or indirectly—positive or negative evaluations of the situation in which the person is involved (Lyons 1980; Calhoun 1984; De Sousa 1987).

However, this conception of emotion as a kind of perceptual experience is far from being unanimously accepted in the literature. Exactly because emotions also have the function of informing us about a situation and because they allow us to evaluate this situation with respect to ourselves (e.g. the fact that I am happy informs me that what has just happened is positive for me), they have often been considered to be cognitive elements. Some philosophers even consider them to be propositional and to consist of beliefs and desires (Marks 1982) or judgments of some kind (Solomon 1980; Neu 2000; Nussbaum 2001): my anger is my belief that I have been treated unfairly together with my desire to be treated fairly; or my anger is the judgment that something unfair has happened to me. In fact, it is because the nature of emotions can be interpreted in this twofold manner as either a particular kind of perception or a particular kind of cognition that the debate between perceptual and cognitive theories of emotions has always been so heated.

Both perceptual and cognitive theories have numerous ramifications and can be differentiated on the basis of various factors. Still, we are not interested in emotion theories in general, but only in how emotions are identified and classified by people. Here, the disagreement between these two views assumes a quite specific form. In the previous section, we discussed what perceptual theories suggest in this respect: people identify, classify and label their emotions on a perceptual basis. Going back to the analysis carried out earlier in this chapter on bodily experiences such as hunger, thirst and pain, the fundamental idea at the bottom of perceptual theories is that we identify and classify emotions as we identify and classify bodily experiences, i.e. on the basis of what they feel like. In other words, the recognition of an emotion from a first person perspective relies on the feelings brought about by the bodily changes that occur when we experience an emotion. Contrary to what we usually think, this theory suggests that it is not the emotion that causes the bodily changes and thus the feelings they give rise to, but rather the bodily changes that cause the emotion since the emotion consists in the feelings we experience due to these changes.

Cognitive theories have been developed in contradistinction to perceptual ones. Given the issue, we are interested in here, the cognitive theories we are going to address are those belonging to the tradition of so-called cognitive psychology.[16] The main thesis they maintain with reference to the issue of how people identify and classify their emotions is that it is impossible to identify clear-cut physiological or perceptual discriminations of the various emotions; in fact, the same bodily changes and bodily feelings occur in very different emotional and even non-emotional states. Emotions such as e.g. anger or fear might indeed correspond to the same pattern of bodily changes and therefore bodily feelings. If we relied on physiological states or on feelings alone, we would not even be able to distinguish an emotion from a bodily experience, e.g. to understand that we are angry and not just experiencing muscular tension and increased blood pressure. What we actually perceive when we have an emotion—this is the main thesis cognitive theories argue

[16]As a matter of fact, the first cognitive theories of emotions can be traced back at last to two millennia ago and especially to the Stoics Seneca and Chrysippus who considered emotions to be judgments about the world and one's place in it. See, e.g., Solomon (2008: 3–16). Moreover, there are also a number of philosophical theories that lean on the cognitive approach but have developed independently from psychological research: among them we should mention at least Solomon (1976), Neu (2000) and Nussbaum's (2001) work.

for—is an undifferentiated state of arousal due to the bodily changes that we experience. What allows us to know that we are experiencing an emotion and to discriminate among the various emotions we experience (i.e. to know which emotion we are experiencing at any time) are the cognitive processes that accompany this arousal, i.e. the knowledge we have of the situation that gave rise to the emotion and our reaction to it (see e.g. Schachter and Singer 1962; Schachter 1970; Clore 1994).[17] What basically distinguishes the perceptual and the cognitive view of emotions is therefore not that cognitive theories deny the very existence of a bodily component of emotions which we become aware of in the form of feelings. The point is rather that cognitive theories consider these bodily components and feelings to be too indeterminate to support identification and classification.

In order to clarify this point from both a historical and a theoretical perspective, it is useful to discuss the famous adrenaline experiment proposed in 1962 by Stanley Schachter and Jerome Singer which supposedly shows that—when a generic state of arousal is induced (even artificially through the injection of adrenaline)—it is possible to produce an arbitrary emotion by conditioning the person with specific cognitions: "given a state of physiological arousal for which an individual has no adequate explanation, cognitive factors can lead the individual to describe his feelings with any of a diversity of emotional labels" (Schachter and Singer 1962: 383; for a discussion which puts this experiment in its historical context see, e.g., Mandler 1975: 88ff).

The main steps in the experiments were the following: the participants in the experiment receive an injection of adrenaline (epinephrine) which causes bodily changes analogous to those corresponding to various emotions (increased blood pressure, heart rate, respiration, muscle tension and the like). The participants were told, however, that they had been injected with a vitamin supplement which might affect their visual capacity. The first group of people was told that the vitamins had some transitory side effects and were informed about the symptoms they were about to experience (increased blood pressure, heart rate, respiration, muscle tension and analogous); the second group was left in the dark about the fact that the injection had such physical effects.[18] After the injection participants were asked to wait in a room for the vitamin to work, then their vision would be tested. Each participant was asked to sit in the waiting room with another person who supposedly had also received a shot. This second person was, however, not a participant but a stooge. In some cases, the stooge engaged in euphoric behaviors (playing with a ball, with paper airplanes and with a hula hoop), while in others s/he was expressively

[17]For a brief overview of how and why cognitive theories diverge from perceptual ones see Christenfeld and Mandler (2012: Chap. 8); for a short overview of the main cognitive theories that belong to this tradition see, e.g., Strongman (2003: Chap. 6).

[18]In the experiment there were actually two more groups, but this is not strictly relevant for the point we would like to make here. One of the additional groups was misinformed about the effect of the injection (they expected completely different symptoms with respect to those they actually experienced); the other group was injected with a placebo that caused no symptoms and was left in the dark about what they were going to experience. See Schachter and Singer (1962: 383).

extremely angry. Then, before proceeding with the vision test, some information on the emotional state of the participants was collected. The results are quite complex, however, the main point might be summarized as follows: the participants who were previously informed about the side effects of the injection weren't really susceptible to the stooge's mood in either the euphoric or the angry condition; by contrast, the participants who did not know that the injection caused physical symptoms like the ones they experienced, reacted in a euphoric or angry way reflecting the stooge's behavior. The interpretation of this result was relatively straightforward: because of the adrenaline injection, participants felt a state of physical arousal; when they were unable to explain these symptoms as physical (e.g. when they did not know that they were a consequence of the injection), they tried to explain them by other means. The alternative explanation that they gave to themselves is that they were experiencing an emotion; however, they did not know what emotion they were experiencing; thus, to decide which emotion they were experiencing, they resorted to the cognitive explanations available to them which in this case were provided by the context, i.e. by the stooge's mood and by how this affected them:

> given a state of physiological arousal for which an individual has no explanation, he will label this state in terms of the cognitions available to him. This implies, of course, that by manipulating the cognitions of an individual in such a state we can manipulate his feelings in diverse directions […] given a state of physiological arousal for which an individual has a completely satisfactory explanation, he will not label this state in terms of the alternative cognitions available (Schachter and Singer 1962: 395–396).

This view does not deny that emotions have a physiological—i.e. a perceptual—component; in fact, a person recognizes that he has a given emotion only to the extent that s/he experiences a state of arousal. Still, the physiological changes that give rise to this arousal are considered to be generic and indeterminate: they cannot be the element on the basis of which people distinguish among their feelings, identify and label them. The element human beings use to carry out this distinction, identification and labeling are the cognitions they have during the arousal and these depend—in turn—on the situation that triggered the emotion. Schachter and Singer's view on emotions belongs to the family of cognitive theories because it maintains that what actually determines the recognition of our emotional states are cognitive elements related to an emotionally significant situation.

Some criticisms have been raised against Schachter and Singer's experiment both with respect to its methodological and theoretical aspects (Plutchik and Ax 1967; Ax 1971; Reisenzein 1983, the latter also discusses a number of unsuccessful attempts to replicate the original experiment). One of the main puzzling elements concerns a discrepancy registered in the anger condition. In fact, while participants in this condition exhibited clear signs of anger in their behaviors (in their gestures and words), in the self-report questionnaire they indicated they felt happy, exactly like the participants in the euphoric condition. According to the authors this happened because they were not willing to admit their irritation to the experimenter.[19] However, this could also be considered as evidence in support of the fact that the arousal the participants experienced was not so generic after all, but—in spite of the upsetting situation they were put into—participants in the angry condition also experienced their feelings as somehow happy (see Prinz 2004a: 71).

Even though the outcome of Schachter and Singer's experiment is indeed controversial, it marks a turning point in the research on emotion and it opens the door to many other cognitive approaches claiming in one way or another that—even though emotions are also accompanied by some feelings—emotions are essentially cognitions; more specifically: cognitions are what determine and qualify any emotional experience. Moreover, the experiment shows that bodily changes and cognitions are two separate factors in emotional experience. In George Mandler's words: "The important contribution that the Schachter experiments have made is the unequivocal statement of a theory of emotion that postulates two independent and interacting factors: physiological arousal and cognitive evaluation" (Mandler 1975: 93). On the other hand, however, it also points out that *cognitive mediation* is the most important factor for determining what emotion an individual is having. In this view, the cognitive factor is indeed a mediation because it links the perception of an emotionally significant situation with a behavior that is appropriate to both the situation and to the thoughts induced by it. This way of explaining emotions on the basis of interactions among perception, thoughts and behavior is entirely consistent with the general approach of cognitive psychology which relies on a functionalist explanatory model (Chap. 1, Sect. 4), according to which thoughts mediate between stimuli and behavioral responses.

Later cognitive approaches to emotion continue along the path Schachter and Singer laid out. They take up and further develop the idea that emotions have a perceptual component which they describe not merely as generic arousal, but in terms of an arousal characterized by a specific valence or hedonic tone (see, e.g., Russell 1980: 1169). Indeed, the arousal corresponding to an emotion does not usually feel 'neutral' to us, but instead feels either positive (pleasant) or negative (unpleasant). Even though valence is a perceptual component of emotions, these theories postulate that valence is primarily a consequence of cognition and is thus secondary. In these views, emotions are some sort of judgments about well-being and about the (positive or negative) relationship that the individual entertains towards his/her environment: they are essentially *appraisal judgments* (Lazarus 1991).

[19]This is a relevant passage in the text where this issue is explained: "Though as the reader will see, the subjects were quite willing to manifest anger when they were alone with the stooge, they hesitated to do so on material (self-ratings of mood and questionnaire) that the experimenter might see and only after the purposes of the experiment had been revealed were many of these subjects willing to admit to the experimenter that they had been irked or irritated. This experimentally unfortunate situation pretty much forces us to rely on the behavioral indices derived from observation of the subject's presumably private interaction with the stooge" (Schachter and Singer 1962: 391).

The general idea underlying these theories is that all emotions include appraisal judgments, i.e. judgments on how a certain situation affects us in some relevant manner (Arnold 1960: 171). If something happens to me which I consider to be very beneficial, then the emotion I will experience is joy. In fact, it is by virtue of appraisal judgments that we apprehend whether a situation is positive or negative for us and for our well-being, while the feelings of arousal and valence that we associate with emotions result from these judgments. What allows us to understand what emotion we are having at a certain time is not the emotional experience per se (the feelings resulting from arousal and valence), but the cognitive component of the emotion, i.e. the appraisal judgment (see, e.g., Lazarus et al. 1970; Lazarus and Smith 1988).

Different theories disagree on how these appraisal judgments actually work and include different "dimensions" on which this appraisal can be measured. More complex theories try to offer more fine-grained distinctions between emotions by identifying various ways in which we 'appraise' situations, i.e. in which we evaluate how something affects our wellbeing, taking into account in which way, how much and for how long we are affected; what is the involvement of our ego; whose merit or responsibility it is that we are in this situation; how easy it is to cope with it; etc. (see, e.g., Lazarus 1991).[20] However, the common denominator of all these views is that—in line with Schachter and Singer's perspective—they rely on a cognitive theory of emotion to explain what emotions are and above all how they are identified: it is not the perceptual component of the emotion that allows us to know what emotion we are experiencing; rather, we identify what emotion we have by virtue of cognitions and in particular by virtue of the appraisal judgments we make with respect to the situation which caused the emotion.

Let's suppose I am angry because my car was damaged by someone. According to this theory, I do not know that I am angry because I feel angry. I rather know that I am angry because I have angry thoughts, i.e. thoughts about how this event affects me (for example concerning the fact that what happened is certainly negative and damaging for me, that the repairs will cost me a lot, that it was unfair for the person who did this not to leave his/her insurance information and the like). Of course, I have these thoughts because I understand the situation I am in (I understand that cars might be damaged, that when this happens you need to repair them, that these repairs cost money, that when you damage somebody else's car you must repay him/her; that cars are insured against accidents etc.). Still, to understand the situation and to know I have these thoughts is not yet sufficient for me to know I am angry; another cognitive element is needed: I must also know that the emotion that goes with such a situation and with such thoughts is anger. Once I have all this knowledge, I can conclude that I am angry.

[20]Different appraisal theories which rely on other dimensions of this judgment have been put forward e.g. by Weiner (1985), Scherer (1984), Roseman (1984), Smith and Ellsworth (1985) and Frijda (1986).

This example of how cognitive theories explain emotions describes first of all what it concretely means to argue that emotions are essentially cognitions, i.e. that they are identified on the basis of cognitions. Secondly, it also points out that considerable knowledge is needed to identify and recognize (classify and label) an emotion: I need to understand the situation I am in, I need to have thoughts appropriate to the situation and I need to know what thoughts and situations go with what emotions or to put it differently, what emotion is appropriate for what thoughts and situations. Last but not least, the example also shows that according to cognitive approaches individuals identify and classify their emotions on the basis of an *inferential* and *third person procedure*. In fact, even though the knowledge a person has about the thoughts that an emotionally significant situation induced in him/her is direct and first-personal, the identification and classification of the emotion per se is certainly not. Individuals do not know what emotion they have because they have a direct experience (a perception) of it; they rather conclude that they have a specific emotion on the basis of the knowledge they have about the situation they are in, their thoughts and about the emotions that go with such situations and thoughts. This is what qualifies this approach univocally as third personal and inferential and what distinguishes it most from the perceptual theories of emotion that on the contrary put forward a view according to which we identify and classify our emotions directly in the first person on the basis of what they feel like.

6 How We Identify Our Own Emotions: A Necessary Mediation

The history of the theory of emotion does not evolve in a linear fashion. Building on the success of cognitive psychology, cognitive views initially prevailed. However, within a different line of inquiry that investigated emotions from a more physiological point of view an alternative approach emerged that was more akin to perceptual theories. This has its historical landmark in Charles Darwin's study in *The Expression of the emotions in man and animals* ([1872] 2009) and it focusses mainly on the physiology as well as on the expression of emotions. In this work, Darwin tried to show that not only do people from different backgrounds express their emotions in a uniform manner, but there are also important similarities in the way humans and animals express emotions, i.e. in their facial expressions, posture and the like. The point of Darwin's study was primarily to find additional evidence for the thesis of the common ancestor that he presents in his previous and more famous book *On the origin of species* ([1859] 2008): in his view, the fact that we exhibit common emotional reactions shows that we share a common evolutionary history. However, his finding and above all the approach that he puts forward are also useful for supporting a different line of argument that is relevant for the research on emotion: *our emotional responses—or most probably a relevant sub-set of them—evolved phylogenetically and are therefore innate and universal* (for a discussion of this thesis also in relation to Darwin see, e.g., Ekman 1999, 2015). Humans also certainly have emotions that are much more complex than those exhibited by animals; however, some of their emotions seem to be more *basic* and these might also be shared by animals and indeed have a biological basis.

The idea at the foundation of this line of research is that humans experience two kinds of emotions which are quite different from each other, i.e. basic level emotions and non-basic (i.e. complex or high level) emotions:[21] Non-basic emotions are uniquely human, require high level cognitive resources and are individually and culturally variable; basic emotions are universal, innate and biologically determined. It is e.g. Paul Ekman who suggests that basic emotions are "a number of

[21]The idea that emotional experience is a complex phenomenon in which we can identify some universal, stable, and simple (originally) elements that then become part of more complex and derived emotional episodes has ancient roots. This position is recognizable already in Plato and Aristotle. In the dialogue Philebus, Plato (1983) identifies two kinds of basic principles that regulate individuals' states of mind and condition their behaviors. These are principles such as pleasure and pain. Pleasure is identified with harmony and with the good functioning of body parts, i.e. with a condition free from damage. Pain is a condition in which the natural cooperation among the body parts or their function is compromised. For this reason, we spontaneously tend towards pleasure and avoid pain. Pleasure and pain regulate emotional states such as desire and fear. Desire is the pull toward harmony—toward the reinstatement of harmony or towards a better harmony—and it is therefore the pursuit of pleasure. On the contrary, fear is the aversion to actual or future pain and to the situations that can provoke it. Aristotle's Rhetoric (2010) suggests a similar view, in which pleasure and pain are considered to be components of other passions and thus function as primitives with respect to them. More complex perspectives on emotional life—that present a more structured view of the fundamental emotions—have been developed by philosophers during the modern period. For example, according to Descartes ([1649] 1985) there are six basic passions (wonder, love, hatred, desire, joy and sadness) from whose composition all other emotional phenomena originate. In Baruch Spinoza's view ([1677] 1994) the primary affects are "nothing but these three [desire; joy or pleasure; and sadness or pain], each one generally being called by a different name on account of its varying relations and extrinsic denominations" (III Def of the Affects XLVIII). The psychological research on emotions inherited this philosophical distinction between primitive and derivative emotions. Indeed, many authors in the field of contemporary (including experimental) psychological research suggest that some of the emotions we experience are basic or primitive, in the sense that they are shared by all human beings if not by all superior animals and constitute the fundamental ingredients for forming more complex emotions. Many lists of supposedly basic emotions have been put forward by different studies. A well-known list of ten basic emotions (interest, joy, surprise, anger, contempt, disgust, distress, fear, guilt and shame) was suggested e.g. by Izard (1971) on the basis of the proposal made previously by Tomkins (1962, 1963). A similar and more notorious list of six emotions (happiness, sadness, fear, surprise, anger, and disgust) was put forward by Ekman et al. (1969) and Ekman (1972). In later writings, Ekman (1994, 1999) expands this list and maintains that all emotions are in a way basic in the sense that what is universal is not specific emotions but emotion families, i.e. general emotional states with shared characteristics that can take different specific forms in different contexts: "each emotion family can be considered to constitute a theme or variation. The theme is composed of the characteristics unique to that family, the variations on that theme are the product of the individual differences, and differences in the specific occasion in which an emotion occurs. The themes are the product of evolution, while the variations reflect learning" (Ekman 1994: 19). Other well-known lists of basic emotions were offered by Oatley and Johnson-Laird (1987) and by Panksepp (2000).

separate, discrete, emotional states" that are culturally universal (i.e. they "occur for all members of the species regardless of culture") and innate (i.e. they "evolved for their adaptive value in dealing with fundamental life-tasks") (Ekman 1999: 46; see also Ekman 1992a, b). According to Ekman basic emotions are "affect programs", i.e. specific evolutionarily determined mechanisms that control emotional behavior and determine what stimuli cause what response (see, e.g., Ekman 2003). Basically, affect programs are automatic appraisal mechanisms: they work very fast, unconsciously and unreflectively and they determine how we react to minimally interpreted stimuli. The kind of appraisal carried out by affect programs is very different from the appraisal assumed by cognitive theories of emotions (see Sect. 5): while the latter consists in a conscious and deliberate judgment made on the basis of certain cognitions (i.e. the knowledge available to the subject), automatic appraisal mechanisms are hypothesized to work without any (or with only minimal) cognitive mediation on the basis of an immediate evaluation of certain characteristics of the stimuli (we are innately predisposed to have certain reactions towards stimuli with specific characteristics that are immediately appraised as positive/negative for us; and we learn to have certain other reactions towards stimuli with other specific characteristics that are immediately appraised as positive/negative for us: see, e.g., Ekman 1999: 51ff). The reactions brought about by these mechanisms are distinctive for specific kinds of appraisal and universal: a specific type of appraisal gives rise to a specific reaction and this reaction is the same for all human beings. This reaction includes specific movements/postures, facial expressions, behaviors, physiological changes and subjective experiences. Taken together, these define our emotional experience for basic emotions. Thus, basic emotions are the result of these automatic appraisal mechanisms and are associated with specific patterns of physiological change; they give rise to analogous behavioral reactions and determine a distinctive subjective experience (for an overview of all these aspects see Ekman 1999); they are characterized by universal facial expressions shared by all cultures and by specific postures/bodily movements (see Ekman 1977; for an overview of facial expressions see, e.g., Ekman 1989, 1993).

The distinction between basic and non-basic emotions assumes an important role and becomes widely accepted in emotion research. Still, not all the authors that embrace this distinction agree on a biological interpretation of it. In fact, some researchers maintain that emotions can be considered to be basic only from a psychological and not from a biological point of view: in this sense, emotions are basic when they are not composed and do not derive from others, but are the primitive building blocks of non-basic emotions (see e.g. Frijda 1986).[22] Emotions that are considered to be psychologically basic might be viewed as innate, but they might also be interpreted as the product of acquisition. If fact, if emotions allow us to deal with fundamental life tasks which are more or less constant for the human

[22]For a recent review on this see e.g. Izard (2007). For a critical overview that challenges both these versions of the theory see e.g. Ortony and Turner (1990).

species, the universality of psychologically basic emotions could result from species-constant learning.

Independently of whether basic emotions are interpreted as biologically or psychologically primitive, as innate or learned, the distinction between basic and non-basic emotions brings the attention of research back to the issue of whether there are at least some basic emotions which do not depend on cognition and are therefore identified without the support of knowledge, but perceptually, i.e. on the basis of the physiological changes and the subjective experiences they are related to (including possibly the perceptions that we have of our facial expressions, postural and behavioral reactions).

The possibility that some emotions are basic and thus independent from cognitive factors has also begun to be reconsidered in the field of cognitive psychology. While Ekman primarily studied basic emotions using psychophysiological measures and intercultural comparison of facial expressions related to emotions,[23] the more cognitively oriented research addresses this issue using a verbal approach and relies on the explicit knowledge that people have of their emotions, i.e. beginning from the way in which the emotional lexicon is used and understood. In particular, the aim of these latter studies is to determine whether different people or different cultures tend to conceptualize emotions in wildly diverse ways on the basis of their background or whether at least some emotions are quite stable from both an intersubjective and an intercultural point of view and can therefore be considered to be universal.

One of the main methods adopted by a number of studies carried out to investigate this phenomenon between the fifties and the eighties was the so-called lexical typing of emotions (on this see e.g. Nowlis and Nowlis 1956; Davidtz 1969; Osgood 1969). The researchers asked a large number of people to define a list of emotions on the basis of their experience. The result was mainly a catalogue of the putative characteristics of all the emotions listed. In a second phase, definitions given by various individuals were compared to check for common denominators for each emotion considered. Moreover, the descriptions of the emotions were presented to a second group of people for a new assessment (people could read only the description of the emotions but did not know what emotion each set of features was supposed to describe). The assignment of this second group of participants was to examine each description in the catalogue and to decide which emotion it should be associated with, i.e. to label the emotions described by the various definitions. The researchers aim was to ascertain whether the description of an emotion could be considered intersubjectively stable, i.e. whether new people were able—on the basis of the definition given by others—to understand which emotions the first group had in mind when giving those descriptions.

---

[23]In some cases, people of different cultures were asked to show how their face would look in certain significant emotional contexts: see, e.g., Ekman and Friesen (1971); in some other cases people were shown photographs of facial expressions and were asked to associate each expression with an emotion selected from a list: see Ekman et al. (1969), Izard (1971) and Ekman (1973, 1989).

The results of these studies were in some respects ambiguous. For a large class of emotions, participants had widely incongruent definitions. Moreover, they did not agree on the appropriate description for the various emotions or their classification. When asked to label an emotion starting from a definition proposed by another person, participants proposed a widely divergent classification. However, in spite of the fact that a large number of emotions turned out to be intersubjectively unstable, some emotions—in particular the emotions that studies on facial expressions had singled out as basic emotions—proved to be quite consistent. This provided further evidence in support of the idea that we should distinguish between basic emotions that are quite independent of cognitions and non-basic (complex or high level) emotions that in contrast largely depend on cognitive factors (see e.g. Oatley and Johnson-Laird 1987, 1992; Oatley 1988; Johnson-Laird and Oatley 1989, 1992).

Since basic emotions do not depend on cognition, adopting this distinction between basic and non-basic emotions means that cognitive approaches to emotion are contaminated and combined with perceptual theories. Still, per se this suggestion does not answer the question of how a person identifies and classifies his/her own basic and non-basic emotions. To address this issue we need to detail the way in which first-person recognition of these emotions actually occurs. The most influential position within this line of research is that proposed by Philip Johnson-Laird and Keith Oatley. In their view, emotions are indeed characterized by a "phenomenological tone", i.e. by a perceptual component that corresponds more or less to what we have called subjective and qualitative experience. As Oatley specifies:

> Emotions also have a distinctive phenomenological tone, of which we might be conscious. Each emotion can typically be felt as different from contrasting emotions and from nonemotions: Sadness feels different form happiness and from states like deductive reasoning or sleepiness. Sometimes we may not be consciously aware of an emotion, though others can see signs of it in our behavior. Sometimes emotions seem inchoate, and we do not quite know how we feel (Oatley 1992: 20).

Although phenomenological tone is always part of any emotion, according to these authors it is generally too indeterminate to be used in identifying an emotion, so we recognize our emotions on the basis of the knowledge we have on what caused them and their effects. However, Johnson-Laird and Oatley propose that *for a small class of basic emotions (they mention happiness, sadness, anxiety or fear, anger, and disgust) which "occur universally in the human species" phenomenological tone is the most important if not the sole factor we need in order to identify what emotion we are experiencing*. The reason is that each basic emotion "has a characteristic phenomenological tone—though no meaning as such, as each is based on a non-propositional signal" (Oatley and Johnson-Laird 1987: 33).[24]

[24]To establish which emotions are basic, Oatley and Johnson-Laird (1987) rely on the studies that investigate the facial expression of emotions. In this article, they refer in particular to Ekman et al. ([1972] 1982).

The same does not apply, however, to non-basic emotions. As the classical cognitive theory suggests, they depend on a cognitive evaluation of the situation (which in the case of complex emotions is considered to be propositional) and specifically on an appraisal judgment, i.e. our evaluation of whether and how the situation affects us and our well-being. Moreover, apart from basic emotions which are the product of biological mechanisms, complex emotions are semantically defined by the conditions that typically cause them. Take for example the social emotion called 'jealousy': jealousy is defined (approximately) as the emotion you experience when you think you have a rival who enjoys or could enjoy something you would like to have (be it an object, a position or someone's love). In order to identify and classify the emotion I am experiencing as 'jealousy' I need to understand the situation I am in and I need to know that the elements of this situation conform to the context for jealousy.

In fact, to identify the emotion I am experiencing as jealousy, I must be aware of many things such as how social relationships work, what my position in the group is, what is considered to be worth pursuing, who is competing with me for what, etc. Moreover, I need to have a concept of 'jealousy' that allows me to interpret my emotion in this situation as 'jealousy': i.e. I need to know that what one typically feels when s/he is competing with someone for something is jealousy. Of course, social emotions have a cultural component. We might well hypothesize that some culture would not have this emotion, at least not in the same form as we have it: a culture could e.g. have instead two different labels to express the emotion that we call 'jealousy', one might be specific for amorous jealousy (J1) and another for all other kinds of jealousy (J2). In this case, a person from this culture will identify amorous jealousy and other kinds of jealousy as two different kinds of emotions applying the definition of jealousy defined by her/his culture. In addition, even though people tend to conceptualize emotions in accordance with the definition given by their culture, the understanding that different people have of socially defined emotions is often partially discrepant. Thus, complex emotions may "vary from culture to culture and from person to person" (Oatley and Johnson-Laird 1987: 47; see also Oatley 1992: §2.3). The same does not apply however for basic emotions because they are (biologically programmed) automatic responses that bring about a univocal qualitative experience (phenomenological tone).

The difference between basic and complex emotions is not that complex emotions do not have a phenomenological tone, since all emotions are characterized by a qualitative component. It lies rather in the fact that—while in the case of basic emotions the perception of the phenomenological tone is sufficient for us to identify which emotion we are experiencing—in the case of complex emotions this is not the case. To be identified, complex emotions also require that people understand the situation that gives rise to them in a propositional manner, have some idea about how the situation affects them and know what emotion is considered to be appropriate (e.g. to be jealous one needs to figure out that s/he has a rival—and this requires both a comprehension of the situation and one's own position in the situation—and understanding that the emotion you might have when you are in a condition of rivalry is jealousy). As Oatley and Johnson-Laird specify: "Many adult emotions are complex in the sense that they are founded on a basic, non-propositional, emotional mode, but have a propositional evaluation which is social and includes reference to the model of the self" (Oatley and Johnson-Laird 1987: 46).

If we go back to the terminology we introduced before, this quotation tells us that Oatley and Johnson-Laird's view of emotions tends to be cognitive and therefore inferentialistic and third-personal because it suggests that emotions are identified and classified on the basis of a cognitive evaluation of the situations which caused them and on the basis of the knowledge we have of the social definition of emotion. However, the description they offer of basic emotions puts a lid on this tendency. Indeed, basic emotions are not conceived as individually and culturally variable and as dependent on cognition, but they are presented as semantic primitives, i.e. as something that is immediately recognized from a perceptual point of view by virtue of its intrinsic characteristics and whose classification does not require any cognitive or definitional elements: "each term denoting a basic emotion is primitive in the sense that it is semantically unanalysable. It refers to a phenomenological primitive that one needs to have experienced in order to grasp the meaning of the terms" (Johnson-Laird and Oatley 1992: 209). In these authors' view, basic emotions are comparable to bodily experiences such as pain (Oatley and Johnson-Laird 1990: 136), because both can be conceptualized on the basis of their phenomenological tone. "Like pain, experiences of basic emotions are phenomenological primitives" (Oatley 1992: 79). We know we are in pain because we can experience and classify the quality of that pain; in the same way, we know we are angry because being angry corresponds to a specific phenomenological tone that we can identify and classify. The difference between basic emotions and bodily experiences such as pain is solely that this last, in contrast to emotions, is characterized also by a specific localization on the body (Oatley and Johnson-Laird 1990: 136).

From the point of view of how humans classify their emotions, the distinction between basic and non-basic emotions suggested by Johnson-Laird and Oatley is of crucial relevance with respect to the debate we began this chapter (Sect. 1) with about referential and inferential competence, and whether it is possible to identify a specifically referential component in our semantic competence with respect to words that do not have external reference, or whether our mastery of this kind of words is only inferential and mediated by language.

If we consider that complex emotions largely rely on cognitive factors and are intersubjectively variable, one could conclude that emotions are analogous to the theoretical terms mentioned by Hartman whose content does not depend on perception at all, but is determined inferentially by the definitions we provide (see Harman 1987: 61). Complex emotions as they are described e.g. in the research mentioned above (that compares the definitions and denominations of emotions given by different people revealing that these depend heavily on subjective and cultural factors) appear to be understood—i.e. classified and conceptualized—on the basis of propositional information only. However, the same does not apply to basic emotions. Indeed, basic emotions are extremely stable since they do not depend on some inferential (cognitive or cultural) factor, but are identified on the basis of the specific sensation or feeling that corresponds to them. This sensation or feeling can be considered to be the referential element in their classification which our capacity to 'perceptually' recognize the emotion we are experiencing relies on.

Using Oatley and Johnson-Laird's clarification, emotions can be positioned along a spectrum that goes from largely referentially based to largely inferentially based. Basic emotions are largely referentially based in the sense that they have a perceptual origin and can also be classified (grouped together) in a nonlinguistic fashion on the basis of their perceptual characteristics. In other words, we can use the sensation that corresponds to basic emotions as a referential basis to identify and classify them. The same is not possible, however, in the case of complex emotions, because their identification and classification depends on the specific, explicit and propositional knowledge that we have of them. The greater the complexity of an emotion, the more its identification and classification depend on propositional and inferential knowledge. Again taking Oatley and Johnson-Laird's perspective, emotional terms can also be hierarchically organized on the basis of whether they rely primarily on referential or on inferential information:

> emotional vocabulary should have a systematic organisation. […] We argue that it is structured like other semantic domains. It is not merely a rag bag of terms that are so heterogeneous as to be not worth analysing […] words denoting emotions should reflect the structure of emotional experience. Hence, some words should denote basic emotions, corresponding to the modes that can be experienced without awareness of evaluations on which they are based. Other words should denote emotions that include propositional information about a cause or object of the emotion. Still other words should denote complex emotions deriving from evaluations of the self (Oatley and Johnson-Laird 1990: 132).

The cognitive theory developed by Nico Frijda follows a similar direction, that also relies on the differentiation between basic and non-basic emotions. This theory is in the spirit of cognitivism since it suggests that all emotions must be understood in terms of complex processes that also include the antecedents (the meaning structure of the event—i.e. how the person interprets it) and the consequences of the emotion (the action readiness—peoples' behavioral disposition). As Frijda summarizes:

> Emotions arise in response to the meaning structures of given situations; different emotions arise in response to different meaning structures. […] Emotions change when meanings change. Emotions are changed when events are viewed differently [… and further …] State of action readiness is a central notion in emotion. All emotions—all states, that is, that one would want to call 'emotions'—involve some change in action readiness (Frijda 1988: 349–351).

Despite their cognitivist structure, emotions in Frijda's view cannot be understood without also considering their intrinsic qualitative dimension and are determined by the so-called "autonomic awareness" and "body awareness" that accompany an emotional experience (see Frijda 1986: Chap. 5). Frijda considers that "autonomic awareness"—which reminds us of the conscious feelings of bodily symptoms corresponding to an emotion described by James and Lange—results from the autonomic response of the peripheral nervous system which causes changes to bodily states such as the heart rate, visceral contractions, etc. Autonomic awareness describes the perception we have of all the physiological reactions that occur beyond our control when we experience an emotion (tachycardia, shakes, blushing, dryness of the mouth, etc.). Moreover, the notion of "body awareness" also characterizes the proprioceptive signals corresponding to the physical outputs of the emotions; these keep track e.g. of information on facial feedback (i.e. feelings derived from changes in facial expression due to an emotion), or information on the muscular tension resulting from the felt action urges deriving from the various emotions.

Even though Frijda does not think that the identification and the classification of emotions can be carried out perceptually on the basis of the qualitative component of emotions *alone*—i.e. *solely based on* autonomic and body awareness—he considers these factors to be fundamental for this aim. In his view, the perceptual (or in our terms, propriosenstitive or 'proprioceptive') dimension is crucial for the identification and for the classification of all our emotions—be they basic or complex—because it determines the *intensity* and the *quality* of the feeling corresponding to each emotion. The difference between basic and non-basic emotions is one of degree: the propriosensitive component is predominant in the identification and classification of basic emotions, while in the case of complex emotions, the inferential aspect is of the utmost importance.

In this respect, Frijda's theory is in opposition to the more orthodox cognitivist view of Mandler (1975). According to Mandler, two states such as e.g. anger and fear are both characterized by comparable states of tachycardia; when people experience emotions such as anger or fear, they are only globally aware of their state of physical arousal, but they do not have any specific perception of e.g. the kind of tachycardia they are experiencing. Thus, the perception of their tachycardia is much too generic to allow them to understand whether they are experiencing anger, fear or any other emotion that results in tachycardia. In order to identify and to classify the emotions they are experiencing, people need to make use of the knowledge they have of the situation and their reaction to the situation, i.e. they need to interpret the situation from a cognitive point of view. In this view, this identification would be carried out on the basis of the non-specific arousal together with a conceptual (inferential) interpretation of the emotion. Frijda thinks that such an extreme cognitivist position must be rejected:

> Mandler's conclusion almost certainly is false. There exist a large variety of different emotionally significant body experiences: trembling; sweating; feeling choked; having a lump in one's throat, feeling oppressed; breathing freely; blushing; feeling one's face flush; feeling one's knees shake or teeth chatter; feeling butterflies in the stomach; having a sinking feeling in the pit of one's stomach; and may more. […] Systematic description of body sensation patterns in emotion has hardly begun (Frijda 1986: 228).

In Frijda's view, while bodily sensations are indeed not sufficient to identify and to classify complex emotions, in the case of basic emotions they might be: "emotions do differ in body experience and […] each of the 'fundamental' emotions, at least, corresponds to a distinct pattern." (Frijda 1986: 229; see also Nieuwenhuyse et al. 1987; Frijda 1987, 1988) Indeed, when people experience an emotion, they do not merely perceive a generic state of arousal, but can rather rely on two kinds of quite specific signals coming from the body and from the autonomic system which co-occur in the experience of that emotion and offer convergent parameters for its identification on a perceptual basis. These are the so-called autonomic and body awareness. Moreover, for Frijda these two kinds of information taken together generate a third, irreducible qualitative state. Indeed, *they give rise to a new ingredient of the emotional state called a "hedonic quality"*. The hedonic quality describes the specific sensation (the specific qualia in Nagel's sense) of the emotion which can be perceived by the subject (Frijda 1986: 242ff). It is this quality that determines, among other things, whether the person perceives the emotion s/he is having as positive or negative.

Taking these three parameters together, a subject can individuate in a non-conceptual (non-inferential) way at least the more basic emotions. The definition of three different parameters that jointly contribute to the identification and classification of emotions is intended as an answer to the cognitivist objection that different emotions give rise to the same bodily changes and that therefore the arousal people perceive in correspondence with emotions is nonspecific and cannot be used for their identification or classification.

A different position with analogous implications has also been suggested more recently by Antonio Damasio (on Damasio's view see also Chap. 2, Sects. 5 and 6). He distinguishes between emotions as mental states (ideas and cognitions whose aim is to activate certain responses and actions) and emotional feelings. "While emotions are actions accompanied by ideas and certain modes of thinking, emotional feelings are mostly perceptions of what our bodies do during the emoting, along with perceptions of our state of mind during that same period of time." (Damasio 2010: 117; see also Damasio 1994, 2000). The awareness of bodily feedback such as the perception of changes in our facial expressions, our gestures and postures, or in the viscera and internal milieu that occur in response to an emotion are also part of emotional feelings. The process through which the felt aspect of emotions is generated is termed 'interoception' by Damasio, and—as we specified previously in this work—is included in the notion of 'proprioception' as we re-defined it on the basis of Gibson's view in terms of propriosensitive information (see Chap. 2, Sects. 3 and 4).

Damasio also distinguishes between basic and complex emotions which he calls universal and social respectively. While basic emotions are common to all cultures and rely on the same bodily feelings (emotional feelings and bodily feedback from emotions), complex emotions have a social and cultural component and also depend on other cognitive factors, including education (Damasio 2010: Chap. 5). This distinction is relevant in Damasio's view because it leads to the idea—quite analogous to Frijda's conclusion—that, while feelings are not per se sufficient to identify complex emotions, they might suffice for the identification of basic emotions. As e.g. Jonathan Cole points out:

> Damasio differentiates between emotions—inner mind states—and feelings—their expression in and on the body. Feelings, in turn, might be closely linked to emotional experience. Thus, basic emotions like anger, happiness, and fear, are felt when the face and autonomic body conform to the pattern for such a state. More subtle emotions, wistfulness, euphoria, for example, are tuned by experience and combine central emotional and cognitive states with subtler feelings in the body. Lastly, he suggests, there are bodily feelings of which we may not be aware. In this model, feelings are felt in the body and the face and fed back to the brain, to be elaborated into emotional experience (Cole and Spalding 2009: 193).

Going back to the terminology of this work, both Frijda and Damasio consider feelings—i.e. qualitative bodily information—as sufficient for making a referential identification of at least some of the emotions we are experiencing: the basic ones. Because of this characteristic, basic emotions can be considered as analogous to perceptual objects in terms of the way they are categorized in the sense that both are identified and classified primarily perceptually on the basis of their features (the former are identified on the basis of their proxytypes, the latter on the basis of their internal qualities). Like concepts referring to concrete objects in the external world, so too concepts denoting basic emotions are formed relying on perceptual (proprioceptive) information and the competence we have with the corresponding words is in both cases primarily referential. The inferential knowledge we have or might regarding these concepts is something extra that is added to the core concept.

In this view, basic emotions are analogous to bodily experiences such as the ones we discussed in the previous sections of this work (hunger, pain, itch, etc.). More precisely, emotions can be located on a continuum that goes from basic emotions that are categorized primarily on the basis of how they feel (i.e. their perceptual/propriosensitive characteristics) to emotions of growing complexity that are categorized widely or primarily on the basis of inferential knowledge. This position should not be mistaken for the (quite trivial) view that emotional states are —like bodily experiences—something we feel. In fact, this suggests something more precise and less predictable, i.e. that *words denoting bodily experiences as well as some (basic) emotional states are learned and semantically mastered like terms denoting observable objects in the external world and that the way we categorize bodily experiences, basic emotions and observable objects in the external world is largely analogous.*

The hypothesis we are defending here is that—like observable objects in the external world and bodily experiences—basic emotions are also identified on a primarily perceptual basis. According to this view, our competence with respect to words referring to basic emotions is grounded in a referential competence which is relatively independent of the inferential competence we have for them and plays the predominant role with respect to their semantic mastery. Both basic emotions and bodily experiences such as pain, itch, and hunger are classified on the basis of internal information detected by our propriosensitive system, i.e. with a specific qualitative feel or phenomenal experience. The difference between them lies in the kind of feel they rely on. Indeed, the phenomenal experience that leads to the identification of emotions has a different quality with respect to that of bodily experiences. The phenomenal character—the *what-it-is-like*—of emotions is not undetermined, but is characterized by a specific quality that allows us not only to distinguish among various emotions but also between emotions and bodily experiences.

As for complex emotions, their identification is also mainly based on inferential factors related to the knowledge we have about the 'context' of the emotion. For this reason, even though complex emotions are also characterized by a specific phenomenal experience, this is not enough to identify them—for recognition and classification, specific inferential information about the emotion is needed. Like all kinds of terms that rely primarily on their inferential component, complex emotions also vary widely between different people and different cultures. Indeed, various people or various cultures might define these emotions differently relying on disparate inferential elements; i.e. on different characterizations of what can cause an emotion and of what consequences should follow from it. Thus, while in the case of basic emotions the stability of the emotional concepts/terms is guaranteed by the referential (perceptual) component—which is common for all human beings—in the case of complex emotions the emotional feeling is indeed not specific enough and various people or cultures might develop different emotional concepts or define a different emotion vocabulary.

7 Emotional Concepts and Their Referential Component: How Feelings Become Schemata

The thesis that we sketched in the previous section and that we are going to discuss in detail more here is that the mastery of terms referring to emotions is also characterized by a referential component of a qualitative character that originates from proprioceptive monitoring of the internal states of the body (see also Chap. 2). In the view we are suggesting, all emotions also have this referential component (i.e. rely on propriosensitive information). However, only in the case of basic emotions is the proprioceptive information predominant for the concept in the sense that it is sufficient to produce a perceptual identification of the emotional feeling. In the same way that perceiving a cat is sufficient to generate a perceptual schema that allows one to perceptually identify cats, the perception (i.e. 'proprioception') of the specific qualitative feeling of a basic emotion is sufficient to generate a schema composed of propriosensitive, qualitatively characterized information that allows one to perceptually identify this emotion. If so, even a creature that does not have any language or propositional knowledge could identify from a non-linguistic point of view the basic emotions it experiences distinguishing e.g. between anger, happiness and fear. In the view we are developing here, *the referential information that supports the conceptual identification and classification of emotions and therefore also our mastery of terms referring to emotions consists in the perception (i.e. 'proprioception') of bodily states and changes of bodily states that occur during an emotional episode as well as the qualitative experience the processing of these states brings about.*[25] In this section we will try to say something more precise on all these aspects.

The first question we will address is: *why do we need to assume that our capacity to conceptually identify and classify emotions and to master emotion terms relies also on a referential component*? To deal with this question, it might be useful to start with something like a mental experiment. Try to imagine that you are watching your neighbor from the window approaching your house and empting a large bag of trash just in front of your door. Imagine (it cannot be difficult!) that you get really angry about this. How do you classify what you are experiencing as 'anger'? On the one hand, it is possible that the identification of the state you experience as 'anger' is carried out relying on the specific knowledge/beliefs that you have about the situation: you believe e.g. that trash is dirty, disgusting and it is not desirable to have it in front of your door or that to empty a bag of trash in front of somebody else's door is a deeply offensive, hostile and rude gesture. On the other hand, the identification of the state you are experiencing could be made on the basis of what you feel when the emotions take place, i.e. when the emotion occurs.

To understand the possible role that these factors might play we need to ask first of all whether we could identify the state we are experiencing as 'anger' even if we did not have any knowledge of the reasons why we are angry (i.e. the fact that trash is dirty, disgusting and it is not desirable to have it in front of your door and that to empty a bag of trash in front of somebody else's door is a deeply offensive, hostile and rude gesture). In the view we are arguing for here, our beliefs on the situation are relevant with respect to the emotion we experience in a twofold way: because they cause the feeling of anger and because they support the interpretation of those feelings as anger. Indeed, the (cultural) belief that some actions are extremely rude and hostile and that it is seriously offensive for us to be victim of such actions is what triggers the emotion (we would not be angry if we instead considered the action to be normal) and this is what helps us to understand our feelings in a propositional manner. However, our first-person knowledge of the fact that we are angry does not rely primarily on these beliefs.

This aspect becomes immediately clear if we turn the situation around. If a person is well-aware that trash is dirty, disgusting and it is not desirable to have it in front of your door and that to empty a bag of trash in front of somebody else's door is a deeply offensive, hostile and rude gesture and has therefore all the knowledge

[25]Some authors suggest that an essential part of our emotion experience consists in the feedback we have from the changes in our facial expression that follow the emotion, e.g. from information about the facial muscles, the tongue, changes in blood flow and temperature etc. (For a strong position on this according to which our emotion experience coincides with our awareness of changes in facial expression see Tomkins 1962; Izard 1977). Even though the awareness people have of the changes in their facial expression is plausibly part of the emotion experience, this cannot be the only or even the main factor in the emotion feeling. Indeed, even people who do not have the ability to make facial expressions (because of a stroke or because they have Moebius's syndrome whose main symptom is facial paralysis), have proven to be capable of feeling emotions normally (see, e.g., Ross and Mesulam 1979; Cole 1997).

s/he needs to get angry when s/he sees the neighbor empting a bag of trash in front of his/her door, but s/he has no physical reaction to this act and no qualitative feeling corresponding to anger, could we still say that s/he is angry? Moreover, could s/he know that s/he is angry?

In a case like this, we could not maintain that this person *feels* angry, since s/he has no feelings (*qualitative experience*) of this kind. As for the issue of whether we could say that s/he knows s/he is angry, at most we could say that this person might know in a purely inferential and intellectual manner that the right feeling s/he to have in this situation is anger. However, this would be a cold and unnatural form of knowledge that does not correspond to the way s/he feels and that is very different from the way in which humans usually 'apprehend' being angry.[26] Thus, if we ask with relation to emotions the same questions raised by Goldman with respect to bodily experiences (i.e. how do I classify my headache?), the answer would be similar, even though probably not identical, in the sense that also in this case the identification cannot be purely functional. We also need a direct, first-person identification based on the intrinsic properties of the emotional states, i.e. their specific feeling.

This kind of view can be supported and further specified using the theory put forward by Lambie and Marcel (2002) which—in spite of still having the spirit of a cognitive approach—aims to develop a first-person account of emotion categorization based on what emotions feel like. According to Lambie and Marcel's interpretation of the available data on emotion experience, each emotion corresponds to a different experience and can be at least in part differentiated on the basis of the phenomenology of this experience alone: "to be in an emotion state *is* to be in a particular phenomenological state, because emotion states are personal-level attitudes that themselves are essentially something it is like to be in or to have" (Lambie and Marcel 2002: 220; see also 233). Lambie and Marcel acknowledge that emotion experience consists of several aspects: they especially consider arousal, action readiness, appraisal, and emotional world (i.e. how an emotion makes us perceive the world and the people it is directed towards: e.g. sad experiences make the world outside appear as empty-depressing-boring, love experiences make the person we love appear beautiful and attractive; anger experiences make the people we are angry with appear hateful and annoying etc.; Lambie and Marcel 2002: 223). However, they also underline that our experience of an emotion is most often holistic and unanalyzable, i.e. that it is a global experience that cannot be broken down in components.

> […] there is evidence that people often experience emotion categorically; that is, they can be aware just of anger or sadness, pure and simple (in nonlinguistic terms). […] It might be argued that such categories are not experiential but only linguistic or that the experience is only categorical when the person is asked to categorize it. We reject these arguments because people often unhesitatingly and without reflection know what emotion they are undergoing (even if they are mistaken) and because they can often do so with no awareness

[26]On this 'detached' and 'unnatural' kind of knowledge see also the next section on Asperger's syndrome.

of the cause (Lambie and Marcel 2002: 224; on this topic see also Oatley and Duncan 1992).

This pure and simple awareness of an emotion consists in the perception of the qualitative feel of the emotion. This feel is not one component of the emotional experience among others such as arousal, action readiness, appraisal, and emotional world. A specific arousal, action readiness, appraisal, and emotional world might rather cause a certain emotional qualitative feel, however this feel is *all that we experience of an emotion*.

> […] the distinctive qualitative feel may be precisely an experience of a particular combination of arousal, action readiness, emotional world and appraisal, and in this sense it is not unanalyzable, even if experienced categorically (Lambie and Marcel 2002: 224).

Thus, the qualitative feel might result from different components of emotions, but it is still the element people rely on to know that they are having an emotion and to classify what emotion they are experiencing. From this point of view, according to Marcel and Lambie, it is a big mistake to identify emotions with the underlying physiological states because this would entail mixing up the causes and effects of emotion.

> Sometimes the question of the *exact content* of experience gets confused with that of *what underlies* the experience. Such a confusion is exemplified by Zajonc and McIntosh (1992). They ask 'What is it that we feel when we feel happy, sad, or disgusted?' (p. 70) and reply that it is changes in brain blood temperature (in the hypothalamus) caused by facial action. In this case, the question seems to be the one about experienced content, but the answer given is appropriate to the question of what underlies it. Although it may be fruitful to investigate how hypothalamic temperature facilitates the release of certain neurotransmitters and how such release is associated with changes in emotion experience, such an investigation can never provide the answer to the question 'What is it that we feel when we feel happy, sad, or disgusted?' Our own brain temperature, qua brain temperature is simply not one of the possible contents of consciousness (we are not aware of our brain states as such), and any answer to the question of what we are aware of when we experience emotion must be in terms of the content of consciousness. Zajonc and McIntosh (1992) may mean that what we are aware of are 'feelings' of emotion, and that these feelings are 'in reality' *equivalent to* perceptions of brain blood temperature, but this answer again is inappropriate to the question of the content of experience. Answers to this question must be in terms of the description under which the awareness occurs, and on this interpretation of Zajonc and McIntosh's claim, this description might be, for example, feeling hot or flushed, having 'anger' or 'anger feeling,' but not 'high brain temperature' (Lambie and Marcel 2002: 227).

In short: the issue of how people can identify and classify their emotions can be approached only starting from the emotion experience as such, while there is no nonconscious informational or neurophysiological states of a purely functional kind that can be relevant to address this problem. In their analysis, Marcel and Lambie try to address this issue more closely by differentiating between "emotion states" and "emotion experiences". Emotion states and emotion experiences both indicate kinds of feelings that we consciously experience. However, they describe respectively first-order and second-order emotion awareness. The emotion state is the feeling we have of the bodily states and changes that occur when we are having an emotion including those that occur as a consequence of the emotion (change in facial expression: the tension of the muscles in the face, the temperature of the face, etc., see Lambie and Marcel 2002: 229). By contrast, the emotion experience is a "second-order introspective awareness or appreciation of one's emotions." (Lambie and Marcel 2002: 219) To put it differently: while the emotion state is of a lower lever and it constitutes so to speak the rough proprioceptive material of the actual emotion experience, the emotion experience describes the fully developed and integrated emotion we have and are aware of when we reflect on what we are experiencing. Emotion states (the rough feeling in the absence of attentional focus) and emotion experience (the second-order emotion experience we have when we pay attention to our feelings and reflect on them) correspond to two separate stages of emotion processing: first order emotion states are not only independent from second order emotion experiences; they are rather a necessary precondition for their development.

In this view, our emotion experiences are related to reflection and depend on/are influenced by attention (on the function of attention with respect to proprioception see also below in this section as well as Chap. 2, Sects. 3–5 and this Sect. 3). Indeed, paying attention to what we are experiencing allows for awareness of the experience and allows us to reflect on it. Attention also influences the quality of the experience we have (in a way, paying attention to our pain creates the pain, even though it was there even before we directed our attention to it). Thus, while emotion states are first-order phenomenal feelings that remain unattended in the background of our experience, emotion experiences describe a kind of feeling that has reached our awareness and which we can reflect on.

The fact that only attention can make us fully aware (reflexively conscious) of our emotions does not imply that the emotional feeling depends entirely on attention. The phenomenology (the rough feel) of the emotion is not created by attention and reflection; rather, it is there before and independently of whether our attentional focus is directed to it: even if we do not pay attention to what we feel, we still experience some feelings in a rather vague and rough manner. However, until we begin to pay attention to them—until we focus on them and reflect on what they are—these feelings stay in the background of our consciousness. The phenomenology of the experience (its feeling) preexists with respect to the attentional focus, but attention is essential for it to reach our full awareness (Lambie and Marcel 2002: 228). As Lambie and Marcel further specify:

> It is true that the way in which we attend to our bodies or to anything else affects the hedonic quality of what is attended, and it also affects other aspects of its phenomenology, such as the way in which it appears to us, how much it matters to us, our sense of ownership of it, and so on. However, not only is it implausible to assume that attention brings into being all aspects of phenomenal experience, there is evidence that it does not (Marcel 1993); that is, there is evidence that the existence of the phenomenology is independent of attention and of awareness, although the precise nature of the phenomenology is not independent of attention […] (Lambie and Marcel 2002: 228).

Going back to the question of whether we could be angry without experiencing any anger, Marcel and Lambie's answer to it is decisively negative. Emotions in humans "do not exist as processes or states independently of phenomenology" (Lambie and Marcel 2002: 220):

> […] to be in an emotion state is to be in a particular phenomenological state, because emotion states are personal-level attitudes that themselves are essentially something it is like to be in or to have. In this sense, quite apart from second-order awareness or interpretation, it would be bizarre to talk of most human emotions, at least for the nonpathological and paradigmatic cases, as separate from phenomenal experience (Lambie and Marcel 2002: 220).

This analysis of emotional feelings and the distinction between emotion states and emotion experiences helps us design an explanation in line with the view we have been presenting on how the categorization of emotions might actually be carried out. Emotion states represent the conditions of possibility for having emotion experiences and therefore the ability to identify and classify them. Like bodily experiences (e.g. pain, itch, thirst, etc.), emotions are also detected through a continuous below awareness mapping of our bodily states and their changes. The result of this mapping is always in the background of our experience without us noticing it. The proprioceptive information we get through this mapping might be compared to the perceptual information we constantly gain from monitoring the external world. We cannot be aware of all the stimuli we perceive though the senses, however, they are the rough perceptual material starting from which perceptual objects are formed. We experience an object, when it becomes salient to us. In the same way, we experience an emotion when changes in bodily states exceed a certain critical threshold and attract our attention.

People are not aware of the individual bodily changes that occur in their body in correspondence to an emotion, what they perceive and are aware of is the emotional experience (in Marcel and Lambie's sense) meant as a unitary and integrated qualitative sensation which originates from the bodily states. Any type of emotion corresponds to a specific pattern of bodily changes that gives rise to a specific sensation of a qualitative kind which is known by acquaintance through proprioceptive awareness. As in the case of internal states, we can hypothesize that people can focus their attention on a single component of these bodily states making that component salient, i.e. they can voluntarily redirect their attention to a single aspect of bodily changes breaking down the original unity of the emotion experience and becoming aware of the quality of this aspect alone. However, what we usually experience is the quality originating from the bodily changes as an ensemble.

Since "the distinctive qualitative feel [of any kind of emotion] may be precisely an experience of a particular combination of arousal, action readiness, emotional world and appraisal" (Lambie and Marcel 2002: 224), Marcel and Lambie claim that all emotions correspond to a specific and unique qualitative experience by virtue of which they can be identified and categorized. In this respect our position diverges from theirs and goes back to that of authors—such as e.g. Frijda, Oatley and Johnson-Laird—who suggested that we need to differentiate between basic and non-basic emotions. Indeed, we agree with these authors that only in the case of certain, particularly simple emotions can identification and classification occur on the basis of the mere qualitative experience. As for complex emotions, the feeling is too indeterminate to be the only element we rely on for emotion recognition. Indeed, in the case of complex emotions, ingredients such as arousal, action readiness, emotional world and appraisal (which in this case would include also the propositional interpretation of the situation, how it will affect us, and how we might cope with it) are separate from the qualitative feel and they contribute to determining the emotion we are experiencing and to making us aware that we are experiencing one emotion instead of another.

When we described categorization in the case of concepts that do not refer to internal experience (see Chap. 3), we pointed out that some of them are primarily referential in the sense that they rely mostly on perceptual information and on perceptual schemata ('cat' would be an example of this kind of concept), while others are primarily inferential in the sense that they are definitional constructs with at most a poor referential basis (e.g. 'quark' would be on the other extreme of the spectrum with respect to 'cat'). In our view, this same distinction applies to emotion words. Basic emotions are considered to be mainly referential in the sense that the mere proprioceptive experience we have of their feel would be sufficient to categorize them: a propositional interpretation of the situation is not needed for the recognition of this kind of emotion. On the other hand, complex emotions are socially defined emotions; to be identified they require a propositional understanding of the ongoing situation.

Thus, even though all kinds of emotions have a referential component and none of them is purely or almost purely inferential such as e.g. 'quark'—i.e. all kinds of emotions correspond to an emotional experience, otherwise they would be only cold thoughts, not emotions in the proper sense—the categorization of complex emotion necessitates third-person, inferential knowledge. Going back to the differentiation we proposed between a perceptual core concept and an inferential (propositional) periphery, we could say that the core concept of all kinds of emotions is given by the emotion experience, while the periphery of the concept is formed by the propositional knowledge we have about the emotion at issue. While in the case of basic emotions the core concept is predominant, in the case of complex emotions the core concepts alone would not be sufficient to support categorization.

Previously (see Sect. 3) when we considered bodily experiences such as pain, hunger and itch, we proposed a hypothesis regarding how referential identification might occur—i.e. what kind of proprioceptive information the core concept might be composed of and how this could be used to categorize such states. Here we will do the same for emotions.

In the view we are considering, bodily experiences and emotions rely ultimately on the same perceptual information, i.e. on bodily sensations associated with a specific qualitative feel which we are directly aware of in the first-person and which are due to specific bodily states and bodily changes (musculoskeletal, visceral, internal milieu, affected by autonomic, limbic, and hormonal systems etc.). This shared referential basis suggests that bodily experiences and emotions are referentially identified and classified starting from the same criteria/properties. For this reason, it can be speculated that they also rely on (at least) the following characteristics of the emotional experience:

(i) specific qualitative sensation;
(ii) somatic localization;
(iii) specific intensity;
(iv) presence/absence of specific concomitant sensations.

In our view, the referential component of emotion concepts (the core concept) consists—as in the case of concepts pertaining to bodily experiences—in the intersection of these four kinds of properties which constitute the schemata that allows one to perceptually identify an emotion. Thus, our referential mastery of words denoting emotions also relies—like the referential mastery of words denoting bodily experiences—on the same intersection of properties. To specify this aspect, let's consider the case of anger again.

(i) Firstly, anger is characterized by a specific qualitative sensation, i.e. a specific what-it-is-like associated among other things with tension and warmth.
(ii) Secondly, anger has a specific localization; even though it is quite widespread and difficult to determine precisely, it could include, among other locations, the muscles, the face, the jaw, the arms and hands and the diaphragm.
(iii) Thirdly, anger is qualified by a specific intensity which also depends on how compelling and serious the situation may be. Intensity allows us among other things to classify the emotion we are experiencing as 'anger' rather than 'fury' or mere 'irritation'.
(iv) Finally, anger is characterized by the presence or absence of specific concomitant sensations such as e.g. the impulse to move as opposed to the sensation of paralysis due to tension.

As we previously suggested in relation to fear, concomitant sensations support, among other things, different classifications of a state that might have diverse names across the various languages. In English, we distinguish for example between 'anger' and 'fury'. The latter is not only more intense than anger, but it is also characterized by an irrepressible impulse to move, which does not necessarily accompany anger or at least not as strongly. Concomitant sensations can also help distinguish among different 'shades' of the same emotion that are not described by specific linguistic labels. For example, it is possible to experience different forms of sadness that are accompanied by stronger or weaker sensations of tiredness and fatigue. Probably a strong concomitant sensation of tiredness and fatigue makes sadness more similar to depression.

As in the case of bodily experiences, also in that of emotional experience the specific quality of the sensation has a predominant importance, at least in the sense that it includes and it also integrates the characteristics described in (ii)–(iii)–(iv). Indeed, what we are aware of in an emotional experience is first of all only the *qualitative feel* as a *gestaltic unity* that also incorporates other aspects. As Lambie and Marcel point out:

> [Emotion experience] consists in the immediate phenomenology of one's physicality and bodily relation to the world […] there are good reasons why this kind of experience is rarely reported (at least faithfully) or reportable. It is often said to be ineffable. It has Gestalt properties and is that aspect of emotion experience that people describe using what seem to be metaphors: 'coiled like a spring', 'ready to snap', 'floating on air', 'an inviting-empty-overwhelming world' (Lambie and Marcel 2002: 224; and also 234 on the 'Gestalt properties').

In spite of its gestaltic, unitary character, emotional experience can—through attentional effort—be decomposed and separated, at least in part, from the properties described in (ii)–(iii)–(iv). In fact, if we drive our attention to an emotion, we can individuate some features of the bodily states associated with it, in particular, the sensations related to the localization of the bodily changes that correspond to the emotion (i.e. the body parts where these changes occur), their specific intensity and the concomitant sensations.

The relevance of attentional focus for the distinction and grasping of specific aspects of any aware experience is widely acknowledged in the literature (see e.g. Prinz 2004a: 209ff; this is also discussed in Marcel 1993; Lambie and Marcel 2002: 228). Even though a specific feel might exist before and independently of the fact that we pay attention to it, when we do so, it changes: it suddenly reaches our awareness (in the sense that it moves from the background to the foreground) and its quality also changes, in the sense that we become aware of its specific locations, intensity and whether it is actually a composite feeling consisting of different concomitant sensations. From this point of view, the overall qualitative aspect of emotions that we perceive can represent something like an 'attentional attractor' which allows people to gain access to further elements of the sensation and to comprehend its mereological composition. In other words, when we pay attention to the emotion experience, we can catch its mereological composition and identify some of its aspects such as its localization, intensity and concomitant sensations. This does not mean that localization, intensity and concomitant sensations exist independently of the qualitative sensations, but rather that—when we focus on the quality of the sensation we are experiencing and analyze it—we can separately identify and describe its parts.

The fact that people can grasp the mereological composition of the states they experience makes their experience of these states more detailed and specific. By virtue of this specificity and degree of detail, this experience turns out to also be a fairly reliable perceptual (i.e. propriosensitive) basis for the development of concepts denoting internal states (bodily experiences as well as emotions). For this reason, it can also support the referential mastery of terms denoting emotions or internal experiences. From this point of view, the classical problem raised by Wittgenstein of how it is possible to safeguard the intersubjectivity of the meanings of words referring to sensations (Wittgenstein [1953] 1986: §244) can be addressed while bypassing extrinsic criteria like the definition of these words by relying on the fact that human beings have similar sensations and that the qualitative element of these sensations is sufficiently detailed and specific to assure a certain intersubjective and even intercultural stability in their identification and classification.

It is because of this qualitative element that people can also differentiate between various kinds of internal states and recognize e.g. that a certain state they are experiencing is a bodily experience rather than an emotion. In line with Goldman's position, we have also argued that the specific quality of the sensation functions as an intrinsic property, i.e. as a referent for a specific kind of state. This is again analogous to external perception: at least in nonpathological cases, people would never mix up the sound made by an object with its visual image because auditory and visual sensations are characterized by unequivocally different qualities. In the same way, the internal states that we perceive through the self-monitoring system also have type-specific qualities: the qualitative experience corresponding to bodily experiences is different from that corresponding to emotions. Thus, it is the intrinsic quality of the states that allows us not only to have referential competence with respect to concepts and words referring to internal states, but also to differentiate among different kinds of internal states and to distinguish for example between bodily experiences and emotions.

8 Some Clinical Evidence on Impaired Mastery of Emotion Terms (I): The Alexithymic Condition

In this chapter, we have tried to highlight the similarity between concepts and words that refer to objects in the external world and concepts and words that refer to internal states. More specifically, we have argued that—just like concepts that refer to objects in the external world—concepts that refer to internal states include a referential component, i.e. a perceptual core concept which supports identification and classification of the corresponding objects. For this reason, semantic mastery of words denoting internal states, just like semantic mastery of words denoting objects in the external world, cannot be based solely on inferential competence, but must also be supported by specifically referential competence. This referential component relies on what we have called 'proprioceptive' or propriosensitive information and consists in the feeling that people experience when they are in the relevant internal state, i.e. the qualitative, aware experience that corresponds to that state. This qualitative experience is due to changes in bodily states and is detected by the constant internal monitoring of information that is generated by our *propriosentitive perceptual system* (which we have described by analogy with the perceptual system (see Chap. 2, Sects. 3–6 and this Sects. 2, 3 and 6).

In the last sections, we specifically addressed the issue of emotion concepts/terms and we suggested that the capacity to conceptualize emotions and to master the corresponding words always requires a referential component, although complex emotions may also require additional information. This 'referential component' consists in the ability to perceptually identify and classify the kinds of feeling that correspond to and qualify each emotion, i.e. the qualitative propriosensitive experience that corresponds to an emotion. The question we raise here is whether our semantic competence with respect to words referring to emotions would be somehow impaired (and how) if a person was unable to detect or to process *propriosensitive information* and thus develop a perceptual core concept to support the categorization of emotions. To address this question, we consider a particular condition called *alexithymia*. In this section, we will primarily examine the general characteristics of this disease. In the following section, we consider how an analysis of this condition might provide information that is essential for a deeper understanding of the relationship between propriosensitive information and semantic competence for emotion.

What is *alexithymia*? Alexithymia is generally defined as a subjective inability to describe emotional experiences in words, and characteristically exhibits four main features: "(1) difficulty identifying and describing subjective feelings; (2) trouble differentiating between feelings and the physical sensations of emotional arousal, (3) limited imaginative processes, and (4) an externally-oriented cognitive style" (Timoney and Holder 2013: 1). This peculiar condition is mostly considered a disease "related to a deficit in the verbal processing of emotional information" (Vermeulen et al. 2006: 66). In fact, its most well-known and debated trait consists in this unique and peculiar linguistic impairment: as e.g. Kym MacLaren explains, alexithymic individuals "are typically unable to give expression to any emotional experience. They can describe the facts of the situation, and they will often ask what 'one' ought to do in such a situation, but they make no reference to how they, themselves, feel about it […] an alexithymic may reason her way to a conclusion concerning her emotional state, using the physical sensations within her body and the behavior of others as premises for this deduction […] the emotion […] is not something expressed in the first person, but rather hypothesized through a third person process of reasoning" (Mac Laren 2006: 140).

The term *alexithymia* was introduced to the lexicon of psychology by Peter Sifneos in the 1970s (Sifneos 1972); it is a neologism derived from ancient Greek which literally means "*no words for emotion*" or "*emotion without words*" (*a = privation/lexis = expression through words/thymos = emotions, feelings or moods*, see Lesser 1981). Sifneos used this word to describe certain specific personality traits that he thought might be the cause of psychosomatic disorders (Sifneos 1973; Nemiah et al. 1976). Using this notion, he intended to develop a single psychological construct to subsume a number of scattered clinical observations on psychosomatic patients that had not previously been considered in relation to each other (von Rad 1983; Gerhards 1988; Krystal 1993; Taylor et al. 1997).[27]

[27]Before Sifneos' research was published, the various symptoms exhibited by psychosomatic patients were accounted for using orthodox psychoanalytic theory. As we pointed out previously, these symptoms were not traced back to a common explanation. The fact that patients found it difficult to describe their own emotions was interpreted as a sign of a psychic conflict originating in mother-infant interactions which led to an interruption in mental development. As a consequence, these individuals had developed an 'infantile personality' (Ruesch 1948) characterized by a pronounced incapacity to access their own internal states and thus to conceptualize and communicate

Sifneos paid particular attention to the link between the reduced or impaired ability to express felt emotions in words exhibited by psychosomatic patients (Nemiah and Sifneos 1970) and other characteristics of their behavior such as a reduced capacity to use their imagination, an uncertain comprehension of words that denote affective and emotional states, and a limited or missing capacity to understand or to actively use metaphors or figurative language to describe affective states. Moreover, these psychosomatic patients also exhibited other specific, related behavioral traits: their communication style appeared elementary, arid and flat; their descriptions of emotions or emotional situations were often limited to descriptions of related external, observable behaviors as if their interior life was 'empty' and devoid of qualitative, subjective and first-personal features (von Rad 1983; Timoney and Holder 2013).

A closer examination of the constellation of symptoms that lead Sifneos to introduce the construct of alexithymia helps us better understand this condition. His analysis began with the various clinical and phenomenological observations reported for patients suffering from 'classical' psychosomatic pathologies such us ulcerative colitis, peptic ulcer disease, asthma, tachycardia, etc. which were often treated using the psychoanalytic method. In fact, during the first half of the last century there was broad consensus on the idea that psychosomatic symptoms constituted a particular class of neurotic manifestations. From this perspective, which relied on orthodox psychoanalytic theory, it was considered that the somatic channel (the manifestation of bodily symptoms) was the main channel through which psychical conflicts that could not be represented and communicated in an explicit and aware manner found expression. In his clinical practice, Sifneos noticed

(Footnote 27 continued)

them appropriately. The infantile personality was also considered the reason why these patients were so heavily dependent on others and had a strong tendency towards imitation and symbiosis (von Rad 1983). People with an infantile personality are not capable of grasping and regulating their own emotional states: as is the case in the syntonic and symbiotic mother–infant interaction, these states must be identified, interpreted and regulated by something or someone external to them. In this sense, the word 'infantile' used to describe the 'infantile personality' was interpreted literally on the basis of the Latin term 'infans', which indicates someone who is not yet capable of speaking and thus cannot autonomously express his/her needs (McDougall 1991: 161). Still, the way psychosomatic patients used language was not only an object of interest for psychoanalytic theory, but also drew the attention in neurophysiological research. Paul MacLean attempted to provide a neurophysiological explanation for the deficits these individuals showed in verbally expressing their emotions. He suggested that such difficulties were due to the compromised transmission of nervous signals related to the somatic aspects of the emotional experience from the ("visceral brain") toward the neocortex ("language brain") (MacLean 1949). However, MacLean's hypothesis could not explain all the complex phenomenology of the linguistic behavior of people suffering from psychosomatic disorders and, in particular, the fact that often they reported generic bodily feelings as if they were their entire and only emotional experience. This last characteristic was the main object of Pierre Marty and Martin de M'Uzan's research, and they interpreted it in terms of 'operatory thought' ('pensée operatoire'), an outward-looking way of thinking that gives rise to a description of affective and emotional experiences which relies on external observable elements (Marty and de M'Uzan 1963): the description of emotional episodes is replaced by a description of the actions, events and situations in which the emotion occurred.

that these patients were not able to describe their emotional states; at best, they would describe such states using very elementary categories such as "happy"/ "unhappy", "sad", "nervous", "frightened" (Sifneos 1967; Nemiah et al. 1976: 431). They were unable to describe emotional content in an adequate and salient manner and often could not answer very simple questions concerning these states (e.g. '*what does it feel like to be sad?*'); frequently it seemed they had not even understood what the word 'feeling' meant (Sifneos 1967). Sifneos also remarked that the psychosomatic patients who exhibited these difficulties in expressing their feelings also manifested specific difficulties in organizing and interpreting their body feelings. In fact, as he and his colleagues noticed, they were "often unable to localize affects in their bodies and appeared unaware of any of the common automatic somatic reactions that accompany the experience of a variety of feelings. If there is a somatic component, it is identical with the symptom of their bodily illness" (Nemiah et al. 1976: 431). Moreover, some of these patients also presented a further linguistic impairment: in particular, they "displayed an undeveloped (or impaired) ability to use the personal pronoun 'I' in emotionally meaningful contexts" (Shands 1958: 501). This suggests that their incapacity to identify emotions and the link between these emotions and the bodily feelings they experienced was accompanied by a personal identity disorder due to the fact that they were not continuously aware of themselves, their bodily feelings and their experience (von Rad 1983: 16, 87).

Sifneos and colleagues' observations were fairly congruent with others made on similar patients by other researchers, working independently, who described such patients as "*emotional illiterates*" with "little awareness of their emotional life" (Freedman and Sweet 1954), who perceive their bodily experience in a detached and mechanical manner: they portray themselves as something like a marionette that—especially in stressful situations—acts in a mostly automatic manner, and whose behavior is not guided by any qualitative awareness of their internal states.[28] To better understand this aspect of the condition, we turn to Henry Krystal's *Integration and Self-healing: Affect, Trauma, Alexithymia* (Krystal 1993), where he presents a systematic reconstruction of the genesis of the alexithymia construct and summarizes the results of a number of relevant studies on psychosomatic patients conducted from the 1960s to the 1980s. In particular, Krystal describes in a very incisive manner why psychosomatic patients have difficulties verbalizing their affects and interpreting their bodily signals and how these difficulties are related to each other and to an anomalous subjective experience.

According to Krystal, although alexithymics are not capable of verbally describing their emotions or even identifying their emotions in a non-verbal way, *they still seem to perceive that something is happening to them in the form of bodily sensations that appear to be 'out of control', dysregulated and not integrated*:

[28]This feature is central one to the definition of the so-called 'Pinocchio Syndrome'; see Sellschopp-Rüpper and von Rad (1977) and von Rad (1983).

> Their reactions are basically somatic consisting in the […] physiological aspects of affects, with minimal verbalization. In addition, their emotions are often undifferentiated; they are vague and unspecific […] The emotions are not experienced as distinct, separate, specific responses of a varied and identifiable nature. Rather, a common distress pattern develops, so that the patients refer to a state of tension or unease [… They also have a diminished capacity to] localize affects in their body and they appear unaware of any of the common automatic somatic sensations that accompany the experience of a variety of feelings (Krystal 1993: 241–243).

In addition to being unspecific, the sensations they have in response to events that typically trigger emotions are also characterized by an extremely low degree of reflexivity: alexithymics "often cannot tell whether they are sad, tired, hungry, or ill" (Krystal 1993: 241). The only sensation they are aware of and are able to grasp and verbalize at least to a certain extent is the physical arousal that usually corresponds to specific affective or emotional events or episodes. However, they are entirely unaware of the qualitative dimension of emotions. They tend to consider themselves and others as mechanical entities—as robots—without subjective qualities. Thus, they show a marked *anhedonia* often associated with a marked *aprosodia*. This lack of reflexivity due to a lack of awareness of the qualitative aspects of their internal life makes them incapable of enjoying gratifying experiences or emotional experiences such as love, happiness and similar feelings. Furthermore, in Krystal's view the emptiness of their emotional language is due to an "absence of human quality" (Krystal 1993: 247). This is often accompanied by a flat and monotonous expository style without rhythm, vitality and participation that makes the interaction with these people "dull, colorless, and boring" even when they are educated and intelligent (Krystal 1993: 247). Krystal's observations in this regard are relevant since they show that—even though these patients are incapable of verbally describing their emotions—they do not exhibit any general impairment of both linguistic and intellectual functioning related to the understanding or the active production of language in the proper sense. In fact, he emphasizes how these patients are often intelligent and how they master the language in an accurate manner: as he points out, they are capable of a "recounting in great and often boring detail, of events in their external environment and their own actions in this setting" (Krystal 1993: 248).

As seems evident, these people do not suffer from a form of *aphasia* in the ordinary sense since their issues are not related to words in general (for an overview, see Code 1989), but only to a specific word-class. And, furthermore, their problems are not of a phonetic nature, related to a distorted pronunciation of sounds due to some motor deficit of the phono-articulatory muscles. Neither do they experience phonological difficulties related to an impairment in the phonological representation of words, or exhibit impaired morphosyntactic abilities. In fact, their linguistic issues are not related to the morphology of words or the syntax of the sentences they use and they don't exhibit any general problems with lexical semantics, either. Instead, their difficulties are selective and relate only to emotion words; moreover, they do not only have issues using (recalling/producing) emotion words, but are also unable to make use of circumlocutions or paraphrases to express them. Their difficulty with verbally expressing emotions is accompanied by a limited imagination and by a dull, impoverished and flat communication style: for example, as we have seen, when patients are asked to describe the emotions they experienced in a given situation, they give the details of the situation or sometimes they describe their bodily states, without mentioning their felt emotions.

Sifneos hypothesized that it might be possible to provide a single account for the complex of apparently unrelated symptoms exhibited by psychosomatic patients and suggested that they might all be different effects of a common cause. His explanation begins with observations concerning the treatment these patients typically received: for psychosomatic patients the psychoanalytic treatment—the so-called 'talking cure'—demonstrated little hope of success (Sifneos 1972–1973, 1974; McDougall 1972; Nemiah 1972, 1977; Krystal 1979). This conclusion was also confirmed by later studies (Krystal 1982; Taylor 1987; Taylor et al. 1997; von Rad 1983: 160ff; Gerhards 1988). According to the main psychoanalytic perspectives at the time (Lesser 1981; von Rad 1983; Timoney and Holder 2013), the onset of psychosomatic symptoms was due to a lack of affective self-regulation which drove too much psychic energy to the bodily organs. Thus, the expected outcome of a psychodynamic therapy was that the patient become more aware of the emotions they experienced so that they could represent and therefore also exercise rational control over those feelings, and eventually achieve increased capacity for emotional regulation. However, most psychosomatic patients did not respond to this treatment: they exhibited difficulties in describing their feelings, moods and/or emotions and, as we have seen, their metaphorical language and their imagination was extremely poor. This made it difficult for them to access the experienced emotions, even indirectly (de M'Uzan 1974, 1977; Apfel and Sifneos 1979). The deep reason for the failure of dynamic treatments could be traced back to the inability of such patients to grasp the semantics of a specific word class which is fundamental for the success of dynamic psychotherapy, i.e. emotion words. Since psychodynamic therapy relies on the role of speech and therefore the function of language to intervene in specific affective blocks, and since these particular patients had difficulties in communicating their affects and emotions through words, they could not adequately share the semantic fields and objects related to emotion terms with the therapist and this necessarily led to the failure of the treatment.

The failure of this treatment brought to light the fact that the difficulties manifested by these patients in verbally communicating their emotions were not just a superficial characteristic associated with their condition that could be treated separately from their main condition. This led to the hypothesis that the inability of those patients to express their emotions through words was not just a superficial problem concerning their linguistic capacity for linguistic terms, but rather the manifest effect of a deeper dysfunction related to processing the information that allows people to 'mentalize' (i.e. internally represent) signals corresponding to emotions and emotional experiences. This made it clear that psychodynamic models were inadequate for treating psychosomatic patients and even for understanding the etiology of the various symptoms that go with psychosomatic diseases such as, among others, the incapacity to verbalize emotions and affects and to correctly interpret bodily feelings related to the experience of emotions and affects. The dynamic approach had to be, therefore, integrated with other perspectives, and, in particular, with research in the fields of developmental psychology and neuroanatomy, which might pinpoint where the specific anomalies observed in the linguistic behavior of psychosomatic patients originated and what they were caused by Nemiah (1977).

Considered as a whole, these observations had important implications for the hypothesis on the nature of *alexithymia* put forward by Sifneos and colleagues in the middle of the last century: alexithymia does not seem to be the consequence of psychosomatic syndromes but rather psychosomatic syndromes appear to be an effect of the alexithymic condition. Indeed, since alexithymic persons cannot correctly understand and represent their emotional and affective experience, they are not able to regulate their emotional life and they are inclined to confuse emotions with dysregulated signals produced by their bodily states: people suffering from alexithymia have a *limited capacity to 'interpret' (i.e. to identify and to classify) the signals coming from their own corporeity*.

The alexithymia construct proposed by Sifneos aimed to identify a specific kind of pathological personality that could explain all the various symptoms observed in psychosomatic patients and that could be considered the origin of all psychosomatic diseases. However, this construct—at least in its 'strong' version—had little success in the clinical field (Gerhards 1988). While early research on alexithymia appeared to sustain Sifneos's hypothesis that alexithymia could be interpreted as a specific pathological condition underlying psychosomatic diseases, later studies weakened this position and showed that the clinical population presenting the characteristic traits of alexithymia might not only include psychosomatic patients but also other people (von Rad 1983; Gerhards 1988; Krystal 1993; Taylor et al. 1997; Timoney and Holder 2013). Alexithymic traits belong to a cluster of symptoms exhibited by various clinical populations including cases involving eating disorders, substance abuse (Taylor et al. 1996; Taylor 2000), somatoform disorders (De Gucht and Heiser 2003; Fuchs and Schlimme 2009), depressive and dissociative episodes, anxiety disorders and post-traumatic stress disorder (Badura 2003). For this reason, alexithymia cannot be considered to be a pathological personality disorder, but is rather a subjective condition that can be described along a *continuum*. From a clinical point of view, high levels of alexithymia might predict or indicate psychological or psychiatric disorders because disorganization in the identification and in the expression of emotions points to a possible problem of emotional self-regulation (Shore 1994; Taylor et al. 1997; Fonagy et al. 1991; Fonagy and Target 1997).

Nowadays, there is broad consensus among researchers that alexithymia is a personality trait with stable characteristics that can be placed along a continuum

(Taylor et al. 1997; Grabe et al. 2001).[29,30] Recent research shows that difficulty with verbally identifying and expressing emotions is widespread also in the non-clinical population (about 10% of the general population exhibits alexithymic traits at various levels and there are no differences in prevalence between men and women). For this reason, and because alexithymia is not classified as a mental disorder in DSM, it has been defined as a "subclinical phenomenon" (Silani et al. 2008: 98). The fact that alexithymic traits are also present in non-clinical populations contributes to its being depicted in a more liberal manner as a subjective condition shared by many people, i.e. as a particular cognitive style also described as "an alexithymic style of emotion processing" (Subic-Wrana 2011: 256).

Research on this condition has outlined a number or relevant aspects both with regard to the specific deficits exhibited by alexythmics and possible causes of the condition. At least two lines of research on this turn out to be particularly relevant for our aims. The first considers how alexithymia affects the mentalizing capacity of individuals, i.e. their ability to understand their own and others' mental states and thus to understand the reasons behind their behavior (see, e.g., Subic-Wrana et al. 2010; Lane et al. 2015a). In fact, when people are not capable of grasping the qualitative dimension of their own emotional experience, they are also unable to correctly interpret the emotions of others and the behaviors these emotions lead to. This suggests that only first-person access to our emotions and to the way they feel can provide us with an adequate comprehension of them (of their nature and effects) and can therefore also support full semantic competence for this word class. A second line of research that we should also take into account addresses the issue of how emotional experience originates from bodily feelings and then reaches awareness. In fact, these studies try to account for some aspects of the alexithymic conditions that are very closely related to the problem of determining how emotions are conceptualized and how emotion words are mastered. People suffering from alexithymia: have difficulty identifying and interpreting their emotions; have difficulty identifying and understanding the emotions of others, both when they are described verbally and when they are presented in the form of visual stimuli (facial expressions, gestures and the like); identify qualitative aspects of their emotions at best in terms of bodily feelings, without any awareness of the reasons why they experience those bodily feelings; describe emotional situations by focusing only on what concretely happened, ignoring the emotional aspects per se; are incapable of describing the qualitative character of their experiences; have difficulty putting themselves in emotional situations (as already mentioned, they are unable to use the

[29]Initially, empirical research supported the idea that it was necessary to distinguish two forms of alexithymia: a primary kind, understood as a dispositional personality trait and a secondary kind, caused by traumatic events and/or illness (Freyberger 1977). For an overview of the empirical research on this see e.g. Krystal (1993), Taylor et al. (1997) and Timoney and Holder (2013).

[30]Initially, empirical research supported the idea that it was necessary to distinguish two forms of alexithymia: a primary kind, understood as a dispositional personality trait and a secondary kind, caused by traumatic events and/or illness (Freyberger 1977). For an overview of the empirical research on this see e.g. Krystal (1993), Taylor et al. (1997) and Timoney and Holder (2013).

pronoun 'I' when describing emotional situations because they do not feel that the reported experience is their own). As we will see in the next section, attempts made in various studies to explain these aspects of alexithymia might support an idea analogous to that put forward by Lambie and Marcel (2002) (this Sect. 7): in order to become a truly emotional experience, the bodily feelings that correspond to an emotion must be integrated and processed at the central level and give rise to a unified, second order and complex qualitative feeling.

9 When Qualitative Information is Missing. Recent Findings on the Alexithymic Condition

Alexithymia is a condition that manifests itself as a problem concerning people's linguistic—in particular, semantic—competence, but it is not per se identifiable as a language disorder. In a recent study, Christiane Wotschack and Gisela Klann-Delius show that—even though alexithymic individuals have clear difficulties in providing a free description of their experienced emotions—their lexical mastery on a formal level seems to be almost intact (Wotschack and Klann-Delius 2013: 520). Their deficits in linguistic mastery specifically and selectively relate to emotion words; moreover, they are not completely unable to use or understand emotion words, but rather that their understanding is anomalous and impaired. According to the description given by Wotschack and Klann-Delius, the knowledge alexithymic individuals have of emotion words seems to be 'disembodied' and 'ungrounded' (in Barsalou's sense; see Barsalou 1999 and Chap. 2, Sect. 1). In order to have well-formed emotion concepts, people need to have perceptual, somatovisceral, and motoric information concerning these emotions and to know what they feel like. As the authors point out, "using language about emotion adequately requires re-experiencing the emotion, encompassing physiological, sensorimotor experiences" (Wotschack and Klann-Delius 2013: 514, 521). Alexithymic individuals appear to lack this kind of information: since they do not have the first-person experience of emotions, they are incapable of re-enacting and re-experiencing them when they speak in order to understand from a referential point of view what emotions are (*feel like*).

Described in this way, the linguistic deficit exhibited by alexithymic individuals appears to be due to a problem related to the perception of emotions. This problem cannot be described in terms of a lack of emotional experience *tout court*: in fact, they are not completely incapable of experiencing emotions in any way and to any degree. The issue is rather that these patients experience emotions in terms of a number of body changes and body feelings, but they do not realize that these feelings are/indicate emotions. Their condition is characterized by a *perceptual dysfunction or distortion*: they perceive the body changes and feelings that accompany emotions (e.g. visceral contractions, increased heart rate, respiration etc.) but their perception remains disorganized and not integrated, so that they do not perceive emotions as emotions with the typical overall qualities of emotions, but rather as body feelings (Thompson 2009). This dysfunctional or distorted perception is the reason why they cannot identify, classify and categorize emotions properly and why they therefore do not have an understanding of these emotion words that is 'perceptually grounded', i.e. that is characterized by a first-person knowledge of how emotions feel.

The fact that they are not capable of integrating the various body feelings that lie at the root of emotional experience is particularly evident if we go back to the reports mentioned in the previous section on how individuals with high levels of alexithymia perceive emotions: they describe emotions as a set of disorganized and dysregulated feelings; they are often unable to identify specific body feelings (hunger, thirst, etc.) on the basis of how they qualitatively feel and they cannot locate such bodily feelings in specific places in their body; sometimes they are even incapable of experiencing emotions as something that *they* feel, i.e. that is part of their personal identity (this is the case when they are unable to use the pronoun 'I' in highly emotional contexts). This interpretation of alexithymia as the effect of poor cognitive organization of body information—i.e. as a problem concerning the perception or the apperception of emotions—is well-grounded in the literature and has led to suggestive descriptions of this disease as a form of *emotional color-blindness* (Krystal 1993: 244) or of *emotional blindsight or blindfeel* (Lane et al. 1997: 838).

Some recent studies focus mainly on the general performance exhibited by alexithymic individuals and suggest that alexithymia is a particular case of *dysfunctional emotional competence* (Denham 1998; Saarni 1999). *Emotional competence* can be summarized as the capacity to *comprehend*, *express* and *regulate emotions*. From this perspective, the analysis of linguistic performance is merely a way to assess the general emotional competence of people as it manifests itself through the vehicle of language. The *comprehension* of emotions implies the capacity to identify, and recognize your own emotions as complex events that involve feelings and it constitutes the precondition for realizing the other two aspects of emotional competence: if we comprehend our emotions, we can label them appropriately and thus express them and control (regulate) them rationally. Moreover, understanding our emotions from a first person perspective appears to be essential for recognizing emotions in others. In fact, many studies show that people with alexithymic traits also have difficulty recognizing the emotions of others both in contexts that involve and that do not involve the use of language (e.g. in reading, in verbal and non-verbal interactions, and when interpreting facial expressions, as well as the looks and the gestures of others; see Lane et al. 1995, 1996, 2000; Vermeulen et al. 2006, 2007; Koponen et al. 2005; Moriguchi et al. 2007; Grynberg et al. 2010, 2012; McDonald and Prkachin 1990).

Below we discuss several recently proposed models which suggest that alexithymia is a condition that goes beyond a problem of lexical semantic disorganization and is rather due to a deficits related to the perception of emotions and the integration of bodily changes and feelings corresponding to emotions. The first two are psychological models which try to describe how emotions are processed, while the third is a philosophical model. We further discuss several other models that attempt a synthesis between the psychological and philosophical versions.

(i) The first model was suggested by Richard Lane at the end of the eighties and it goes by the name of *LEA* ("*Levels of Emotion Awareness*"). The aim of this model is primarily to investigate the epigenesis of our emotional experiences, i.e. how we become gradually aware of our emotions (in Lane's view emotional experience and emotional awareness are one and the same thing, while the first person has privileged authority in the identification of his/her own emotions exactly because only s/he can experience them, i.e. be aware of them: see Lane and Schwartz 1987, 1992). This model relies on the idea that we are fully aware of our emotions when we are able to identify, distinguish and recognize the various components included; these components form actual, independent 'levels' of our emotional experience that are hierarchically organized (Lane and Schwartz 1987).

Lane's model is based on Piaget's genetic epistemology and uses Piaget's notion of a cognitive schema as something people use to organize patterns of information. In Piaget's view, the schemas primarily organize motoric information to improve the perceptual capacities of an individual, to gradually acquire knowledge about the external world and finally to develop abstract concepts. According to Lane, we use analogous schemas for organizing the information we have about our internal environment: these allow us to gain knowledge about what happens in our own body. Thanks to these schemas we progressively improve our perceptual capacities: we become increasingly better able to identify the states and the events that happen inside us, to distinguish between them and thus we also improve our linguistic capacity to describe and express these states and events (Lane and Schwartz 1992: 4–6).

In Lane's account, the levels of awareness of our emotions we gradually develop (thanks to which we become better able to identify and discriminate among them) is due to our capacity to organize various kinds of internal information of increasing complexity in a congruent manner: (1) *bodily sensations*; (2) *action tendencies*; (3) *single emotions*; (4) *blends of emotions*; (5) *blends of blends of emotions*. (1) The schemas that organize the perception and the awareness of the bodily sensations include patterns of sensomotoric information: they are mainly perceived as individual somatic sensations—i.e. as states or events internal to the body—that are still not causally related to any other events. (2) At a higher level of organization, these patterns of sensomotoric information broaden to include an awareness of the causal relationships between specific bodily sensations and particular action tendencies as well as of the hedonic tone of the sensations and an appraisal of the related situation. (3) On the basis of these elements, we can become aware of a single emotion that is identified on the basis of the pattern of the bodily sensations that goes with it, by its hedonic tone, the appraisal of the situation that caused it and the overall quality that characterizes the emotion (how it feels). (4)–(5) Once we have the capacity to identify a single emotion on the basis of what it feels like, we can become aware of emotions at growing degrees of complexity by identifying the specific quality and the specific bodily sensations that characterize any single emotion that composes the complex (i.e. the blend). When people become capable of identifying blends of blends of emotions, they also develop a descriptively accurate emotional lexicon through which they can express their emotional experience using complex shades of meaning, metaphors and figurative language.

One important aspect of Lane's model concerns the fact that—when we develop more complex schemas at higher levels—the information and the organization available at lower levels is not deleted or 'overwritten':

> The hierarchical nature of the model is such that each subsequent level transforms but does not eliminate function at lower levels. In general, just as each new level constitutes an expansion in the complexity of the representation of emotions, expressed as greater differentiation and integration preceding levels become more attenuated and manifest greater differentiation and integration of function as well (Lane and Schwartz 1992: 4).

In Lane's model, we acquire completely developed emotional competence when we are fully aware of our emotional experience in all its aspects (and at all levels of organization), starting from the most basic bodily components and continuing on up to the most complex aspects related to the overall quality of the (blends of) emotions (see Lane and Schwartz 1987, 1992; Lane et al. 1997).

In light of this model, the impairment alexithymic individuals show in their emotional competence cannot be due to a deficit concerning the capacity to verbally express their emotions but must be rather explained in terms of problems with that emotional awareness and development that allow us to identify and recognize emotions on the basis of how we perceive them (what they feel like). In fact, recently, Lane suggested that alexithymia be considered a form of "*affective agnosia*", i.e. a deficit in the ability to mentally represent our emotions, i.e. to categorize them on the basis of their overall qualities, "which typically requires the integration of information from multiple interoceptive and somatic sensory modalities as well as conscious/unconscious appraisals of one's situation" (Lane et al. 2015b: 595).[31] According to Lane, "the concept of agnosia highlights the importance of mental representation of the emotional meaning of one's perceived interoceptive/somatic emotional reactions states as a prerequisite for both understanding one's experience and naming it" (Lane et al. 2015b: 600).

(ii) The second psychological model which also supports an interpretation of alexithymia as a deficit in the perception of emotion was put forward by Wilma Bucci and is known as *Multiple Code Theory* (Bucci 1997b). According to Bucci's

[31]According to Taylor, Bagby and Parker, Peter Sifneos also arrived at the hypothesis that alexithymia is a form of 'emotional agnosia' (see Sifneos 1967), however, he never developed this intuition in a systematic manner. Taylor, Bagby and Parker also report that in his later writings he also suggested that alexithymia could be seen as a form of 'feeling aphasia' (see Sifneos 1994, 1996). This hypothesis relies on Damasio's definition of aphasia as an impairment in the ability "to convert the sequence of nonverbal mental representations that constitute thought in symbols and grammatical organizations that constitute language" (Damasio 1992: 531). In Taylor, Bagby and Parker's view, if Sifneos had developed his intuitions about alexithymia being a form of 'emotional agnosia' or 'feeling aphasia' in a systematic manner, he would have come to the conclusion that alexithymia in, in fact, a form of affective agnostic aphasia', i.e. a particular kind of naming impairment (anomia) due to "deficient cognitive activity" that "involves more than labeling emotion with words" (Taylor et al. 2016: 1007).

theory, people process information at three different levels which also describe the ways in which they organize and understand their experience. Bucci uses a lexicon driven by the psychoanalytic tradition and suggests that these levels aim to achieve a *symbolization* of our experience, i.e. an internal representation of the various components of our experience; his notion of symbol is, however, quite close to Lane's idea of schema. In the perspective she proposes, human experience—including, in particular, emotional experience—is processed at three different levels which are distinct but related to each other (Bucci 1997a, b). (1) The first level is defined as *sub-symbolic and non-verbal*: at this level, information is processed in parallel and includes sensory and motor stimuli as well as qualitative sensations triggered by sensory or motoric stimuli (e.g. the sensations evoked by listening to a musical piece, the sensations evoked by recognizing that the facial expression of someone in front of us is aggressive, friendly or whatever, the sensation of discomfort provoked by a chaotic situation, etc.). This level does not have access either to verbalization or to other forms of awareness mediated by symbols such as iconic representations, images and the like. (2) The second level is *symbolic but non-verbal* and includes information that has a symbolic format of a figurative nature: the first-person is aware of this information in the form of images which can be translated into words. (3) The third level is *symbolic and verbal* and it includes all that is explicitly codified by means of language. In normal conditions these three levels are related to each other in a bidirectional manner. In fact, the use of language presupposes the availability of sensory and motor stimuli and the qualitative sensations evoked by such stimuli which 'ground' the language from a perceptual point of view throughout the process of mediation at the second level.

In Bucci's model, the anomalies observed in the use of language by alexithymic individuals are due to something like *asymbolia*: they have no symbols for somatic states because their language has no connections or only weak connections with the first subsymbolic, non-verbal level due to malfunction or poor functioning at the middle level which should have performed the mediation function providing preliminary organization of information from the first level and making it available for verbalization. This is why alexithymic individuals provide such disorganized descriptions of their emotions: they include deregulated somatic sensations; they confuse emotions with actions or action tendencies; they do not coordinate all the elements in a well-constructed manner including by using images or metaphors. This interpretation of alexithymia as a particular form of *asymbolia* presupposes that alexithymic individuals are not able to *integrate* the level of bodily signals with the consequent qualitative sensations at the language level. Alexithymic individuals do perceive these bodily signals and the qualitative sensations they give rise to, but they do so only in a vague, disorganized and undifferentiated manner because the signals are not integrated at the higher level of symbolic processing.

(iii) These two models are also congruent with a hypothesis on alexithymia developed by the philosopher Jesse Prinz based on his theory concerning how emotions are cognitively processed. Prinz suggests that the alexithymic condition should be interpreted as a recognition deficit and in particular as a form of *associative agnosia* analogous to the better-known *visual associative agnosia* (Prinz

2004a: 216–218). The term “visual agnosia” usually indicates “disorders of visual object recognition following brain damage” (Farah 2004: 1). In the literature it is widely acknowledged that there is great diversity among agnostic patients as far as their abilities and deficits are concerned. Even though there is no standard taxonomy for agnosia (see, e.g., Farah 2004: 1–10), many attempts have been made to classify different forms of agnosia on the basis of underlying abilities and deficits. A particularly useful classification was suggested already in 1890 by Heinrich Lissauer who distinguishes between “*apperceptive agnosia*” and “*associative agnosia*”: while in *apperceptive agnosia* the object recognition fails because there is some impairment in visual perception (even though there is no visual field defect), in *associative agnosia* perception does not appear to be impaired in any way, but object recognition just does not occur (Lissauer 1890; see also Shallice and Jackson 1998). Associative agnostics are unable to name the object they see or to accomplish categorization tasks in a non-verbal fashion (such as picking out pictures depicting only objects of the same type—e.g. chairs—from a variety of pictures). However, usually they are able to copy pictures of objects in an exceptionally precise manner; even though they do so in a rather unusual way, focusing on each line separately as if they were drawing meaningless shapes (Farah 2004: 69ff). As Prinz also suggests, this syndrome is analogous to the alexithymic condition which might be interpreted as “an inability to recognize emotions.”

> People with alexithymia have difficulty saying how they are feeling, because they cannot recognize how they are feeling. This does not mean they lack emotions. People with alexithymia often report an unusual number of unexplained bodily ailments […]. One explanation is that they are experiencing the body perturbations that comprise normal emotions but they fail to recognize those perturbations as emotions. Like associative agnosics, who experience shapes without comprehension, alexithymics may experience emotions without comprehension (Prinz 2004a: 217).[32]

Indeed, the problem alexithymic individuals suffer from seems to lie in the transmission and processing of internal signals (i.e. propriosensitive signals). More specifically, the bodily signals they receive are not well-integrated, thus they do not perceive emotions as a rich, unitary experience which is clearly, qualitatively characterized but also made up of various somatic components each with a specific localization and intensity. The alexithymic condition appears to be due to specific problems of ‘*body-reading*’ (concerning the identification and classification of bodily signals) which leads to specific problems of *mind-reading* both in the first

[32]The explanation that Prinz offers for alexithymia as associative agnosia differs somewhat from the one we develop here. In fact, Prinz links his interpretation of alexithymia to his theory of consciousness, the so called AIR (Attended Intermediate-level Representations) theory based on Jackendoff’s view (Prinz 2004a: 208ff). According to this theory, sensory systems are hierarchically organized into three main levels of processing: experience becomes conscious at the intermediate level, while the recognition of this occurs at the higher level. Therefore “damage to early or intermediate-level centers should disrupt experience, and damage to high-level centers should impair recognition” (Prinz 2004a: 217). We do not rely on this distinction among the levels; the interpretation of alexithymia we argue for here instead relies on the hypothesis of emotional processing we described in the previous sections of this chapter.

and in the third person. Thus, the apparently exaggerated attention that people with alexithymia pay to their bodily signals serves to compensate for deficits in the integration of bodily information which prevents them from capturing the qualitative aspects of the various emotions.

Prinz gives a rather univocal interpretation of alexithymia as associative agnosia. However, based on his analysis and the conclusions drawn by Lane, more recently several authors have suggested that this condition should instead be considered a particular form of *anomia* (inability to express emotions, feelings and affects) which can be due to *either associative or to apperceptive agnosia* (Taylor et al. 2016). In their view, considering alexithymia in terms of anomia is useful firstly because it helps account for the linguist deficits exhibited by alexithymic individuals who cannot appropriately express their emotions. Secondly, and most importantly, by distinguishing between associative and apperceptive agnosia allow us to explain why alexithymic traits *can be located along a continuum that goes from low to high*. In fact, alexithymic traits are much higher when alexithymia takes the form of an apperceptive agnosia than when it takes the form of an associative agnosia. In the first case, alexithymic individuals are unable to perceive and to integrate their somatic sensations which are experienced as disorganized and disconnected. In the second case, the somatic sensations are integrated but for some reason are not recognized as emotions: "an individual will experience coherent patterns of his or her entire bodily state, and also manifest facial and other bodily expressions of emotion, but may not recognize these experiences as emotional experiences and use their emotional meaning" (Taylor et al. 2016: 1007–1008).

The idea that alexithymia is, in one way or another, a form of agnosia suggests that—in order to develop full semantic competence with respect to emotion words—the body feelings that go with emotions must be integrated and must give rise to an overall quality on the basis of which the emotion can be identified and recognized. This conclusion is perfectly in line with the hypothesis we put forward previously in this chapter: that processing emotions relies on propriosensitive information which is the kind of perceptual information people rely on to identify and recognize them referentially.

Recent neurocognitive research on the constitution of emotional representations and emotional awareness is also congruent with the view we argue for here (this research was already discussed previously: on this see also Chap. 2, Sects. 5 and 6). This research supports our interpretation of alexithymia as based on the relationship between emotional experience and semantic competence.

Indeed, many recent studies in the field focus on and underline the relevance of what they call '*interoceptive awareness*'. This is a particular form of awareness related to the perception of signals coming from different bodily systems and captured in different modal ways and "it comprises sensing the physiological condition of the body as well as the representation of the internal state within the context of ongoing activities" (Herbert and Pollatos 2012: 693). This way of interpreting interoception is an extension of how it is usually understood—as the perception of visceral sensations: "The concept of 'interoception' was classically restricted to visceral sensations, but recent neuroanatomical and neurophysiological results indicate that sensations related to the ongoing physiological condition of all organs of the body—muscles, joints, teeth, and skin as well as the viscera—are processed together" (Craig 2004: 239). This concept of interoception also significantly overlaps with the notion of 'proprioception' we introduced (relying on Gibson's view) as a blend of what is usually called proprioception (in a narrow sense: the internal perception of the body position and movements) and interoception, which—in our hypothesis—is obtained through the constant activity of the propriosensitive perceptual system (Chap. 2, Sects. 3–6).

In fact, recent neuroanatomical research underpins the hypothesis that interoceptive awareness is produced by an *interoceptive neural network*. This network comprises "somatosensory and somatomotor cortices, the insular cortex, cingulate cortex (ACC), and prefrontal cortices" (Herbert and Pollatos 2012: 694) and is responsible for the coordination and integration of all our bodily feelings, as well as for the modulation of the sensory information coming from the various bodily parts. It is the activity of this network that makes us aware of our bodily feelings and emotional states (also from a qualitative point of view). Interoceptive awareness therefore consists in a very complex representation of our internal states which comprises sensations at various levels of complexity, from the very general to the very specific (Tsakiris et al. 2007, 2011).

This new perspective on interoception suggests that it produces a "topologically organized map of the body that provides a basis for both stimulus identification and localization" (Craig 2010: 564). Indeed, this generates an integrated representation of the general somatic condition that gives rise—among other things—to the qualities (the *qualitative salience*) of emotional experience. In this view, emotions are something like a "global emotional moment" which includes both the mapping of somatic activation and the corresponding qualitative feel (see, e.g., Craig 2009). This view supports the hypothesis we have put forward in this work that, to be able to conceptualize emotions, people must be aware of their global quality and must be able to drive their attention to specific aspects of the interoceptive signal to 'feel' the traces of its individual constitutive elements (localization, intensity, concomitant sensations).

In recent times, the role of the insula within the so called *interoceptive neural network* has been investigated and its importance with respect to emotional awareness has been stressed. As mentioned above, the insula is considered to be the hub where the integration of information connected to mapping general somatic states takes place; it generates a global representation of the 'functioning' of the bodily form which in turn gives rise to a qualitative awareness of bodily states. This lends support, among others, to the idea that alexithymia is a form of agnosia and that this condition is a dysfunction due to problems with the transmission of information on various body states. More specifically, it suggests that alexithymia can be related to insular function, since the insula is the most important component of the network in which interoceptive feelings are integrated and give rise to interoceptive awareness. According to some research using neuroimaging techniques, "alexithymics demonstrate lower reactivity in brain regions associated with emotions" while they also "show an amplified activity in areas considered to be involved in physical sensation", for example, in pain. This supports the hypothesis that alexithymia consists in a "somatosensory amplification" (Kano and Fukudo 2013: 1–3) in which the global qualitative components are missing. From this perspective, alexithymia must be interpreted as a problem related to *interoceptive awareness* in which people show diminished interoceptive awareness of a global kind (Herbert et al. 2011), while exhibiting "hyperinteroceptive arousal" related to stimuli of a physical kind (e.g. pain). From this perspective, the semantic deficits exhibited by alexithymic individuals can be traced back to the fact that such individuals are not able to categorize emotions because they cannot perceive and distinguish all the features that are necessary in this process, i.e. the global quality of emotions as well as the localization of the bodily changes that accompany emotions, specifically emotional intensity and the concomitant sensations (see Sect. 7).

Thus, the linguistic problem alexithymic individuals exhibit is actually a conceptual problem: they are not able to identify and classify interoceptive signals to create a conceptual representation of the emotion they are experiencing; since they do not have any emotion concepts, they cannot link these to words. If so, the basic problem of alexithymic individuals is that they cannot conceptually grasp their emotions as 'global emotional moments (or experiences)': they only perceive the unstructured bodily feelings that give rise to the emotion; they do not have an integrated and global representation of the overall quality of the emotions, but are only aware of bodily sensations and feelings. In other words: *people suffering from alexithymia cannot conceptually identify and classify their emotions because they cannot adequately process interoceptive signals in order to obtain an integrated representation of the global quality of their emotional experience. They only perceive the unstructured bodily signals that in normal cases gives rise to this global experience. Thus, their linguistic difficulties are due to a categorization deficit.* This is also the reason why they pay so much attention to their bodily states and why they describe their reactions to emotional situations in terms of bodily states, while they never speak of emotions.

In our opinion, one of the most relevant studies supporting this interpretation of the nature of alexithymia has been offered by Giorgia Silani and colleagues (Silani et al. 2008). This research relies on Lambie and Marcel's (2002) differentiation, already considered in detail (see Sect. 7), between an 'emotion state' (a first order emotional experience consisting of the feeling we have of the bodily states and changes that occur when we experience an emotion) and an 'emotion experience' (a second order emotional experience consisting of the fully developed and integrated emotion we have and are aware of when we reflect on it). In line with Craig's hypothesis, Marcel and Lambie consider that the interoceptive cortex plays a crucial role in second order emotions. In their study, they correlate the self-reported degree of awareness of own and other emotions in individuals with a range of alexithymia scorings and neural activity. They have established that self-reported poor awareness of own and others' feelings is strongly associated with a reduced response in interoceptive cortex.

> Correlational analyses showed that the level of activity in this region correlated significantly and negatively with scores on both alexithymia measures. The higher the alexithymia score the lower the activity. These data suggest that in the condition most relevant to alexithymia, i.e., when the individual's own emotional state must be consciously represented, activity in the insula is associated with the degree to which alexithymia is reported. This brain response was assessed when participants introspected on their inner experience in the presence of unpleasant stimuli, and this is consistent with the proposal that the function of the insula is to provide a representation of bodily states that enables conscious awareness of feelings. We suggest it is a lack of conscious awareness of these representations that characterizes alexithymic symptoms in emotional awareness, i.e., an impairment in Lambie and Marcel's second-order emotional experience (Silani et al. 2008: 105–106).

This view is congruent with the hypothesis we are suggesting here: alexithymics do not comprehend emotions because—even though they are aware of the bodily states that usually give rise to emotions (that constitute the rough material for emotions)—they do not have access to the integrated, unitary and qualitatively characterized experience of emotions. They perceive their emotional states as disorganized physiological signals, and they almost completely lack any awareness of the qualitative dimension of their experience, including its hedonic tone (i.e. pleasantness or unpleasantness). In other words, alexithymia appears to be a "deficit in the subjective awareness and cognitive processing of affects." (Taylor et al. 1997: 5): the impairment exhibited by people with alexithymia seems to be due to their inability to integrate the proprioceptive signals coming from their body to obtain a global, conscious and unitary qualitative experience.

The qualitative character of experience typically results from processing bodily signals and, as we have maintained in previous sections, is essential for the first-person to identify what s/he is experiencing. Since alexithymic persons are not capable of processing the bodily signals they perceive in an integrated manner and representing their salient qualitative properties, they cannot identify and classify the states they are experiencing. In light of these considerations, the difficulty exhibited by alexithymic persons in expressing their emotions with words can be explained in terms of an impairment of their capacity to process internal 'proprioceptive' information and to form the conceptual representations people need to grasp and to master the corresponding words in their language. Specifically, alexithymics cannot form *the perceptual core representation of emotional concepts that allows them to categorize emotions*.

This interpretation of alexithymia helps us support and further refine the hypothesis that concepts denoting internal states are also characterized by a perceptual component and that our capacity to master words denoting internal states also includes referential competence which relies on this perceptual (i.e. proprioceptive) information. First of all, the fact that alexithymic people are incapable of grasping and mastering emotion words shows that—when 'proprioceptive' information is not correctly processed and does not support referential competence—semantic competence is impaired and people are not capable of mastering words denoting internal states. Thus, in order to support our semantic competence, our conceptual representations of internal states must also include a perceptual (i.e. 'proprioceptive'/propriosensitive) component. If this component is missing, people cannot identify the internal state that they are experiencing. Secondly, alexithymia also demonstrates that at least two steps are required for information processing related to the development of core concepts for internal states. On the one hand, proprioceptive information must be detected (i.e. the signal must be captured): this seems to be the case in alexithymia, since alexithymic people perceive some signals from their bodies. On the other hand, however, the signal must be correctly processed—i.e. integrated and transformed into qualitative aware information. If this does not happen, the signal remains undetermined and the state cannot be identified and classified, i.e. conceptually represented.

This conclusion is consistent with the more general hypothesis we sustain in this work: words referring to internal states (bodily experiences as well as emotions) are analogous to words denoting observable objects in the external world; both rely on concepts consisting of a perceptual (referential) core. Specifically, to master emotion words, people need to form a corresponding conceptual representation based on a perceptual (referential) core which includes information on the bodily and qualitative experience underlying the corresponding emotional experience. From this point of view, alexithymia can be interpreted as a condition in which bodily, propriosensitive information cannot be adequately processed and integrated to constitute the referential core concept that is indispensable for the identification and the classification, i.e. for the categorization, of emotions. Since alexithymics are not able to master words referring to both basic and complex emotions, we can conclude that this referential core concept is essential for comprehending and using all kinds of emotion words, including those that also require propositional (i.e. inferential) knowledge in order to be adequately mastered. If our hypothesis is correct, to categorize emotions of any kind we first need to form a perceptual schema that includes information about the general quality of the emotion experience and its corollary characteristics: intensity, localization and concomitant sensations.

10 Some Clinical Evidence on the Mastery of Emotional Terms (II): Autism Spectrum Disorder

Further evidence for the relevance of internal information to people's semantic competence with respect to words denoting emotions is provided by so-called *Autism Spectrum Disorder* and especially by people suffering from Kanner's Syndrome and from what was once called Asperger's Syndrome (Kanner 1943; Asperger 1944, 1968; Wing 1981), i.e. a form of high functioning autism that is characterized by a high level of cognitive functioning (social skills are compromised but general intelligence is high and formal linguistic capacities are almost intact).[33] Individuals suffering from these syndromes are interesting for our work

[33]Kanner's Syndrome (also called 'early infantile autism', 'early autism' or 'Kanner's autism') and Asperger's Syndrome (or Asperger's Disorder) are no longer listed separately in the clinical

because they combine two favorable characteristics. On the one hand, they exhibit high-level linguistic capacities; the main evidence for this is that some of them are authors of relevant autobiographic material on their condition (indeed, most of the following analysis will be based on this work). On the other hand, like alexithymic individuals, people suffering from these syndromes also have a limited comprehension of words denoting emotional states and, more generally, words related to internal states.

As in the case of alexithymia, autistic people do not seem to grasp the qualitative aspects of their emotional experience. The commonalities that characterize the linguistic competence of these people are so pronounced that some research in the literature has specifically investigated the alexithymic traits of people with Autism Spectrum Disorder (ASD) (Bird et al. 2010, 2011; Bird and Cook 2013; Hill et al. 2004; Fitzgerald and Bellgrove 2006; Samson et al. 2012). In fact, it is useful to recall that—apart from alexithymia (which is considered a personality trait and a subclinical phenomenon)—ASD is a condition that is been included in the taxonomy of the DSM for a long time. Its etiology (which in the past was erroneously attributed to a problem of intrapsychic development due to relational issues, e.g. due to an inadequate mother/child affective relationship as in the "refrigerator mothers" hypothesis put forward by Bettelheim 1967), is nowadays traced back to issues of a genetic nature. These result in a generalized impairment of the individuals' social and especially *emotional capacities*.

Both the behavioral and the linguistic deficits exhibited by patients with ASD are considered to be the effect of a lack of first-person experience with respect to their interior life: because such people have difficulties accessing their internal states, they have also difficulties in interpreting and in communicating about them, as well as in understanding the states of others (Frith 1989). This picture of ASD is very well summarized by Simon Baron-Cohen's term "*mindblindness*", which conveys the idea that these individuals are functionally blind toward their own and others' mental states. Recent research that analyses these aspects of ASD considers alexithymia to be a condition mainly related to ASD: it focuses on whether ASD patients' incapacity to read their own and others' minds is accompanied by alexithymic traits and thus deals primarily with overlaps between ASD and alexithymia.

ASD and alexithymia give rise to extremely similar linguistic impairments; common symptoms also include so-called *operatory thinking*, which we already

(Footnote 33 continued)

taxonomy of the Diagnostic and Statistical Manual of Mental Disorders edited by the American Psychiatric Association starting from the V edition of 2013 (APA 2013: 31–32, 50–59). All of the characteristics previously described as belonging to either Kanner's or Asperger's syndromes are now included as varying degrees along a dimensional continuum of the so-called 'Autism Spectrum Disorder' (ASD). For a differentiation see the fourth edition of the Diagnostic and Statistical Manual of Mental Disorders (APA 1994: 66–78. Even though the term Asperger's Syndrome is no longer used as a clinical description, we sometimes use this label, since we are specifically interested in the cognitive characteristics of the patients once called 'Aspergers'. In fact, 'Asperger's Syndrome' indicates a form of high functioning autism which is characterized by a high level of verbal and cognitive functioning.

mentioned with respect to alexithymia: i.e. the tendency to describe objects, actions and events that characterize the context in which an emotion takes place as if they were the emotion itself and an incapacity to grasp the qualitative dimension of emotions. However, in spite of this affinity, these two conditions should be differentiated at least in some respects. First of all, while alexithymic individuals exhibit a poor capacity to use their imagination to describe emotions (i.e. to rely on iconic and pictorial representations to conceptualize emotions), people suffering from ASD do not manifest this trait (Bird and Cook 2013). Secondly, apart from alexithymics, high-functioning people suffering from ASD—*the variety that was previously called Asperger's (a term we keep using here because it is well-known and comprises the two aspects of the condition that interest us: ASD and high functioning*) have a manifest interest in their interior life (Happè 1991). They seem to be aware that their 'interior life' is different from that of normal developing people and this makes them 'curious' and it raises their interest in understanding the differences. Because of these differences and since alexithymia is a sub-clinical condition while ASD is a clinical condition, it is useful to distinguish between them, at least as far as possible, and to discuss the linguistic impairment of high-functioning people suffering from ASD separately.

A large amount of autobiographic literature written by people with Asperger's syndrome proves they are very interested in their interior life and also capable of using figurative and iconic language to describe it. Among other things, this literature provides us with relevant information on how they experience themselves and on how they understand words denoting emotions. These texts confirm that people with Asperger's have good general linguistic competence. However, as e.g. Uta Frith points out, even though autistic writers show extensive self-knowledge and describe their inner feelings and bodily sensations with extraordinary vividness, their stories are characterized by something like a sort of an "absent self" (Frith 2003: 218, 2008: 102). More specifically, she notes that they exhibit a deficit in mentalizing (in understanding what others are thinking/feeling) accompanied by a lack of awareness about the *qualitative dimension* of their own and others' interior life. People with Asperger's used different forms of introspection than those used by non-autistic people; their introspection is always visual, concrete and is 'limited' in terms of awareness of their own and other's mental life. Some examples will help illustrate Uta Firth's point more precisely.

The first case that helps us understand how the interior experiences of Asperger patients are limited and how this limitation affects their understanding of emotional language is the autobiography of Temple Grandin (1995), a well-known Professor of animal sciences at Colorado State University, who is affected by Asperger's syndrome. In her book, Grandin tries to elucidate in detail the way in which she perceives, conceptualizes and communicates emotion and more generally emotional and social situations. There are two main aspects that emerge from her description: (i) specific problems with the comprehension of the qualitative aspect of emotion words; and (ii) a more general difficulty related to understanding the meaning of all kinds of words that do not have any observable referent in the external world. We will consider this last issue in the next chapter (which is devoted to the discussion of abstract words), while here we will focus on the first one. With respect to emotions, Grandin tries to dispel the prejudice that people with autism do not have emotions and to examine instead the particular characteristics of the emotional life they do have:

> Some people believe that people with autism do not have emotions. I definitely do have them, but they are more like the emotions of a child than of an adult. My childhood temper tantrums were not really expressions of emotion so much as circuit overloads. When I calmed down, the emotion was all over. When I get angry, it is like an afternoon thunderstorm; the anger is intense, but once I get over it, the emotion quickly dissipates. I become very angry when I see people abusing cattle, but if they change their behavior and stop abusing the animals, the emotion quickly passes (Grandin 1995: 89).

And further:

> My emotions are simpler than those of most people. I don't know what complex emotion in a human relationship is. I only understand simple emotions, such as fear, anger, happiness, and sadness (Grandin 1995: 91).

The broad description Grandin offers of her interior life reveals that she experiences emotions in the form of an indiscriminate physical excitement. She perceives emotions as polarized between pleasant and unpleasant states. Unpleasant states are most prevalently related to agitation and are described mostly in physical terms as stiffening, tachycardia, intense perspiration, increased heartbeat, loss of balance, and the wish to avoid a situation (which can be interpreted as an escape instinct). Pleasant states are mostly related to containing and discharging agitation and are associated with somatic states like relaxing and the reduction of agitation obtained through moving the body (e.g. jumping for joy). The impression one gets is that Grandin is capable of identifying and distinguishing the hedonic tones corresponding to positive and negative emotions on the basis of their physiological symptoms and has learned to classify them into a few broad groups which she then gives a name ('fear', 'anger', 'happiness', and 'sadness').

The understanding of the semantics of emotion words exhibited by Grandin appears strongly related to physical sensations and specifically to physical excitement and does not seem to be characterized by any globally qualitative dimension. This may be the reason why Grandin is sure that she is able to experience primary emotions (i.e. basic emotions that are more strongly associated with somatic components and that are more directly elicited by external conditions), while she is just as sure that she does not experience complex emotions, above all complex emotions related to interpersonal relationships: "complex emotional relationships are beyond my comprehension […] I have learned how to understand simple emotional relationships […] however, emotional nuances are still incomprehensible to me […] I realized that other people are guided by their emotions during most social interactions. For me, the proper behavior during all social interactions had to be learned by intellect." (Grandin 1995: 91–92, 97). Thus, in high functioning autistic individuals the most compromised emotions seem to be the complex ones, for which qualitative aspects and nuances carry the biggest weight. If we go back to Richard Lane's model of the *Levels of Emotional Awareness*, then it appears that in the case of people suffering from ASD what is impaired are levels 4 and 5 of emotion awareness where individuals learn to identify *blends of emotions* and *blends of blends of emotion*. The lower levels of emotion awareness (body sensation, action tendencies, single emotions) appear to be intact, even though, since they are not integrated into higher levels of processing, they are perceived in an oversimplified manner (Grandin 1992; Grandin and Scariano 1986).

As in the case of people suffering from alexithymia, the inability of individuals with Asperger's Syndrome to understand emotions seems to be due to the fact that people are not aware of any global and qualitatively characterized emotional experience: "I know I am missing something when other people swoon over a beautiful sunset. Intellectually I know it is beautiful, but I don 't feel it" (Grandin 1995: 91). Basic emotions correspond to a rather clear pattern of somatic activation and this can be used as a perceptual basis (referent) for the emotion. Indeed, Grandin seems to experience basic emotions in terms of bodily sensations with a positive or negative hedonic tone, depending on whether they tend toward relaxation or tension. These sensations are the rough material she relies on to form at least some very rudimentary emotional concepts which she uses to understand what emotions are. More complex emotions that are less directly linked to specific bodily changes are just inaccessible to her: "I do not experience the full range of emotions. My first inkling that my emotions were different came in high school, when my roommate swooned over the science teacher. Whatever it was she was feeling, I knew I didn't feel that way toward anyone" (Grandin 1995: 97).

The fact that Temple Grandin has a certain awareness—at least at some level—of her emotional experience makes her capable of verbally describing the situations in which emotions occur. However, her focus is either in the bodily states or, more often, especially when she describes emotions that others have while she does not, on the external situation/context that gives rise to the emotion. What is completely missing from Grandin's story is the qualitative dimension of emotions: she does not grasp what it is like—i.e. how it feels—to have an emotion experience. This is the reason why she needs to develop a compensation strategy that allows her to speak about emotions, even though she does not know exactly what they are, since she is not able to use emotional information to guide her interactions with others in social contexts. To compensate for her lack of feelings, Grandin adopts a rigorously behavioristic strategy: she observes how other behave in various situations and can describe in detail how they react and what they do. This allows her to behave more or less adequately in social situations and to have apparently normal linguistic competence in terms of using words related to emotions. As a matter of fact, however, she has no idea what emotions are: e.g. she knows how people in love behave, the types of interactions they engage in, what they say, etc.; in short, she has detailed intellectual (propositional and *inferential*) knowledge about 'love', but she has no direct (*referential*) knowledge of love (on this see Grandin and Barron 2005: 40).

In such cases, inferential knowledge compensates for the lack of referential access to the emotional feel. This strategy is similar to that adopted by some individuals with alexithymic traits who use external indicators as support to orient themselves in a world of emotions. They act like colorblind people who must rely on elements other than color to know how to behave in a colorful world and to overcome their impairment: to understand for example when others see the traffic light as red, yellow or green and thus what is the right time to stop or to move. The capacity they develop to obviate their difficulty does not, however, change the fact that they lack a perception that others have and that would make their world more lively and vivid (see, e.g., Krystal 1993; Mac Laren 2006).

The strategy used by Grandin is analogous to this since her linguistic mastery is only apparently compliant with typically developing people. In reality, she does not fully master the meaning of emotion words and her 'appropriation' of the semantics of these words is only extrinsic and impersonal: she is able to use emotion words in a way that is inferentially adequate, but she cannot directly grasp and represent the qualitative component of the corresponding emotion experience. Because it is only an attempt to imitate the linguistic behavior of other speakers, her use of these words is always uncertain and it is likewise accompanied by uncertain beliefs about the nature of emotions and the appropriate way to behave in relation to them. The superficial adequacy of her linguistic competence in describing emotional experiences is only 'fake', in the sense that her use of words is guided by extrinsic instead of by intrinsic criteria.

Another case that helps us understand the limitations of emotions experienced by people with Asperger's and how this limitation affects their capacity to conceptualize emotions and to use words related to emotions is reported by Peter Hobson, an English psychologist and psychoanalyst. Hobson notes that it is clear that Asperger patients have difficulties understanding interior states including emotions and emotion related states and that this difficulty has relevant implications for their language comprehension. However, in his view it is extremely difficult to capture the exact nature of their deficits as well as to map the exact kinds of word meanings that seem to present difficulties for them (Hobson 1993: 165ff). Hobson describes the situation of a patient who has a persistent difficulty in understanding the words 'friend' and 'friendship'.

> Now this individual had a number of preoccupations, but foremost amongst these was his inability to grasp what a 'friend' is. He would ask again and again: 'Are you a friend?', 'Is he a friend?', and so on. The ward staff made every effort to teach him the meaning of the word 'friend', they even found someone to act as a 'befriender' to accompany him on outings to the local shopping centre. All this was to no avail—he seemed unable to fathom what a 'friend' is (Hobson 1993: 5).

This person tries hard to acquire knowledge about friendship. However, in spite of his broad knowledge concerning the definitions and descriptions of the meaning of the word, he continues to miss the deep sense of what a friendship/friend is. Thus, Hobson wonders why these terms are tricky for the patient who does not otherwise exhibit particular conceptual difficulties. The conclusion he comes to is that to understand these words one needs to 'experience' friendship from a first-person point of view.

> As children, most of us can be taught the meaning of the word 'friend' because we experience something of what it is like to engage with others in ways that are fitting between friends. We know what it is to have friends, to be a friend, to enjoy doing with friends those things that are the stuff of friendship. One cannot *really* know what a friend is, simply by 'observing' as one who stands outside and watches behavior (Hobson 1993: 5–6).

Using the terminological distinction we introduced earlier, we might say that—even though this patient masters 'friend' quite precisely from an inferential point of view (since he has collected detailed information on what a friend is)—he has a deficit in his referential competence because he is not capable of recognizing whether a person he meets is a friend or not. The reason he lacks this kind of competence seems to lie in the fact that he cannot experience friendship directly: i.e. he seems incapable of having the emotional experiences that characterize friendship.

The case of this last patient shows—among other things—that the class of terms whose correct mastery requires internal information of an emotional kind is larger than we implicitly hypothesized in the previous section. In fact, this class of words not only comprises emotion words, but also terms which are, like friendship, indirectly related to emotions and whose meaning is a mixture of external perception, internal emotional experience and linguistic information. To this last group belongs cultural information that people generally acquire about word definitions—in this case, for the definition of friendship. In typically developing people, we can assume that this information is perfectly integrated assuring a full (referential and inferential) mastery of this term and analogous words. On the contrary, Hobson's patient lacks the 'internal' (emotional) information that would allow him to know who 'feels like' a friend. He tries to make up for this lack of direct experience by collecting additional linguistic (inferential) information about friendship. However, this kind of information does not help him to develop the capacity to identify friends and recognize them. In this sense, internal (emotional and qualitative) information seems to be essential in developing full semantic mastery not only for emotion words but also for words related to emotions, and no other kind of information can make up for the lack of this internal information.

What exactly is the 'internal' information that Asperger patients lack, or, in other words, why are they unable to develop actual referential competence with respect to emotion words? The answer to this question lies in the fact that their internal experience of emotions is too limited to enable the referential identification and classification of emotions. Thus, the deficit is certainly conceptual and not linguistic. The real issue is, however, to determine the possible cause of this deficit.

It is possible to speculate broadly about what Asperger patients really feel, but if we remain anchored in the evidence we have collected with special reference to Grandin's report, at least two aspects seem clear: on the one hand, since autistics can distinguish between the hedonic tone of at least a few very basic emotions, we should conclude that they have conscious access to some qualitative information. However, on the other hand, since they cannot access the full range of emotions that typically developing people experience, it is clear they are not able to perceive most of the qualities that characterize our emotion experience. A plausible explanation of this discrepancy is that they have access only to those hedonic qualities that are directly dependent on bodily states (i.e. the pleasantness or unpleasantness of such bodily states).

According to the hypothesis we suggest here, people who suffer from ASD might exhibit difficulty experiencing and thus categorizing emotions properly for two different reasons. On the one hand, it is possible that the bodily changes which characterize emotional experience are not detected precisely enough, i.e. the signal is too weak or the self-monitoring process that tacitly maps our bodily states doesn't work properly (Moseley et al. 2015). On the other hand, even though such bodily information is detected correctly, it is possible that it is not properly processed and integrated to produce a global qualitative experience (Lombardo et al. 2007). Although both hypotheses are compatible with the evidence we have presented, the second seems more plausible, if we consider that such patients correctly experience basic emotions that are directly related to bodily states and their issues largely concern complex emotions which depend more heavily on the processing and *integration* of all bodily sensations and related feelings, i.e. an *impairment with the integration of all relevant propriosensitive information*. This last conclusion is also compliant with a well-accepted view in the literature according to which Asperger's Syndrome is due to, among other things, *weak central coherence*, i.e. difficulty integrating signals coming from the various perceptual modalities, including proprio- and interoceptive signals, into a global representation.

References

American Psychiatric Association—APA. (1994). *Diagnostic and statistical manual of mental disorders* (IV ed.). Washington, DC: APA Press.

American Psychiatric Association—APA. (2013). *Diagnostic and statistical manual of mental disorders* (V ed.). Washington, DC: APA Press.

Apfel, R. J., & Sifneos, P. E. (1979). Alexithymia: Concept and measurement. *Psychotherapy and Psychosomatica, 32*(1–4), 180–190.

Aranyosi, I. (2003). Physical constituents of qualia. *Philosophical Studies, 116*(2), 103–131.

Aristotle. (2010). In E. M. Cope & J. E. Sandys (Eds.), *Rhetoric*. Cambridge: Cambridge University Press.

Arnold, M. B. (1960). *Emotion and personality*. New York: Columbia University Press.

Asperger, H. (1944). Die, Autistichen Psychopathen‘ im Kindesalter. *Archiv für Psychiatrie und Nervenkrankheiten, 117*, 76–136.

Asperger, H. (1968). Zur Differentialdiagnose des kindlichen Autismus. *Acta Paedopsychiatrica, 35*(4), 136–145.

Ax, A. F. (1971). Review of 'Neurophysiology discovers the mind'. *Contemporary Psychology, 16*, 365–367.

Badura, A. S. (2003). Theoretical and empirical exploration of similarities between emotional numbing in posttraumatic stress disorder and alexithymia. *Anxiety Disorders, 17*(3), 349–360.

Barsalou, L. W. (1999). Perceptual symbol systems. *Behavioral and Brain Sciences, 22*(4), 577–660.

Bermond, B. (2007). The emotional feeling as a combination of two qualia: A neurophilosophical-based emotion theory. *Cognition and Emotion, 22*(5), 897–930.
Bermudez, J. (2008). *Thinking without words*. Oxford: Oxford University Press.
Bettelheim, B. (1967). *The empty fortress. Infantile autism and the birth of the self.* New York: The Free Press.
Bigelow, J. C., & Pargetter, R. (1990). Acquaintance with qualia. *Theoria, 61*(3), 129–147.
Biggs, S., Matthen, M., & Stokes, D. (2015). Sorting the senses. In D. Stokes, M. Matthen, & S. Biggs (Eds.), *Perception and its modalities* (pp. 1–19). Oxford: Oxford University Press.
Bird, G., & Cook, R. (2013). Mixed emotions: The contribution of alexithymia to emotional symptoms of autism. *Translational Psychiatry, 3,* e285. doi:10.1038/tp.2013.61.
Bird, G., Press, C., & Richardson, D. C. (2011). The role of Alexithymia in reduced eye-fixation in Autism Spectrum conditions. *Journal of Autism and Developmental Disorders, 41*(11), 1556–1564.
Bird, G., Silani, G., Brindley, R., Withe, S., Frith, U., & Singer, T. (2010). Empathic brain responses in insula are modulated by levels of alexithymia but not autism. *Brain. A Journal of Neurology, 133*(Pt5), 1515–1525.
Brosch, T., Pourtois, G., & Sander, D. (2010). The perception and categorization of emotional stimuli: A review. In J. De Houwer & D. Hermans (Eds.), *Cognition and emotion. Reviews of current research and theories* (pp. 66–98). New York: Psychology Press.
Bucci, W. (1997a). Symptoms and symbols: A multiple code theory of somatization. *Psychoanalytic Inquiry, 17*(2), 151–172.
Bucci, W. (1997b). *Psychoanalysis and cognitive science. A multiple code theory*. New York, London: The Guilford Press.
Buck, R. (1993). What is this thing called experience? Reflections on the neuropsychology of qualia. *Neuropsychology, 7*(4), 490–499.
Buddensiek, F. (2008). Stoa und Epikur: Affekte als Defekte oder als Weltbezug? In H. Landweer & U. Renz (Eds.), *Klassische Emotionstheorien. Von Plato bis Wittgenstein* (pp. 63–93). Berlin: De Gruyter.
Calhoun, C. (1984). Cognitive emotions? In C. Calhoun & R. C. Solomon (Eds.), *What is an emotion?* (pp. 327–342). Oxford: Oxford University Press.
Cannon, W. (1927). The James–Lange theory of emotions: A critical examination and an alternative theory. *American Journal of Psychology, 39*(1–4), 106–124.
Cannon, W. (1929). *Bodily changes in pain, hunger, fear and rage* (2nd ed.). New York: Appleton.
Christenfeld, N. J. S., & Mandler, G. (2012). Emotion. In D. K. Freedheim & I. B. Weiner (Eds.), *Handbook of psychology. Volume 1: History of psychology* (pp. 157–175). New York: Wiley.
Cicero, M. T. (2002). In M. R. Graver (Eds.), *Cicero on the emotions: Tusculan disputations 3 and 4*. Chicago: Chicago University Press.
Clore, G. (1994). Why emotions require cognitions. In P. Ekman & R. J. Davidson (Eds.), *The nature of emotion* (pp. 181–191). New York, Oxford: Oxford University Press.
Code, C. (Ed.). (1989). *The characteristics of aphasia*. London, New York, Philadelphia: Taylor and Francis.
Cole, J. (1997). *About face*. Cambridge, MA: MIT Press.
Cole, J., & Spalding, H. (2009). *The invisible smile. Living without facial expression*. Oxford: Oxford University Press.
Craig, A. D. (Bud). (2004). Human feelings: Why are some more aware than others? *Trends in Cognitive Science 8*(6):239–241.
Craig, A. D. (Bud). (2009). How do you feel—Now? The anterior insula and the human awareness. *Nature. Review of Neuroscience 10*(1):59–70.
Craig, A. D. (Bud). (2010). The sentient self. *Brain Structure Function 214*(5–6):563–577.
Crick, F., & Koch, C. (1990). Towards a neurobiological theory of consciousness. *Seminars in the Neuroscience, 2,* 263–275.
Crick, F., & Koch, C. (1991). Understanding awareness at neuronal level. *Behavioral and Brain Sciences, 14*(4), 683–685.

Crick, F., & Koch, C. (2003). A framework for consciousness. *Nature Neuroscience, 6*(2), 119–126.

Damasio, A. R. (1992). Aphasia. *The New English Journal of Medicine, 326*(8), 531–539.

Damasio, A. R. (1994). *Descartes' error: Emotion, reason and the human brain*. New York: Putnam Publishing.

Damasio, A. R. (2000). *The feeling of what happens*. London: Heinemann.

Damasio, A. R. (2010). *Self comes to mind. Constructing the conscious brain*. New York: Vintage Books.

Darwin, C. ([1859] 2008). *On the origin of the species*. Oxford: Oxford University Press.

Darwin, C. ([1872] 2009). *The expression of the emotions in man and animals* (Anniversary ed.). London: Harper.

Davitz, J. R. (1969). *The language of emotions*. New York: McGraw-Hill.

De Gucht, V., & Heiser, W. (2003). Alexithymia and somatization. A quantitative review of the literature. *Journal of Psychosomatic Research, 54*(5), 425–434.

de M'Uzan, M. (1974). Psychodynamic mechanisms in psychosomatic symptoms formation. *Psychotherapy and Psychosomatics, 23*(1–6), 103–110.

de M'Uzan, M. (1977). Zur Psychologie der psychosomatischen Kranken. *Psyche, 31*(4), 318–332.

De Sousa, R. (1987). *The rationality of emotions*. Cambridge, MA: MIT Press.

Dellantonio, S., & Pastore, L. (2010). Introspektion. In H. J. Sandkühler (Ed.), *Enzyklopädie Philosophie* (pp. 1156–1160). Hamburg: Felix Meiner Verlag.

Denham, S. A. (1998). *Emotional development in young children*. London, New York: The Guilford Press.

Descartes, R. ([1649] 1985). The passions of the soul. In J. Cottingham, R. Stoothoff & D. Murdoch (Eds.), *The philosophical writings of René Descartes* (Vol. 1, pp. 325–404). Cambridge: Cambridge University Press.

Dixon, T. (2003). *From passions to emotion. The creation of a secular psychological category*. Oxford: Oxford University Press.

Duffy, E. (1941). An exploration of emotional phenomena without the use of the concept of emotion. *Journal of General Psychology, 25*(2), 283–293.

Ekman, P. (1972). *Emotions in the human face*. New York: Pergamon Press.

Ekman, P. (1973). Cross-cultural studies on facial expressions. In P. Ekman (Ed.), *Darwin and facial expression: A century of research in review* (pp. 169–222). New York: Academic Press.

Ekman, P. (1977). Biological and cultural contributions to bodily and facial movement. In J. Blacking (Ed.), *Anthropology of the body* (pp. 34–84). London: Academic Press.

Ekman, P. (1989). The argument and evidence about universals in facial expressions of emotion. In H. Wagner & A. S. R. Manstead (Eds.), *Handbook of social psychophysiology* (pp. 143–164). Chichester: Wiley.

Ekman, P. (1992a). Are there basic emotions? A reply to Ortony and Turner. *Psychological Review, 99*(3), 550–553.

Ekman, P. (1992b). An argument for basic emotions. *Cognition and Emotion, 6*(3), 169–200.

Ekman, P. (1993). Facial expression of emotion. *American Psychologist, 48*(4), 384–392.

Ekman, P. (1994). All emotions are basic. In P. Ekman & R. J. Davidson (Eds.), *The nature of emotion. Fundamental questions* (pp. 15–19). Oxford: Oxford University Press.

Ekman, P. (1999). Basic emotions. In T. Dalgleish & T. Power (Eds.), *The handbook of cognition and emotion* (pp. 45–60). New York: Wiley.

Ekman, P. (2003). *Emotions revealed. Recognizing faces and feelings to improve communication and emotional life*. New York: Times Books.

Ekman, P. (Ed.). (2015). *Darwin and facial expression. A century of research in review*. Los Altos, CA: Malor Books.

Ekman, P., & Friesen, W. V. (1971). Constants across cultures in the face and emotion. *Journal of Personality and Social Psychology, 17*(2), 124–129.

Ekman, P., Friesen, W. V. & Ellsworth, P. ([1972] 1982). What emotion categories or dimensions can observers judge from facial behavior? In P. Ekman (Ed.), *Emotion in the human face* (2nd ed., pp. 39–55). Cambridge: Cambridge University Press.

Ekman, P., Sorenson, E. R., & Friesen, W. V. (1969). Pan-cultural elements in facial displays of emotions. *Science, 164*(3875), 86–88.

Farah, M. J. (2004). *Visual agnosia* (2nd ed.). Cambridge, MA: MIT Press.

Fitzgerald, M., & Bellgrove, M. A. (2006). The overlap between alexithymia and Asperger's syndrome. *Journal of Autism and Developmental Disorders, 36*(4), 573–576.

Fonagy, P., Steele, M., Steele, H., Moran, G. S., & Higgit, A. (1991). The capacity for understanding mental states: The reflective self in parents and child and its significance for security of attachment. *Infant Mental Health Journal, 12*(3), 201–212.

Fonagy, P., & Target, M. (1997). Attachment and reflective function: Their role in self organization. *Developmental Psychopathology, 9*(4), 679–700.

Freedman, M. B., & Sweet, B. S. (1954). Some specific features of group psychotherapy and their implications for selection of patients. *International Journal of Psychotherapy, 4,* 355–368.

Freyberger, H. (1977). Supportive psychotherapeutic techniques in primary and secondary alexithymia. *Psycotherapy and Psychosomatics, 28*(1–4), 337–342.

Frijda, N. H. (1986). *The emotions*. New York: Cambridge University Press.

Frijda, N. H. (1987). Emotions, cognitive structure and action tendency. *Cognition and Emotion, 1* (2), 115–144.

Frijda, N. H. (1988). The laws of emotion. *American Psychologist, 43*(5), 349–358.

Frith, U. (1989). A new look at language and communication in autism. *British Journal of Disorders and Communication, 24*(2), 123–150.

Frith, U. (2003). *Autism: Explaining the Enigma*. London: Wiley-Blackwell.

Frith, U. (2008). *Autism. A very short introduction*. New York: Oxford University Press.

Fuchs, T., & Schlimme, J. E. (2009). Embodiment and psychopathology: A phenomenological perspective. *Current Opinion in Psychiatry, 22*(6), 570–575.

Fürst, M. (2004). Qualia and phenomenal concepts as basis of the knowledge argument. *Acta Analytica, 19*(32), 143–152.

Galati, D. (2002). *Prospettive sulle emozioni e teorie del soggetto*. Torino: Bollati Boringhieri.

Gallagher, S. (2007). Phenomenological and experimental contributions to understanding embodied experience. In T. Ziemke, J. Zlatev, & R. M. Frank (Eds.), *Body, language, and mind. Volume 1: Embodiment* (pp. 271–293). Berlin: De Gruyter.

Gerhards, F. (1988). *Emotionsausdruck und emotionales Erleben bei psychosomatisch Kranken. Eine Kritik des Alexithymiekonzepts*. Opladen: Westdeutscher Verlag.

Goldman, A. I. (1993a). *Philosophical applications of cognitive science*. Colorado, Oxford: Westview Press.

Goldman, A. I. (1993b). Consciousness, folk psychology, and cognitive science. *Consciousness and Cognition, 2*(4), 364–382.

Goldman, A. I. (1993c). The psychology of folk psychology. *Brain and Behavioral Sciences, 16* (1), 15–28.

Goldman, A. I. (2006). *Simulating minds. The philosophy, psychology, and neuroscience of mindreading*. Oxford: Oxford University Press.

Gopnik, A. (1993). How we know our minds: The illusion of first-person knowledge of intentionality. *Behavioral and Brain Sciences, 16*(1), 1–14.

Gopnik, A., & Meltzoff, A. N. (1998). *Words, thoughts, and theories*. Cambridge, MA: MIT Press.

Gordon, R. (1987). *The structure of emotions*. Cambridge: Cambridge University Press.

Grabe, H. J., Spitzer, G., & Freyberger, H. J. (2001). Alexithymia and the temperament and character model of personality. *Psychotherapy and Psychosomatics, 70*(5), 261–267.

Grandin, T. (1992). An inside view of Autism. In E. Schopler & G. B. Mesibov (Eds.), *High functioning individuals with autism* (pp. 105–126). New York: Plenum.

Grandin, T. (1995). *Thinking in pictures and other reports from my life with autism*. New York: Vintage Books.

Grandin, T., & Scariano, M. (1986). *Emergence: Labelled autistic*. Navato, CA: Arena Press.

Grandin, T., & Barron, S. (2005). *Unwritten rules of social relationships. Decoding social mysteries through the unique perspectives of autism*. Arlington, TX: Future Horizons Inc.

Graver, M. R. (2007). *Stoicism and emotion*. Chicago, London: Chicago University Press.

Grice, H. P. (1962). Some remarks about the senses. In R. J. Butler (Ed.), *Analytical philosophy* (pp. 36–54). Oxford: Blackwell.

Grynberg, D., Chang, B., Corneille, O., Maurage, P., Vermeulen, N., Berthoz, S., et al. (2012). Alexithymia and the processing of emotional facial expressions (EFEs): Systematic review, unanswered question and further perspectives. *PLoS ONE, 7*(8), e42429.

Grynberg, D., Luminet, O., Corneille, O., Grèzes, J., & Berthoz, S. (2010). Alexithymia in interpersonal domain: A general deficit in empathy? *Journal of Personality and Individual Differences, 49*(8), 845–850.

Gunther, Y. H. (Ed.). (2003). *Essays on nonconceptual content*. Cambridge, MA: MIT Press.

Happè, F. G. E. (1991). The autobiographical writings of three Asperger's Syndrome adults: Problem of interpretations and implications for the theory. In U. Frith (Ed.), *Autism and Asperger's syndrome* (pp. 207–242). Cambridge: Cambridge University Press.

Harman, G. (1987). (Nonsolipsistic) conceptual role semantics. In E. Lepore & B. Loewer (Eds.), *New directions in semantics* (pp. 55–81). London: Academic Press.

Harman, G. (1990). The intrinsic quality of experience. *Philosophical Perspectives, 4,* 31–52.

Herbert, B. M., Herbert, C., & Pollatos, O. (2011). On the relationships between interoceptive awareness and alexithymia: Is interoceptive awareness related to emotional awareness? *Journal of Personality, 79*(5), 1149–1175.

Herbert, B. M., & Pollatos, O. (2012). The body in the mind: On the relationship between interoception and embodiment. *Topics in Cognitive Science, 4*(4), 692–704.

Hill, E., Berthoz, S., & Frith, U. (2004). Cognitive processing of own emotions in individuals with autistic spectrum disorders and their relatives. *Journal of Autism and Developmental Disorders, 34*(2), 229–235.

Hobson, R. P. (1993). *Autism and the development of mind*. London: Psychology Press.

Izard, C. E. (1971). *The face of emotion*. New York: Appleton-Century-Crofts.

Izard, C. E. (1977). *Human emotions*. New York: Plenum Press.

Izard, C. E. (2007). Basic emotions, natural kinds, emotion schemas, and a new paradigm. *Perspectives on Psychological Science, 2*(3), 260–280.

James, W. (1884). What is an emotion? *Mind, 9*(34), 188–205.

James, W. (1890). *The principles of psychology* (Vol. 1). New York: Henry Holt & Co.

Johnson-Laird, P. N., & Oatley, K. (1989). The language of emotions: An analysis of a semantic field. *Cognition and Emotion, 3*(2), 81–123.

Johnson-Laird, P. N., & Oatley, K. (1992). Basic emotions, rationality, and folk theory. *Cognition and Emotion, 6*(3–4), 201–223.

Kanner, L. (1943). Autistic disturbance of affective contact. *Nervous Child, 2,* 217–250.

Kano, M., & Fukudo, S. (2013). The alexithymic brain: The neural pathways linking alexithymia to physical disorders. *BioPsychoSocial Medicine, 7*(1), e1. doi:10.1186/1751-0759-7-1.

Koponen, S., Taiminen, T., Honkalampi, K., Joukamaa, M., Viinamaki, H., Kurki, T., et al. (2005). Alexithymia after traumatic brain injury: Its relation to magnetic resonance imaging findings and psychiatric disorders. *Psychosomatic Medicine, 67*(5), 807–812.

Krystal, H. (1979). Alexithymia and psychotherapy. *American Journal of Psychotherapy, 33*(1), 17–31.

Krystal, H. (1982). Alexithymia and the effectiveness of psychoanalytic treatment. *International Journal of Psychoanalytic Psychotherapy, 9,* 353–388.

Krystal, H. (1993). *Integration and self-healing. Affect, trauma, alexithymia*. Hillsdale, NJ: The Analytic Press.

Lambie, J. A., & Marcel, A. J. (2002). Consciousness and the varieties of emotion experience: A theoretical framework. *Psychological Review, 119*(2), 219–259.

Landweer, H., & Renz, U. (Eds.). (2008). *Klassische Emotionstheorien. Von Plato bis Wittgenstein*. Berlin: De Gruyter.

Lane, R. D., Ahern, G. L., Schwartz, G. E., & Kasziak, A. W. (1997). Is alexithymia the emotional equivalent to blindsight? *Biological Psychiatry, 42*(9), 834–844.

Lane, R. D., Hsu, C.-H., Locke, D. E. C., Ritenbaugh, C., & Stonnington, C. M. (2015a). Role of theory of mind in emotional awareness: Implications for conceptualization and measurement. *Consciousness and Cognition, 33,* 398–405.

Lane, R. L., & Schwartz, G. E. (1987). Levels of emotional awareness: A cognitive-developmental theory and its applications to psychopathology. *American Journal of Psychiatry, 144*(2), 133–143.

Lane, R. D., & Schwartz, G. E. (1992). Levels of emotional awareness: Implication for psychotherapeutic integration. *Journal of Psychotherapy Integration, 2*(1), 1–18.

Lane, R. D., Sechrest, L., Reidel, R., Brown, V., Kaszniak, A., & Schwartz, G. (1995). Alexithymia and nonverbal emotion processing deficits. *Psychosomatic Medicine, 57*(1), 84–85.

Lane, R. D., Sechrest, L., Riedel, R., Shapiro, D. E., & Kaszniak, A. W. (2000). Pervasive emotion recognition deficit common to alexithymia and the repressive coping style. *Psychosomatic Medicine, 62*(4), 492–501.

Lane, R. D., Sechrest, L., Reidel, R. G., Weldon, V., Kaszniak, A. W., & Schwartz, G. E. (1996). Impaired verbal and non verbal emotion recognition in alexithymia. *Psychosomatic Medicine, 58*(3), 203–210.

Lane, R. D., Weihs, K. L., Herring, A., Hishaw, A., & Smith, R. (2015b). Affective agnosia: Expansion of the alexithymia construct and a new opportunity to integrate and extend Freud's legacy. *Neuroscience and Biobehavioral Reviews, 55,* 594–611.

Lange, C. G. ([1885] 1912). The mechanism of the emotion. In B. Rand (Ed.), *The classical psychologists* (pp. 672–684). Boston: Houghton Mifflin.

Lange, C. G. ([1885] 1922). The emotions. A psychophysiological study. In K. Dunlap (Ed.), *The emotions* (Vol. 1, pp. 32–90). Baltimore: Williams & Wilkins Company.

Lazarus, R. S. (1991). *Emotion and adaptation*. New York: Oxford University Press.

Lazarus, R. S., Averill, J. R., & Opton, E. M., Jr. (1970). Towards a cognitive theory of emotion. In M. B. Arnold (Ed.), *Feelings and emotions* (pp. 207–232). New York: Academic Press.

Lazarus, R. S., & Smith, C. A. (1988). Knowledge and appraisal in the cognition-emotion relationship. *Cognition and Emotion, 2*(4), 281–300.

Lesser, I. M. (1981). A review of alexithymia concept. *Psychosomatic Medicine, 43*(6), 531–543.

Lewis, C. I. (1929). *Mind and the world-order. Outline of a theory of knowledge*. New York, Chicago: Charles Scribner & Sons.

Lindsley, D. B. (1950). Emotion and the electroencephalogram. In M. L. Raymert (Ed.), *Feelings and emotions. The Mooseheart Symposium* (pp. 238–246). New York: McGraw-Hill.

Lindsley, D. B. (1951). Emotion. In S. S. Stevens (Ed.), *Handbook of experimental psychology* (pp. 473–516). New York: Wiley.

Lindsley, D. B. (1970). The role of nonspecific reticulo-thalamo-cortical systems in emotions. In P. Black (Ed.), *Physiological correlates of emotions* (pp. 147–188). New York, London: Academic Press.

Lissauer, H. (1890). Ein Fall von Seelenblindheit nebst einem Beitrage zur Theorie derselben. *Archiv für Psychiatrie und Nervenkrankheiten, 21,* 222–270.

Lombardo, M. V., Barnes, J. L., Wheelwright, S. J., & Baron-Cohen, S. (2007). Self-referential cognition and empathy in autism. *PLoS ONE, 2*(9), e883. doi:10.1371/journal.pone.0000883.

Lyons, W. (1980). *Emotion*. Cambridge: Cambridge University Press.

Mac Laren, K. (2006). Emotional disorder and the mind-body problem: A case study of alexithymia. *Chiasmi International, 8,* 139–155.

MacLean, P. D. (1949). Psychosomatic disease and the "visceral brain": Recent developments bearing on the Papez theory of emotion. *Psychosomatic Medicine, 11*(6), 338–353.

Macpherson, F. (2011). *The senses. Classical and contemporary philosophical perspectives*. Oxford: Oxford University Press.

Mandler, G. (1975). *Mind and emotion*. New York: Wiley.

Mantegazza, P. (1885). *La physionomie et l'expression des sentiments*. Paris: s.n.

Marcel, A. J. (1993). Slippage in the unity of consciousness. In G. R. Bock & J. Marsh (Eds.), *Experimental and theoretical studies of consciousness. Ciba Foundation Symposium* (Vol. 174, pp. 168–180). Chichester: Wiley.

Marconi, D. (1987). Two aspects of lexical competence. *Lingua e Stile, 12*(3), 385–395.

Marconi, D. (1991). Understanding and reference. *Semiotiques, 1*(1), 9–25.

Marconi, D. (1995). On the structure of lexical competence. *Proceedings of the Aristotelian Society. New Series, 95,* 131–151.

Marconi, D. (1997). *Lexical competence*. Cambridge, MA: MIT Press.

Marks, J. (1982). A theory of emotion. *Philosophical Studies, 42*(1), 227–242.

Marty, P., & de M'Uzan, M. (1963). La "pensée opératoire". *Revue Française de Psychanalyse, 27,* 1345–1356.

McDonald, P. W., & Prkachin, K. M. (1990). The expression and perception of facial emotion in alexithymia: A pilot study. *Psychosomatic Medicine, 52*(2), 199–210.

McDougall, J. (1972). The anti-analysand in analysis. A clinical portrait and a theoretical concept. *Revue Francaise de Psychoanalyse, 36*(2), 167–184.

McDougall, J. (1991). Reflections on affects: A psychoanalytic view of alexithymia. In J. McDougall (Ed.), *Theaters of the mind. Illusion and truths on the psychoanalytic stage* (pp. 147–179). London, NewYork: Routledge.

Moriguchi, Y., Decety, J., Ohnishi, T., Maeda, M., Mori, T., Nemoto, K., et al. (2007). Empathy and judging other's pain: An fMRI study of alexithymia. *Cerebral Cortex, 17*(9), 2223–2234.

Moseley, R. L., Shtyrov, Y., Mohr, B., Lombardo, M. V., Baron-Cohen, S., & Pulvermüller, F. (2015). Lost for emotion words: What motor and limbic brain activity reveals about autism and semantic theory. *Neuroimage, 104,* 413–422.

Mosso, A. (1881). *The circulation of the blood in the human brain* (C. N. Fabbri, Trans.). London: s.n.

Mosso, A. ([1884] 1886). *Fear* (E. Lough & F. Kiesow, Trans.). London: s.n.

Muller, H. (2008). Why qualia are not epiphenomenal. *Ratio, 21*(1), 85–90.

Musacchio, J. M. (2005a). Why do qualia and the mind seem nonphysical. *Synthese, 147*(3), 425–460.

Musacchio, J. M. (2005b). The ineffability of qualia and the world-anchoring problem. *Language Science, 27*(4), 403–435.

Nemiah, J. C. (1972). The psychosomatic nature of anorexia nervosa. *Advances in Psychosomatic Medicine, 7,* 316–321.

Nemiah, J. C. (1977). Alexithymia. Theoretical considerations. *Psychotherapy and Psychosomatics, 28*(1–4), 199–206.

Nemiah, J. C., Freyberger, H., & Sifneos, P. E. (1976). Alexithymia. A view of the psychosomatic process. In O. W. Hill (Ed.), *Modern trends in psychosomatic medicine* (Vol. 3, pp. 430–439). London: Butterworths.

Nemiah, J. C., & Sifneos, P. E. (1970). Psychosomatic illness. A problem in communication. *Psychotherapy and Psychosomatics, 18*(1), 154–160.

Neu, J. (2000). *A tear is an intellectual thing: The meanings of emotion.* Oxford: Oxford University Press.

Niedenthal, P. M. (2008). Emotion concepts. In M. Lewis, J. M. Haviland-Jones, & L. Feldman Barrett (Eds.), *Handbook of emotions* (3rd ed., pp. 587–600). New York, London: The Guilford Press.

Nieuwenhuyse, B., Offenberg, L., & Frijda, N. H. (1987). Subjective emotion and reported body experience. *Motivation and Emotion, 11*(2), 169–182.

Nowlis, V., & Nowlis, H. H. (1956). The description and analysis of moods. *Annals of New York Academy of Sciences, 65*(4), 345–355.

Nussbaum, M. C. (2001). *Upheavals of thought. The intelligence of the emotions.* Cambridge: Cambridge University Press.

O'Callaghan, C. (2008). Seeing what you hear: Cross-modal illusion and perceptions. *Philosophical Issues, 18*(1), 316–338.

O'Callaghan, C. (2015). Not all perceptual experience is modality specific. In D. Stokes, M. Matthen, & S. Biggs (Eds.), *Perception and its modalities* (pp. 132–165). Oxford: Oxford University Press.

Oatley, K. (1988). Plans and the communicative function of emotions: A cognitive theory. In V. Hamilton, G. H. Bower, & N. H. Frijda (Eds.), *Cognitive perspectives on emotion and motivation. NATO advanced science institute. Series D* (Vol. 44, pp. 345–364). Dordrecht: Kluwer.

Oatley, K. (1992). *Best laid schemes. The psychology of emotions*. Cambridge: Cambridge University Press.

Oatley, K., & Duncan, E. (1992). Incidents of emotion in daily life. In K. T. Strongman (Ed.), *International review of studies of emotion* (Vol. 2, pp. 249–293). Chichester: Wiley.

Oatley, K., & Johnson-Laird, P. N. (1987). Towards a cognitive theory of emotions. *Cognition and Emotion, 1*(1), 29–50.

Oatley, K., & Johnson-Laird, P. N. (1990). Semantic primitives for emotions: A Reply to Ortony and Clore. *Cognition and Emotion, 4*(2), 129–143.

Oatley, K., & Johnson-Laird, P. N. (1992). Terms of emotion: The inferences that can be drawn. *Revista de Psicologia Social, 7*(1), 97–104.

Ortony, A., & Turner, W. (1990). What's basic about basic emotions? *Psychological Review, 97* (3), 315–331.

Osgood, C. E. (1969). On the why's and wherefore's of E, P, and A. *Journal of Personality and Social Psychology, 12*(3), 194–199.

Panksepp, J. (2000). Emotions as natural kinds within the mammalian brain. In M. Lewis & J. Haviland-Jones (Eds.), *Handbook of emotions* (2nd ed., pp. 137–156). New York: Guilford Press.

Peacocke, C. (2001). Does perception have a nonconceptual content? *The Journal of Philosophy, 98*(5), 239–269.

Picciuto, V., & Carruthers, P. (2015). Inner sense. In D. Stokes, M. Matthen, & S. Biggs (Eds.), *Perception and its modalities* (pp. 277–294). Oxford: Oxford University Press.

Plato. (1983). In R. H. Waterfield (Ed.), *Philebus*. Penguin Classics.

Plutchik, R., & Ax, A. F. (1967). A critique of 'determinants of emotional state' by Schachter and Singer (1962). *Psychology, 4,* 79–82.

Prinz, J. J. (2002). *The conscious brain. How attention engenders experience*. Oxford: Oxford University Press.

Prinz, J. J. (2004a). *Gut reactions. A perceptual theory of emotions*. Cambridge, MA: MIT Press.

Prinz, J. J. (2004b). The fractionation of introspection. *Journal of Consciousness Studies, 11*(7–8), 40–57.

Prinz, J. J. (2010). When is perception conscious? In B. Nanay (Ed.), *Perceiving the world* (pp. 310–332). Oxford: Oxford University Press.

Raftopoulos, A., & Müller, V. C. (2006). The phenomenal content of experience. *Mind and Language, 21*(2), 187–219.

Raftopoulos, A. (2009). *Cognition and perception. How do psychology and neural science inform philosophy?* Cambridge, MA: MIT Press.

Reisenzein, R. (1983). The Schachter theory of emotion: Two decades later. *Psychological Bullettin, 94*(2), 239–264.

Rorty, A. (Ed.). (1980). *Explaining emotions*. Los Angeles: California University.

Roseman, I. (1984). Cognitive determinants of emotion. A structural theory. *Review of Personality and Social Psychology, 5*(1), 11–36.

Ross, P. W. (2001). Qualia and the senses. *The Philosophical Quarterly, 51*(205), 495–511.

Ross, E. D., & Mesulam, M. (1979). Dominant language functions of the right hemisphere? Prosody and emotional gesturing. *Archives of Neurology, 36*(3), 144–148.

Ruesch, J. (1948). The infantile personality. *Psychosomatic Medicine, 10*(3), 134–144.

Russell, B. (1910–1911). Knowledge by acquaintance and knowledge by description. *Proceedings of the Aristotelian Society. 11*, 108–128.

Russell, B. (1912). *The problems of philosophy*. Oxford: Oxford University Press.

Russell, J. A. (1980). A circumplex model of affect. *Journal of Personality and Social Psychology, 39*(6), 1161–1178.

Russell, J. A. (2003). Core affect and the psychological construction of emotions. *Psychological Review, 110*(1), 145–172.

Saarni, C. (1999). *The development of emotional competence*. London, New York: The Guilford Press.

Samson, A. C., Huber, O., & Gross, J. J. (2012). Emotion regulation in Asperger's Syndrome and high-functioning autism. *Emotion, 12*(4), 659–665.

Schachter, S. (1970). The assumption of identity and peripheralist-centralist controversies in motivation and emotion. In M. B. Arnold (Ed.), *Feelings and emotions* (pp. 111–121). New York: Academic Press.

Schachter, S., & Singer, J. (1962). Cognitive, social and physiological determinants of emotional state. *Psychological Review, 69*(5), 379–399.

Scherer, K. R. (1984). Emotion as a multicomponent process: A model and some cross-cultural data. *Review of Personality and Social Psychology, 5,* 37–63.

Scherer, K. R. (2009). Emotion theories and concepts (psychological perspectives). In D. Sander & K. R. Scherer (Eds.), *The Oxford companion to emotion and the affective sciences* (pp. 145–151). Oxford: Oxford University Press.

Schmidt, E. (2015). *Modest nonconceptualism. Epistemology, phenomenology, and content.* Berlin: Springer.

Schoemaker, S. (2002). Introspection and phenomenal character. In D. J. Chalmers (Ed.), *Philosophy of mind. Classical and contemporary readings* (pp. 457–472). Oxford: Oxford University Press.

Sellschopp-Rüpper, A., & von Rad, M. (1977). Pinocchio—A psychosomatic syndrome. In W. Bräutigam & M. von Rad (Eds.), *Towards a theory of psychosomatic disorders. Alexithymia—pensée operatoire—psychosomatisches Phänomen* (pp. 357–360). Basel: Karger.

Shallice, T., & Jackson, M. (1998). Lissauer on agnosia. *Cognitive Neuropsychology, 5*(2), 153–192.

Shands, H. C. (1958). An approach to measurement of suitability for psychotherapy. *Psychiatric Quarterly, 32*(3), 500–521.

Sherrington, C. S. (1906). *The integrative action of nervous system*. Yale: Yale University Press.

Shore, A. N. (1994). *Affect regulation and the origin of self. The neurobiology of emotional development*. London, New York: Routledge.

Sifneos, P. E. (1967). Clinical observations on some patients suffering from a variety of psychosomatic diseases. *Acta Medica Psychosomatica, 7,* 1–10.

Sifneos, P. E. (1972). *Short-term psychotherapy and emotional crisis*. Cambridge, MA: Harvard University Press.

Sifneos, P. E. (1972–1973). Is dynamic psychotherapy contraindicated for a large number of patients with psychosomatic disease? *Psychoterapy and Psychosomatics 21*(2):133–136.

Sifneos, P. E. (1973). The prevalence of 'alexithymic' characteristics in psychosomatic patients. *Psychotherapy and Psychosomatics, 22*(2–6), 255–262.

Sifneos, P. E. (1974). A reconsideration of psychodynamic mechanisms in psychosomatic symptom formation in view of recent clinical observations. *Psychotherapy and Psychosomatics, 24*(2), 151–155.

Sifneos, P. E. (1994). Affect deficit and alexithymia. *New Trends in Experimental and Clinical Psychiatry, 10*(4), 193–195.

Sifneos, P. E. (1996). Alexithymia: Past and present. *American Journal of Psychiatry, 153*(7), 137–142.

Silani, G., Bird, G., Brindley, R., Singer, T., Frith, C., & Frith, U. (2008). Levels of emotional awareness and autism: A fMRI study. *Social Neuroscience, 3*(2), 97–112.

Slaby, J. (2008). James: von der Physiologie zur Phänomenologie. In H. Landweer & U. Renz (Eds.), *Klassische Emotionstheorien. Von Plato bis Wittgenstein* (pp. 549–567). Berlin: De Gruyter.

Smith, B. C. (2015). The chemical senses. In M. Matthen (Ed.), *The Oxford handbook of philosophy of perception* (pp. 314–352). Oxford: Oxford University Press.

Smith, C. A., & Ellsworth, P. C. (1985). Patterns of cognitive appraisal in emotion. *Journal of Personality and Social Psychology, 48*(4), 813–838.

Solomon, R. C. (1976). *The passions: The myth and the nature of human emotions*. Garden City: Anchor Books.

Solomon, R. C. (2008). The philosophy of emotions. In M. Lewis, J. M. Haviland-Jones, & L. Feldman Barrett (Eds.), *Handbook of emotions* (pp. 3–16). New York, London: The Guilford Press.

Solomon, R. L. (1980). The opponent-process theory of acquired motivation: The costs of pleasure and the benefits of pain. *American Psychologist, 35*(8), 691–712.

Spence, C., & Bayne, T. (2015). Is consciousness multisensory? In D. Stokes, M. Matthen, & S. Biggs (Eds.), *Perception and its modalities* (pp. 95–132). Oxford: Oxford University Press.

Spinoza, B. ([1677] 1994). Ethics. In E. Curley (Ed.), *A Spinoza reader: The ethics and other works* (pp. 85–265). Princeton: Princeton University Press.

Strongman, K. T. (2003). *The psychology of emotion from everyday life to theory*. New York: Wiley.

Subic-Wrana, C. (2011). Emotion regulation and mentalization in somatoform disorders. In I. Nyklicek, A. Vingerhoets, & M. Zeelenberg (Eds.), *Emotion regulation and well-being* (pp. 245–260). Berlin: Springer.

Subic-Wrana, C., Beutel, M., Nkebel, A., & Lane, R. L. (2010). Theory of mind and emotional awareness deficits in patients with somatoform disorders. *Psychosomatic Medicine, 72*(4), 404–411.

Taylor, G. J. (1987). *Psychosomatic medicine and contemporary psychoanalysis*. Madison (CT): International University Press.

Taylor, G. J. (2000). Recent developments in alexithymia: Theory and research. *Canadian Journal of Psychiatry, 45*(2), 134–141.

Taylor, G. J., Bagby, R. M., & Parker, J. D. A. (1997). *Disorders of affect regulation. Alexithymia in medical and psychiatric illness*. Cambridge: Cambridge University Press.

Taylor, G. J., Bagby, R. M., & Parker, J. D. A. (2016). What's in the name of alexithymia? *Neuroscience and Biobehavioral Reviews, 68,* 1006–1020.

Taylor, G. J., Parker, J. D. A., Bagby, R. M., & Bourke, M. P. (1996). Relationships between alexithymia and psychological characteristics associated with eating disorders. *Journal of Psychosomatic Research, 41*(6), 561–568.

Thalberg, I. (1977). *Perception, emotion and action*. New Haven: Yale University Press.

Thompson, J. (2009). *Emotionally dumb. An overview on alexithymia*. Adelaide: Soul Books.

Timoney, L. R., & Holder, M. D. (2013). *Emotional processing: Deficits and happiness. Assessing the measurement, correlates, and well-being of people with Alexithymia*. Berlin: Springer.

Tolman, E. C. (1922a). A new formula for behaviorism. *Psychological Review, 29*(1), 44–53.

Tolman, E. C. (1922b). Concerning the sensation quality. A behavioristic account. *Psychological Review, 29*(2), 140–145.

Tolman, E. C. (1923). A behavioristic account of the emotions. *Psychological Review, 30*(3), 217–227.

Tomkins, S. S. (1962). *Affect, imagery, consciousness. Volume 1: The positive affects*. New York: Springer.

Tomkins, S. S. (1963). *Affect, imagery, consciousness. Volume 2: The negative affects*. New York: Springer.

Tsakiris, M., Jimenez, A. T., & Costantini, M. (2011). Just a heartbeat away from one's body: Interoceptive sensitivity predicts malleability of body-representations. *Proceedings of the Royal Society. Biological Science, 278*(1717), 2470–2476.

Tsakiris, M., Schütz-Bosbach, S., & Gallagher, S. (2007). On agency and body-ownership. Phenomenological and neurocognitive reflections. *Consciousness and Cognition, 16*(3), 645–660.

Tye, M. (2002). Visual qualia and visual content revisited. In D. J. Chalmers (Ed.), *Philosophy of mind. Classical and contemporary readings* (pp. 447–456). Oxford: Oxford University Press.

Tye, M. (2003). *Consciousness and persons unity and identity*. Cambridge, MA: MIT Press.

Tye, M. (2005). Nonconceptual content, richness and fineness of grain. In T. Gendler & J. Hawthorne (Eds.), *Perceptual experience* (pp. 504–530). Oxford: Oxford University Press.

Tye, M. (2006). The thesis of nonconceptual content. *European Review of Philosophy, 6,* 7–30.

Vanheule, S., Desmet, M., Megank, R., & Bogaerts, S. (2007). Alexithymia and interpersonal problems. *Journal of Clinical Psychology, 63*(1), 109–117.

Vermeulen, N., Luminet, O., & Corneille, O. (2006). Alexithymia and the autonomatic processing of affective information: Evidence from the affective priming paradigm. *Cognition and Emotion, 20*(1), 64–91.

von Rad, M. (1983). *Alexithymie. Empirische Untersuchungen zur Diagnostik und Therapie psychosomatisch Kranker*. Berlin, Heidelberg, New York: Springer.

Watson, J. B. (1913). Psychology as the behaviorist views it. *Psychological Review, 20*(2), 158–177.

Watson, J. B. ([1924] 1931). *Behaviorism*. London: Kegan Paul, Trench, Trubner & Co.

Watson, J. B., & Morgan, J. J. B. (1917). Emotional reactions and psychological experimentation. *American Journal of Psychology, 28*(1), 163–174.

Watson, J. B., & Rayner, R. (1920). Conditioned emotional reactions. *Journal of Experimental Psychology, 3*(1), 1–14.

Watson, D., & Tellegen, A. (1985). Towards a consensual structure of mood. *Psychological Bulletin, 98*(2), 219–235.

Weiner, B. (1985). An attributional theory of achievement motivation and emotion. *Psychological Review, 92*(4), 548–573.

Wing, L. (1981). Asperger's syndrome: A clinical account. *Psychological Medicine, 11*(1), 115–129.

Wittgenstein, L. ([1958] 1986). *Philosophical investigations*. Oxford: Basil Blackwell.

Woodward, R., & Sherrington, C. S. (1904). A pseudoaffective reflex and its spinal path. *Journal of Physiology, 31*(2), 234–243.

Wotschack, C., & Klann-Delius, G. (2013). Alexithymia and the conceptualization of emotions: A study of language use and semantic knowledge. *Journal of Research in Personality, 47*(5), 514–523.

Wundt, W. (1890). *Grundriss der Psychologie*. Leipzig: Engelmann.

Zajonc, R. B., & McIntosh, D. N. (1992). Emotions research. Some promising questions and some questionable promises. *Psychological Science, 3*(1), 70–74.

Chapter 6
The 'Proprioceptive' Component of Abstract Concepts

Abstract In this chapter, we address the issue of whether the mastery of abstract words requires only inferential knowledge and thus, if the concepts that support the mastery of abstract words include only linguistic information. We start by differentiating the notions of 'abstract' and 'general' which are often erroneously confused. We then identify a strict definition of abstract, as contrasted with 'concrete', that applies to words or concepts whose referent cannot be experienced by the senses. We argue that abstract words/concepts would be better described as theoretical, because they are usually conceived as structured sets of inferential knowledge expressed linguistically; that is, as small theories. Pointing out parallels with Carnap's analysis of this issue in philosophy of science, we hypothesize that words/concepts denoting non-observable entities are not all 'equally theoretical', because their link to sensory experience can be stronger or weaker. We revive the distinction, inspired by Quine, between theoretical and intertheoretical concepts/words. This distinction relies on the fact that the former—unlike the latter—retains a strong, although indirect, connection with perception. In Quine's discussion, perception is understood uniquely in terms of observability, i.e. of external sensory experience. Here we argue, however, that bodily, 'proprioceptive' (i.e. propriosensitive) experience can also serve to referentially ground theoretical (i.e. abstract) concepts/words. We frame this issue using the example of the theoretical concept 'freedom' and Lakoff's hypothesis that this concept is developed on the basis of bodily information. We contrast this with the example of 'democracy' which more closely resembles an intertheoretical concept/word. Furthermore, we show that one of the classical views put forward in psycholinguistic research to explain how abstract concepts are mentally represented—i.e. Paivio's *Dual Coding Theory*—points in the same direction as our analysis. The same is true of Barsalou's work suggesting that we use internal information to understand at least some abstract words. To sustain this position, we put forward two lines of evidence: the first comes from psycholinguistic studies while the second examines deficits of semantic competence exhibited by people with Autism Spectrum Disorder. On the basis of our analysis, we put forward a classification that distinguishes between *different kinds of concreteness* and *different degrees of abstraction*: concepts/words referring to body experiences and basic emotions are described as analogous to

S. Dellantonio and L. Pastore, *Internal Perception*, Studies in Applied Philosophy, Epistemology and Rational Ethics 40, DOI 10.1007/978-3-662-55763-1_6

concrete concepts/words because they are grounded in perceptual (i.e. propriosensitive) experience, while abstract concepts/words are considered more or less abstract depending on whether they are intratheoretical (and rely entirely on inferential information) or theoretical (and are partially grounded in perceptual—or more often in propriosensitive perceptual—information). In the last section of the chapter we consider two scales that have been used in psycholinguistic research to measure the degree of concreteness versus abstractness of words and we show that —used conjointly—they can provide a measure of the internal versus external grounding of specific words.

1 Abstract Terms: How Should 'Abstraction' Be Interpreted

At the beginning of the previous chapter we recalled that semantic competence for words like 'cat' which have an external observable referent can be explained in terms of the conjunction between referential and inferential competence (Marconi 1997). A person can be considered fully competent with respect to e.g. 'cat' if s/he is able to report that cats are 4-legged, domestic animals with hair, tails and pointed ears and if s/he can at the same time identify cats in the external world and distinguish them from other kinds of things. However, in the case of words that do not have an external perceivable referent, it is not clear whether it is possible to speak specifically about referential competence or whether our semantic mastery consists uniquely in a linguistic capacity to associate certain words with others. This chapter deals with this question.[1]

Words/concepts that do not have a concrete, directly observable referent in the external world are most commonly called 'abstract' and, less frequently, 'theoretical'. We will mainly use the term 'theoretical'. There are two main reasons why we prefer the term 'theoretical' to the term 'abstract'. (i) First, we wish to avoid the intrinsic ambiguity of the term 'abstract'. (ii) Second, we try to stress the essentially 'theoretical' nature of this group of words, which stems from the fact that their meaning depends largely on an inferential network, i.e. on a model which determines how they should be used and what they should be applied to. In the following part of this section we will more closely examine each of these reasons in turn.

[1]A number of studies carried out on particular types of words or concepts (e.g. social concepts, scientific concepts, formal or mathematical concepts, natural kinds, artifacts etc.) show that there are important differences in the information they rely on and in the way this information is acquired and used. Thus, it would surely be useful to address the issue of semantic competence separately for all these classes. However, the distinction between words/concepts that do and [words that]? do not have an observable (concrete) reference in the external world seems to apply across the board. As e.g. Jesse Prinz points out at the beginning of his book on concepts: "The human conceptual repertoire ranges from the sensory to the abstract" (Prinz 2002: 3).

(i) The word 'abstract' has been used in a number of different ways both in the philosophical and the psychological literature. As a result, its use is widely variable and therefore ambiguous. First of all, the opposition between abstract and concrete can be interpreted in a purely linguistic sense, excluding any implications regarding the nature of their respective referents. As for example Gideon Rosen summarizes: "Traditional grammar distinguishes the abstract noun 'whiteness' from the concrete noun 'white' without implying that this linguistic contrast corresponds to a metaphysical distinction in what these words stand for" (Rosen 2012). In addition, the discussion on the opposition between abstract and concrete steadily swings between two different notions of abstract: (a) the one equates abstract with general and (b) the other defines concepts or words denoting 'things' that cannot be experienced by the senses as abstract.[2] Our analysis deals with this second sense of abstraction but before proceeding to a more precise definition, we first address some issues regarding the interpretation of abstraction in terms of generality.

(a) The equation between abstract and general has a long story in the philosophical debate on categorization. The contemporary version of this debate runs along lines similar to those that characterized discussions in the 17th century.[3] Regardless of the specific positions embraced by various authors, the issue of abstraction is defined in terms of the problem of understanding how humans come to form general ideas that can represent different particular instances. According to John Locke, for example, the idea of a right triangle is a specific idea, while the corresponding abstract idea is the general idea of triangle which includes all kinds of triangles despite their specific characteristics.

> […] when we nicely reflect upon them, we shall find that general ideas are fictions and contrivances of the mind, that carry difficulty with them, and do not so easily offer themselves as we are apt to imagine. For example, does it not require some pains and skill to form the general idea of a triangle, (which is yet none of the most abstract, comprehensive, and difficult); for it must be neither oblique nor rectangle, neither equilateral, equicrural, nor scalenon; but all and none of these at once. In effect, it is something imperfect, that cannot exist; an idea wherein some parts of several different and inconsistent ideas are put together. It is true, the mind, in this imperfect state, has need of such ideas, and makes all the haste to them it can, for the conveniency of communication and enlargement of knowledge; to both which it is naturally very much inclined. But yet one has reason to suspect such ideas are marks of our imperfection; at least, this is enough to show that the most abstract and general ideas are not those that the mind is first and most easily

---

[2]There is at least one further major debate on the meaning of 'abstract' that we will not consider here. This debate concerns the metaphysical nature of certain specific objects. In its contemporary version, this debate starts with Frege's argument that numbers and thoughts are neither material beings nor ideas in the mind, because—if they were—they wouldn't be objective, and must instead belong to a timeless 'third realm'. Frege never used the word 'abstract'; however, 'abstract' is used in the later tradition to refer to objects that—like mathematical ones—are supposed to exist apart from both the mental and the physical (for an overview on this see e.g. Rosen 2012).

[3]In general terms this debate during the 17th century recalls an older philosophical issue which played an important role especially during medieval times concerning the generality of thought as opposed to the particularity of the world and more specifically the question of how thought in its generality can match (i.e. be true of) the world in its particularity (Robinson 1999).

> acquainted with, nor such as its earliest knowledge is conversant about (Locke [1690] 1999: Book IV, Chap. VII, Sect. 9).

According to Locke, abstract (general) ideas are complex to master and are learned therefore only after the specific ones. Since any triangle we can figuratively picture in our mind is specific and has specific characteristics that do not apply to triangles in general (e.g. it is right angle, scalene etc.), a general idea of triangle might result only from the composition of parts of specific ideas. As David Hume states, discussing the same example: "[…] it is a principle generally received in philosophy that everything in nature is individual, and that it is utterly absurd to suppose a triangle really existent, which has no precise proportion of sides and angles" (Hume [1740] 2009: 44). Considered from this perspective, the problem of abstraction clearly coincides with that of categorization in general, i.e. with the problem of explaining how several particular ideas can be subsumed in the same general idea.

> It is evident, that in forming most of our general ideas, if not all of them, we abstract from every particular degree of quantity and quality, and that an object ceases not to be of any particular species on account of every small alteration in its extension, duration and other properties. It may therefore be thought, that here is a plain dilemma, that decides concerning the nature of those abstract ideas, which have afforded so much speculation to philosophers. The abstract idea of a man represents men of all sizes and all qualities; which it is concluded it cannot do, but either by representing at once all possible sizes and all possible qualities, or by, representing no particular one at all (Hume [1740] 2009: 41).

The issue of abstraction is addressed in this same way by Berkeley who uses the example of MAN.[4] George Berkeley suggests that the abstract idea of MAN is obtained from the observation of different 'things' that exhibit a certain resemblance to each other: we can form the idea of different men and then focus selectively only on what is common to them, while all their differences and specificities are overlooked:

> For example, the mind having observed that Peter, James, and John resemble each other, in certain common agreements of shape and other qualities, leaves out of the complex or compounded idea it has of Peter, James, and any other particular man, that which is peculiar to each, retaining only what is common to all; and so makes an abstract idea wherein all the particulars equally partake, abstracting entirely from and cutting off all those circumstances and differences, which might determine it to any particular existence. And after this manner it is said we come by the abstract idea of man or, if you please, humanity or human nature […] (Berkeley [1710] 1999: 10).

In these discussions, abstraction is understood as a generalization process through which one subtracts and adds specific properties to a particular idea in order to obtain a general idea which subsumes various particular instances. Thus, abstraction indicates basically the process of concept formation, while concepts are considered to be abstract when they are general.

---

[4]In this respect, it is appropriate to recall that it was Hume who went back to Berkeley's example and to the way Berkeley dealt with abstraction and not the other way around.

The contemporary debate on concepts and categorization implicitly assumes that all concepts are general in the sense defined by empiricism. Even individual concepts (names) like 'Peter', 'James', and 'John'—so-called token concepts that allow us to identify the same instances in different times and under different perspectives (see e.g. Jackendoff 1985: Chap. 5)—are general concepts in the sense that they do not keep track of the detail of single experiences (they don't keep track e.g. of Peter's dress, or of his hairstyle on a specific occasion) and in this way they allow us to identify this person on different occasions (regardless of outfits, perspectives, etc.). The fact that all concepts are considered general does not mean however that we can dispense with the equation (or the confusion) of general with abstract. Indeed, concepts can be more or less general, for example 'poodle' is considered less general than 'dog' and 'animal' is considered more general than both 'poodle' and 'dog' and *the level of generality of a concept is often equated (and confused) with its level of abstraction*. Words like 'abstract' and 'abstraction' are often used to describe generalization procedures or scales of generality as implied by the statement e.g. "the intermediate level of abstraction is privileged" (meaning that concepts of medium generality are the most used and those which are learned first), "the basic level is the highest level of abstraction that can be captured by mental images"; (Prinz 2002: 10, 163), "concepts that are more abstract than basic-level concepts" (Lakoff 1987: 302); "it could have been that the most abstract categories [in the hierarchy] were easiest to use, because they have fewer features" or "the ability to form categories at different levels of abstraction and the willingness to include the same objects in categories at two different levels" (Murphy 2004: 325), etc.

(b) However, in spite of the fact that 'abstract' and 'general' are still often used interchangeably, this characterization of abstractness as generality is neither the only nor the most important one in the present debate on categorization. In fact, the label 'abstract' is not only or mainly used in opposition to 'particular'; rather, *abstractness is more often defined in contrast to concreteness on the basis of the kind of things a concept or a word refers to: a concept or word is considered concrete if the class of objects it refers to can be observed (perceived) through the senses; it is abstract otherwise*. For example, Katja Wiemer-Hastings, Jan Krug and Xu Xu report: "Concrete and abstract nouns are commonly defined with reference to perceivability: Concrete entities are considered to be physical entities with characteristic shapes, parts, materials, etc., whereas abstract entities lack physical attributes (e.g. Crystal 1995). […] If an entity cannot be perceived, it is abstract" (Wiemer-Hastings et al. 2001: 1134–1135).[5] According to this criterion a concept like e.g. MAN as discussed by Berkeley and Hume is certainly concrete instead of abstract since it has an observable (perceivable) reference in the external world

[5]This article also mentions the ambiguous status of some word classes that are considered abstract by some authors, but as different from both abstract and concrete concepts by other authors. The most important among these is the class of emotion words, which we analyzed extensively in the previous chapter and will discuss again later in this chapter (see Sect. 5). For other definitions of abstract see e.g. Paivio ([1986] 1990) and Barsalou (1999, 2003).

(men). In contrast, concepts like TRUTH, IMPOSSIBLE, BELIEF, DEMOCRACY, FREEDOM, ART, WILL etc. which represent things that are not perceptually experienced are considered to be abstract. As e.g. Jesse Prinz suggests, DEMOCRACY is an abstract concept because its referent cannot be directly perceptually experienced: because a democracy cannot be seen, heard, smelled, or tasted (Prinz 2002: 167).

This differentiation between generality and abstraction allows us to identify some relevant features of the concepts/words belonging to this class. First of all, generality is necessarily a relational condition: all concepts are general, thus the distinction among concepts as regards their generality can only be a matter of degree and can be established only relative to others. For example, the sequence POODLE, DOG, MAMMAL, ANIMAL describes a hierarchal scale of increasing generality. As this taxonomy already makes clear, abstract concepts like those mentioned before (e.g. TRUTH, IMPOSSIBLE, BELIEF, DEMOCRACY, FREEDOM, ART, WILL etc.) do not belong to any hierarchal scale of this kind. Thus, while more general concepts/words are defined starting from less general concepts/words and retain, so to speak, some concrete roots, abstract concepts/words are not the result of a generalization procedure and cannot be traced back to any scale of ascending generality. The equation and the confusion between these two uses of the notion of 'abstraction' might be due to the fact that concepts/words with a low degree of generality are always concrete. And the coincidence between low generality and concreteness might lead one to think that the converse equation applies and that very general concepts are abstract. But general concepts are still concrete as long as they (directly or indirectly) refer to instances that can be perceived through the senses.

Concepts/words like e.g. ANIMAL, FURNITURE and VEHICLE are quite general, but this does not mean that they are abstract. In fact, if 'abstract' is defined in opposition to 'concrete', then a concept/word is considered to be abstract only if it denotes something that cannot be perceived by the senses. But the instances denoted by concepts/words like ANIMAL, FURNITURE and VEHICLE—i.e. animals, pieces of furniture and means of transport—can be perceived by the senses and are therefore not abstract. In the case of concepts/words like ANIMAL, FURNITURE and VEHICLE the issue at stake is still the one put forward by Hume—how can we form an idea (we would say, a representation) that includes all possible instances that belong to a concept without resembling any specific one of them? Put differently, in the case of general concepts/words the problem is how to build (more) general concepts/words starting from (more) specific ones or from individual experiences of particular instances. *The question raised by abstract concepts/words is quite different and concerns the problem of explaining how it is possible to define new concepts at least apparently from scratch through a procedure other than generalization.*

This opposition between abstract and concrete concepts/words results from what we called (Chap. 3, Sect. 7) the "standard picture of meaning" which suggests that natural languages are composed of two different kinds of words: "a set of words, known as the observation vocabulary, whose meaning are fixed by their relations with observable properties of the environment" and another set whose meaning is "fixed by a network of inferential or other relations to the meanings of other words, including those belonging to the observation vocabulary" (Cruse 2000: 52). The observational vocabulary is usually identified with the set of concrete words/concepts, while abstract words/concepts are generally considered those whose meaning/content is artificially produced through language and specifically through verbal inferential relationships. Still, despite appearances, the opposition between concrete and abstract is not dichotomic since it is possible to differentiate between more and less abstract words/concepts on the basis of 'the distance' of a word from the observational level. While words/concepts denoting instances that are perceivable with the senses are univocally concrete, words/concepts that do not directly refer to perceivable instances are considered to be more or less abstract depending on whether the inferential relations that define them rely more or less directly on the observation vocabulary.

(ii) Our attempt to avoid as much as possible the word 'abstract' and to use (in a sense that still needs to be specified) 'theoretical' instead, is motivated by the desire to both avoid the ambiguity of the term abstract and to account for the inferential nature of abstract words which makes them analogous to non-rigorously defined theories of limited extent. In fact, the definition that emerges from the standard picture of meaning identifies abstract words/concepts as small theories, that is, as structured sets of knowledge expressed linguistically.

The idea of 'theoretical' we appeal to as well as the standard picture of language it invokes go back to Rudolf Carnap and to logical empiricism which distinguishes between two distinct types of language in science: observational language and theoretical language. "The observation language uses terms designating observable properties and relations for the description of observable things or events. The theoretical language, on the other hand, contains terms which may refer to unobservable events, unobservable aspects or features of events [...]" (Carnap 1956: 38) The theoretical terms (or concepts)[6] are therefore defined in opposition to the observation terms (or concepts) as those terms/concepts whose reference is not something we can perceive with our senses, but are theoretical constructs that depend on a specific theory of the world. According to Carnap, in any area of knowledge we can identify basic concepts and organize all other concepts starting from these using a coherent constitution system that derives non-basic concepts from basic ones through explicit definitions (Carnap [1928] 2003). The distinctive trait of non-basic terms is that their meanings undergo relevant variations over time which depend on the theoretical contexts that determine them. Indeed, according to Carnap and to logical empiricism the content of many theoretical concepts that cannot be directly linked to observable concepts relies uniquely on a definition

[6]An important difference between Carnap's position and our own is that Carnap considers the notions of 'concept' and 'term' as interchangeable. Carnap is not interested in understanding how concepts are represented in the individual mind, but only addresses the question of how some theoretical concepts/words are defined in the field of a theory: see e.g. Carnap ([1928] 2003), Verein Ernst Mach ([1929] 2012) and Carnap (1956). These three writings also allow us to appreciate Carnap's view that theoretical concepts/terms evolve over time.

which itself can change radically over time (the examples these theorists rely on are mainly taken from the field of physics but the same considerations apply to all kinds of theoretical concepts). The new concepts are still labeled using the same terms as the original ones, but the modifications made to their definitions lead to a radical transformation of their content and structure.[7]

As many philosophers of science point out (for a review see e.g. Van Fraassen 1980: §2), even though the distinction between observational and theoretical terms is always one of degree and not absolute (no term is entirely observational or entirely theoretical), theoretical terms are different from observational ones because their meanings cannot be directly traced back to any observable element, but are built inferentially on the basis of indirect evidence in the field of a given theory. As e.g. Willard van Orman Quine states (with relation to sentences instead of single terms): "Unless pretty firmly and directly conditioned to sensory stimulation, a sentence *S* is meaningless except relative to its own theory; meaningless intertheoretically" (Quine [1960] 2001: 24). This means essentially that the meaning of a sentence—and as a consequence also of a word—which is not directly connected to experience is of an inferential kind, in the sense that it is determined by the relations of a definitional character that it entertains with other sentences and terms belonging to the same theory or theoretical complex.

The distinction between observational and theoretical terms is important not only for addressing the issue posed by philosophy of science concerning how to identify the inferential relationships that contribute to constituting theoretical concepts in order to specify their content. It is also crucial with respect to a theory of semantic competence since the different nature of these word classes leads to the postulate that they are learned and mastered differently. This is for instance the position of the psychologist Allan Paivio, who proposed one of the most important models to explain why observational and theoretical words—which he explicitly identifies with concrete and abstract words respectively—are mastered (i.e. are cognitively processed) differently: "the observational-theoretical distinction becomes psychologically real when interpreted in terms of the correlated difference between concrete and abstract terms" (Paivio [1986] 1990: 11; see also Clark and Paivio 1989).

Carnap as well as many other philosophers of science do not distinguish between the notion of concept and that of word in the way we do (see Chap. 3, Sects. 1 and 2) because they are not interested in analyzing how individuals come to master observational and theoretical words, but only consider how the meaning of the observational and theoretical vocabulary is fixed in the field of a theory. However, if we apply the observational-theoretical distinction to the issue of semantic mastery, we will need to distinguish between two kinds of concepts (i.e. conceptual representations)—observational and theoretical—that allow people to master two

[7]According to logical empiricists in the transition from old to new theories, theoretical concepts lose their initial naïve traits, due to the influence of everyday experience, and acquire an entirely new characterization. In particular, they become mathematically tractable (Verein Ernst Mach [1929] 2012: §3.2).

kinds of words: those belonging to the observational and to the theoretical dictionaries respectively.

From the perspective of the debate we are considering, words are social and intersubjective constructs. In order to master them people must develop corresponding internal concepts that allow them to categorize things in accordance with the application rules of words. While the meaning of observational words depends quite closely on the observable proprieties of the objects they denote, the meaning of theoretical terms is more free-floating and depends on how a word is defined in relation to others in a certain social and theoretical context. Because the meaning of observational words depends more closely on perceivable properties, we can assume that concepts corresponding to those words can be developed more easily on the basis of sensory experiences and that they will fairly reliably support people's semantic competence with respect to the corresponding words. By contrast, the acquisition of concepts corresponding to theoretical words will largely depend on inferential information that people need to acquire. For this reason, we might expect that an individual's mastery of theoretical terms will in general be less precise and accurate than their mastery of observational words and that may sometimes even be mistaken.

In order for a concept to support semantic mastery of a theoretical word, it must include (at least some part of) the inferential information which the meaning of the corresponding word is composed of. In other words, to master a theoretical word a person must know at least in part the theory that supports its definition and the inferential characterization of the word and this knowledge must be part of his/her concept. If the theory a person develops is mistaken (i.e. if the concept a person has developed leads him/her to categorize things according to different criteria with respect to the application rules of the corresponding words), his/her use of the word will be misleading too. Often people's knowledge of the inferential relationships defining a word will be only partial or unilateral since the social definition of theoretical words is usually very complex and therefore not suitable for non-experts. Take for example the word 'freedom' (which we will discuss in more detail later). The definition given for this word by political theories is quite complex. Different theories—take e.g. liberalism and socialism—provide very divergent definitions. However, although the common-sense knowledge of this word does not include much of this theoretical content, it still relies on some of its simpler elements. Thus, the concept of freedom most people have will not include precise knowledge of any political theory on freedom, but will still include aspects of some theory which have become common knowledge. This is the reason why everyone's concept of freedom has political connotations: some people think e.g. that everyone should have access to education or the medical treatment of her/his choice without charge while others think that people should be free to spend their money as they like without being forced to pay taxes to support others (we will address this aspect in more detail later in this chapter: see Sects. 3 and 4).

This idea of the kind of semantic competence people have with respect to theoretical words closely resembles the view put forward by Hilary Putnam when he introduces the notion of stereotypes (Putnam 1975: 247–252). According to

Putnam, we should make a sharp distinction between the linguistic mastery of experts and that of common persons. Experts are in a privileged position with respect to other language users since they grasp the causal chain that led to establishing a certain linguistic meaning and they therefore have a complete understanding of this meaning. Speakers who do not know the causal processes that led to establishing this meaning only have a partial and superficial understanding of the word and what it denotes. However, the basic level of competence required for the correct use of a word is not that of an expert; less sophisticated criteria suffice. The rules that common speakers apply in the use of words are simplified versions of the scientific criteria experts use. While experts look for criteria that are as univocal and specific as possible, common people use words according to *stereotypical* rules that prescribe only vague and intuitive principles for their application. This suggests that—even though to have an adequate degree of competence with respect to a word the structure and the content of individual concepts must correspond at least to some degree to that of their social meanings—a full match is not necessary.

In this chapter, we do not aim at determining whether and when a person has full semantic competence with respect to a theoretical word, but rather to explain how an individual can come to master (in a better or worse manner) a theoretical term. Our goal is primarily to define what information the concepts supporting this mastery consist in and how this information is structured. We interpret 'abstraction' as referring to a theoretical product. Abstract words/concepts are not obtained through some kind of generalization procedure; they differ from general concepts/words since they are constructed on a linguistic basis through relations with other words/concepts. The issue we are going to raise here is what kind of information theoretical concepts must include in order to allow for mastery of theoretical words.

2 Theoretical and Intratheoretical Terms in Science and in Ordinary Usage

What does our semantic competence with respect to words that do not have an observable reference look like? As we already mentioned (Chap. 5, Sect. 1), the suggestion implicit in Marconi's hypothesis with respect to this issue is that people have exclusively inferential competence for terms that do not refer to concrete, observable objects (which we call observation words): "the distinction between the two aspects of semantic competence, inferential and referential, is not relevant to all words in the lexicon of a natural language" (Marconi 1997: 64). Marconi embraces Gilbert Harman's point of view that: "it is possible that there are certain sorts of theoretical terms, like 'quark', that play no role in perception at all, so that the content of the concepts they express is determined entirely by inferential role" (Harman 1987: 61). A similar suggestion is also made by Susan Carey when she says that: "While there certainly is also a causal story to be told about how the content of the concept *electron* is determined, it will have a very different flavor from the explanation for how the content of round is determined. I shall argue […] that inferential role plays a part in the causal process through which conceptual representations [i.e. theory laden representations] pick out their referents" (Carey 2009: 8). These positions refer to an idea of semantic competence which complies with the mentioned standard picture of language according to which "our ability to apply words may be based on the ability to apply other, more common words" that can be "cognitively basic" because they refer to observable objects in the external world (Marconi 1997: 61).

The common suggestion emerging from these considerations is that—while observational terms are characterized by a referential component which is relatively independent from the inferential definition—in the case of words that do not denote concrete, i.e. directly observable objects in the external world, it doesn't make sense to speak of a referential competence independent or relatively independent from the inferential characterization of the words. The knowledge of theoretical terms like e.g. 'electron' or 'quark' relies primarily on linguistic definitions of an inferential kind and the referential capacity to identify something as an 'electron' or 'quark' depends entirely on the inferential information that defines what electrons and quarks are. In other words, as for theoretical terms the referential side of our semantic competence seems to be entirely mediated by inferential competence and the content of the corresponding concepts does not seem to include any direct referential component but to be of a linguistic and inferential kind only. However, a closer look at the examples of 'electron' and 'quark' discussed e.g. by Harman and Carey shows that not all theoretical terms are 'equally theoretical', and there are relevant differences among them.

Let us consider the case of 'electron' together with the (analogous but more intuitive) case of 'cell'. 'Cell' and 'electron' are—exactly like 'quark'—scientific terms defined in the field of biology and physics respectively that have a specific inferential characterization.[8] All these terms are theoretical, because they were developed in the field of a specific theory and do not refer to objects that are directly observable *without the use of specific instruments*. However, they are not 'equally theoretical': in fact, electrons and cells can be observed, even though only with the support of instruments, while the existence of quarks can be tested only through indirect procedures. Cells can be observed by simply using a microscope, electrons can be visualized on photographic plates (nuclear emulsions), while quarks in isolation cannot be observed. The existence of quarks can be tested *only indirectly* through experiments in which some particles are hit by others; since the behavior of the particles in the hit (the so called 'scattering of particles') can be explained only by assuming the existence of particular entities which must have specific characteristics, physicists have come to the conclusion that these entities exist and have the hypothesized characteristics. In other words, the difference between 'cell' and

[8]The inferential characterization of a theoretical concept can be of a verbal kind only or can also include mathematical models. This is irrelevant for the aims of our analysis.

'electron' on the one hand and 'quark' on the other can be traced back to the fact that—while in cases like 'cell' and 'electron' is it possible to observe what objects correspond with the word provided that we can set up controlled situations and resort to specific instruments—in cases like 'quark' the object denoted by the word results merely from a mathematical and reasoning procedure. Thus, theoretical terms like 'cell' or 'electron' *do have a directly observable referent*, even though this may not be appreciated by everyone in daily life, since a microscope, nuclear emulsions and an appropriate camera obscura is required to bring it to light. On the contrary, the referent of words like 'quark' is entirely determined inferentially via theory.

This analysis shows that theoretical terms might be different from each other and that we need to distinguish (at least) two kinds of theoretical terms (and correspondingly two different kinds of concepts that support their mastery). On the one hand, we have terms/concepts like 'cell' and 'electron' which—in spite of their theoretical nature—retain a referential component which is relatively independent from their inferential characterization. On the other hand, we have terms/concepts like 'quark' which are *intratheoretical* in the sense that their content is entirely mediated by and dependent on the theory that defines them. We will differentiate between these two kinds of terms by using the label 'intratheoretical' for the latter, and keeping the name 'theoretical' for the former.

The difference between these two cases is not a moot point: while in the case of theoretical terms like 'cell' and 'electron', theories and calculations are only needed for interpreting observations, but don't produce the reference of the word, in the case of *intratheoretical* terms like 'quark' the referent (the empirical evidence) is entirely created by the theory and by its mathematical apparatus. The *ratio essendi* of intratheoretical terms is completely internal to the theory itself. In fact, these terms cannot be taken out of their context without being neutralized. Furthermore, while in the case of theoretical terms there is an extra-theoretical condition we can go back to in order to confirm or falsify the inferential relationships that contribute to determining the meaning of the term, in the case of intratheoretical terms there is nothing external or different from the theory itself we can resort to for confirming the correctness of the inferential structure.

This difference between theoretical and intratheoretical terms has clear consequences for the stability of the meaning of the term itself. Theoretical terms like 'cell' or 'electron' are indeed relatively independent and therefore relatively stable with respect to the theory thanks to which they were discovered and defined. If, at some point, this theory should change or be falsified in favor of another, the existence of the entity corresponding to the word would not be challenged. The new theory will rather need to include this entity even though it is possible that its definition and theoretical (inferential) description will now be different. On the other hand, intratheoretical terms like 'quark' derive entirely from the theory that led to their discovery and definition since the existence of the objects these terms refer to depends on this theory and on its procedures of indirect verification. Thus, the meaning of these terms will vary pursuant to the theory: different theories will lead to the identification of different references. The stability of the meaning will therefore depend on the stability of the theory. A theory change would lead to an entirely new definition of its intratheoretical terms and to the identification of correspondingly different references.[9]

Even though the examples of theoretical/intratheoretical terms we discussed come from the fields of biology and physics, natural or hard sciences are not the only contexts in which theoretical and intratheoretical terms can be defined. Many theoretical/intratheoretical terms are formed in the 'soft(er) sciences' such as the social or human sciences and it is plausible to assume that a number of others are formed in common sense theories (we could say 'folk theories') that our everyday knowledge of the world relies on. Most often it happens that theoretical/intratheoretical terms are defined both by science and by common sense. This is the case, for example, with terms like freedom, democracy, courage, friendship, beauty, chaos, hope, hostility, devotion, suffrage, veto, axiom, deduction, etc. which describe notions defined both by specific scientific disciplines (political science, logic, psychology, philosophy, theology) and by the folk theories which each of us develop.[10]

For common speakers, folk theories are certainly the most important element for the determination of semantic competence with respect to theoretical terms. This means that the inferential competence common speakers have with these words is generally not of a technical kind, and does not draw directly on knowledge of a scientific kind, but relies mainly on everyday knowledge and beliefs (which are often simplified versions of scientific theories). On the basis of the considerations presented at the end of the last section, we could say that—while the competence experts have with theoretical terms relies directly on scientific knowledge which is precisely and univocally defined—the competence non-experts exhibit is based on folk-theories which are much vaguer and more superficial. In addition, since the

[9]The fact that a scientific concept might have or lack a referential component and therefore be classified as theoretical or intratheoretical should not be confused with the issue of whether people can form some kind of image of the entities specific scientific concepts are supposed to denote. In the examples we have considered, 'electron' is e.g. more imaginable than 'quark' since—while it is difficult to visualize what quarks might be—a number of people associate the concept of electron with a metaphorical picture in which electrons are envisioned as planets in a solar system. According to our present theories, this metaphor is inaccurate since it suggests that electrons follow a specific orbit. In fact, people that associate electrons with this picture generally do not have precise inferential competence with respect to 'electron'. This metaphor is precisely due to the fact that when electrons were discovered, their behavior seemed to be analogous to that of the planets in the solar system. Non-experts tend to rely on these metaphorical pictures for their understanding of theoretical terms. Even though these pictures appear to make theoretical terms easier to visualize, they also often lead to incorrect mastery.

[10]The examples we mention have been selected more or less randomly mainly from philosophical/psychological studies on abstract concepts. Some of them will be discussed more specifically later in this chapter. However, to be sure that we have only chosen words that are commonly considered as univocally abstract we controlled their abstractness ratings on the basis of a psycholinguistic database—the so called MRC psycholinguistic database—which includes a measure for concreteness vs. abstractness. For a more detailed explanation of what this database is, on where to find it and on the nature of the ratings it includes see Sect. 5 of this chapter.

social and human sciences are often characterized by a larger number of alternative hypotheses and by greater disagreement between the varying points of view, the meanings of theoretical terms defined by these sciences are in general less uniform and less shared than those of theoretical terms defined in the natural sciences. For this reason, the folk-theories that derive from the social and human sciences are not only simplified, but also a composite of different hypotheses which might differ or even be inconsistent amongst themselves. As a result, the concepts that people develop in order to master theoretical words may be blends derived from different elements of their social meanings and turn out to be not only highly individual but also lacking in internal uniformity or consistency inconsistent. However, in spite of the many differences in the way sciences of various kinds or folk-theories define their theoretical terms and in the way experts and non-experts master theoretical terms, there is no reason to think that the distinction we made with respect to natural sciences between theoretical and intratheoretical terms does not apply to folk-theories.

According to the definition we provided previously, the difference between intratheoretical and theoretical terms is that the former depend in larger measure on inferential definitions and are therefore more instable, while the latter have a 'semantic core' of a referential kind and are therefore relatively more independent of their definitions—i.e. from their inferential periphery (on this aspect see also Chap. 3)—which can itself vary radically across time. Theoretical words rely on their inferential characterization more than concrete words, however they are still less contingent upon this characterization than intratheoretical words. Even though theoretical terms are found in various theoretical contexts which might differ from each other and give rise to very different definitions of these terms, there is a large, so to speak, 'cross-contextual' agreement on their *basic* meaning, in the sense that some fundamental aspects of the meaning are consistent across various contexts. This agreement on basic meaning is not due to the definition of the word, since this can change, but to some other aspects of the word that are stable because they have a perceptual origin. For this reason, we can expect that the mastery of these terms will rely also on competence of a referential kind. By contrast, in the case of intratheoretical terms any variation or drift in definition might significantly alter or even upset a previous perspective. Intratheoretical terms depend entirely on their inferential definition; thus, they remain stable within their context, but any modification of their theoretical context can radically transform them. As a consequence, the mastery of these words relies entirely or almost entirely on the knowledge of their definition.

This differentiation between theoretical and intratheoretical terms can be related to the classical debate on abstract concepts. If we use 'abstract' to indicate concepts that do not directly consist of any sensory information or to words whose meaning does not rely on any property that we can directly experience through the senses, then theoretical terms are only partially abstract, since they do keep some form of direct connection with perception. On the contrary, intratheoretical terms are truly and entirely abstract in the sense that they are definitional constructs of an inferential kind and their relationship with referential information is only indirect and derived from other terms that are inferentially linked to and define them. Even though we do not have a general criterion that allows us to univocally distinguish which terms are theoretical (i.e. only partially abstract) and which intratheoretical (i.e. fully abstract),[11] we can give an example of these two classes of words and discuss why they are likely to be mastered in different ways and why they might rely on concepts consisting in different kinds of information. We will consider 'democracy'[12] as an example of an intratheoretical term in ordinary usage while we will take 'freedom'[13] as an example of a theoretical term in ordinary usage.

DEMOCRACY—Let us first consider the case of democracy. As we explained previously, intratheoretical terms rely entirely on the theory that identifies them and therefore on their definition in the scope of this theory. For this reason, they are quite unstable and their definition may change even radically from one theory to another, while their mastery is uniquely or almost uniquely inferential and depends on the knowledge of other words that define them. The term 'democracy' fulfills these criteria and exhibits both (i) an unstable meaning across history and above all (ii) a strongly definitional characterization that makes the meaning of this word and the content of the corresponding concept strongly dependent on other words and links to other concepts.

(i) The word 'democracy' comes from ancient Greek and specifically from the conjunction of '*demos*', i.e. people and '*kratos*', i.e. force or power in a coercive sense and thus indicates a form of government in which people hold power. However, in the Greek culture that shaped this word, the notion of 'demos' by no means assumed the positive sense that characterizes it in contemporary culture. 'Demos' denoted indeed an indistinct set of persons which included also poor and uneducated people whose force derives from quantity rather than quality. This notion was opposed to that of *aristoi* which indicated the best men, i.e. the ones who would guarantee a good (a wise) form of government. Since common people were considered to have only an inferior form of intelligence, their government

[11]Later we will discuss two constructs developed in the field of psycholinguistics to measure the degree of abstractness of words and we will show that they support our differentiation between abstract words as theoretical and intratheoretical.

[12]Prinz also discusses the case of the concept democracy. However, according to his classification this concept is of a 'social kind', which needs to be differentiated from theoretically derived concepts like 'electron'. For the aim of this work—which is mainly to consider the function and the weight of referential and inferential information in the mastery of specific word classes—it is not relevant to distinguish among concepts/word classes as long as they do not exhibit differences with respect to their internal composition, i.e. with respect to the kind and proportion of referential and inferential information they consist in. For this reason, we need to consider whether there are concepts that consist entirely of inferential information or whether there are concepts that consist primarily of inferential information. However, it is not relevant to distinguish in what context they have been developed—if they have been developed in the field of an exact science or in the field of a social theory or even in the folk-theoretical context of a folk theory. See Prinz (2002: 166ff).

[13]The case of freedom has been introduced and discussed in a particularly detailed manner by Lakoff. The following analysis takes into account Lakoff's position (mainly Lakoff 2006), even though this will be directly addressed only in the next section.

would necessarily be incompetent and deplorable. For this reason, democracy was held to be a bad form of government as contrasted with aristocracy which was considered a good form of government in which the best men wielded power as peers and took decisions on the basis of their superior wisdom. Thus, a more correct way to convey the original sense of democracy is something like a 'dictatorship of the masses'.

This negative sense of democracy was due to the fact that the ancient Greeks had a view of human kind and of the hierarchies among humans which was very different from the contemporary view: in their culture, inequality was deemed part of the natural order of things. This view changed in modern times when the idea took root that rights and dignity are inherent to all human beings. At this point the 'demos' began to be thought of as a set of individuals with rights and dignity. This change in the notion of an individual led to a radical revision of the meaning of democracy which came to be considered no longer as a form of misgovernment but rather as a form of good government. Even though the label 'democracy' remained the same, the content of the concept changed radically since it was linked to a different view of people and to equality amongst people (for a detailed description of the controversial history of this concept see e.g. Canfora 2006).

(ii) With respect to the question of how our term democracy is defined and of how we come to master it, 'democracy' is related to the way a state is organized and to the way it takes its decisions. Specifically, democracy is that form of government in which each citizen contributes to reaching decisions. There are many ways in which democratic participation can be realized: public assemblies that all are invited to attend are just one possibility. However, our common idea of democracy is of a representative kind of governance, in the sense that decisions are taken by a few who are delegated to represent everyone. The delegation is realized through universal suffrage, i.e. via the vote. For this reason, we define as democratic those states that have free elections open to all adult citizen and—in our more concrete representation of democracy—this is connected to the possibility of voting and to the concrete act of voting.

Thus, our knowledge of what democracy is does not rely on a specific reference, but is rather due to the fact that we have learned specific inferential information, i.e. that we acquired a definition according to which democracy is a form of government in which all citizens contribute to making decisions by voting for their representatives. Democracy doesn't refer to anything observable in the world; however, the inferential information through which we define democracy (and understand what democracy is) is 'less abstract' than democracy itself and it allows us to have a more concrete representation of what democracy implies. Knowing this definition, we can link the concept of democracy with that of delegation and thus also with the vote and with the concrete act of voting. Thus, it is plausible to assume that our mastery of democracy will depend on the inferential competence we have with respect to this word while reference is provided *indirectly* by the words that inferentially qualify democracy and which allow us to identify as democratic the states in which people vote.

The great instability of the meaning of democracy which even underwent a change of valence from very negative to very positive and its strongly definitional characterization lead us to think that this word can be qualified as intratheoretical in the sense defined previously. In the case of democracy our referential competence (our capacity to identify e.g. a specific system as a democracy) depends entirely on the definition we learned and the connection between this word and observation is only mediated by other words which define it. This strong dependence on the definition does not mean that 'democracy' is fully disconnected from perceptual experience. 'Democracy' is rather only *indirectly* linked to experience through the words we use to define it and through their reference. It is these (and only these) that anchor the word to a concrete dimension. Since, however, definitions can change, even this concrete anchoring is variable.

FREEDOM—Even though 'democracy' might seem quite close and quite similar to 'freedom', the case of 'freedom' is actually very different since this term exhibits semantic stability across time due to an independent or relatively independent component of meaning. Even though the word 'freedom' has been used in quite different ways depending on the inevitable tensions among different claims about freedom—and on who was claiming the freedom of doing what at the expense of whom—*its meaning has remained relatively stable across time and importantly, all the various senses of this word have been characterized by the same strong positive connotation.*

The stability of the meaning of this word across time is not due to the fact that it is defined in a univocal and unequivocal manner. On the contrary, unlike democracy, the exact definition of freedom is quite confused and subject to dispute. We all know—or believe we know—what 'freedom' means and use this word quite naturally as it were perfectly clear what everyone has in mind when s/he thinks of freedom. However, when we try to determine more precisely exactly what freedom consists of (and what exactly we should fight for in order to pursue freedom), it becomes clear that the idea of freedom different people have in fact varies widely. In concrete terms, the situations that different political parties qualify as expressions or violations of freedom are often divergent, if not even contradictory, so that it is almost a truism to state that the idea of freedom each person holds varies widely and largely depends on his/her social and political opinions. Different political, moral and social views lead to different definitions of the word 'freedom', and when people adhere to a specific view they inherit the corresponding concept of freedom. As e.g. George Lakoff suggests: "When a hard-core conservative uses the word 'liberty' and applies it to 'economic liberty' and 'religious liberty', he has in mind the conservative version of the word's meaning. And when a progressive uses the same word—that the freedom to marry is a matter of a gay person's liberty—he has in mind a progressive version of the concept's meaning" (Lakoff 2008: 178–179).

The notion of 'freedom' is indeed considered a cornerstone by all or almost all political parties across the spectrum from the far-right to the far-left. The sense attributed to the word by right oriented politics is (more or less deeply) liberalistic: freedom is thought of in terms of economic freedom and the free market; e.g. in terms of the freedom to make profits and to keep one's earnings without being forced to give up part of them in taxes; freedom to produce and to be competitive in the market without any restrictions regarding environmental protection, minimum wage and the like; freedom to make commercial transactions (buy and sell) any kind of good including mass media, artworks and natural treasures, etc.; In left-oriented politics, the word invokes the opposite ideals: it invokes the freedom everyone should enjoy to live with dignity—to have a fair salary, a healthy environment, the right to medical care, access to multiple sources of free information and to common goods such as artwork or natural treasures. These radical dissimilarities between the way left and right parties use the word 'freedom' show that this word assumes meanings which are also radically different from each other depending on the traditions they rely on. For this reason, the concepts that people develop to master the word can also differ profoundly depending on whether a person learns a definition that is more right or left oriented. In this sense, the social variability of the meaning implies a subjective variability of the concepts people develop which is contingent on the way a specific person interiorizes the various available definitions.

Nevertheless, in spite of the fact that the concepts of freedom people have may differ widely, freedom is not entirely variable and relative, but—beyond the disagreement about specifics aspects—it is possible to detect a kind of common denominator of freedom, on which there is unanimous agreement. Indeed, as e.g. Lakoff's cognitive analysis of the concept of freedom points out, the sense of this concept is not entirely variable, but characterized by an "*uncontested core*" of "*central cases*". These central cases describe a number of paradigmatic situations, that *everybody*—independent of his/her social, political and moral views—considers to correspond to or conversely to violate the ideal of freedom (see Lakoff 2006: 10–14 and 28–31; see also Lakoff 2008: 178–180). The existence of this uncontested core explains not only why deep down everyone believes not only that they know for sure what 'freedom' means, but also that their concept of freedom has something in common with the concept of freedom of others, even though different people might rely on different definitions of the term. In other words, even though the inferential dimension of the word can change even radically in the transition from one theoretical context to another, the impression we have is that the common core concept does not change.

The hypothesis that the various concepts of freedom share a common core is also supported by another factor concerning the feelings everybody has about freedom. In fact, our everyday experience shows that people always have an emotional, visceral attachment to their own concept of freedom, regardless of how they conceptualize freedom and the world view they associate with it. Everybody thinks that freedom is something fundamental they cannot give up, something also worth fighting for. Everybody's concept of freedom is related to an emotional response, in the sense that the prospect of a restriction of freedom causes always and for everyone—regardless of the view s/he has on politics, morality and society—a *visceral* negative emotional response (Lakoff 2006: 30–31, 2008: 180–181). Since this emotional attachment is independent from the inferential, variable component of the concept, it is plausible to assume that it is connected with some other part of the concept that must be shared by everyone. If this is true—and we will further support this conclusion in the next section—this would offer further support for the idea that our concepts of freedom must have some common, stable core.

Relying on the analysis we carried out in the previous sections, the concurrent presence of a variable component of inferential information (which depends on which definition of the word the subject has interiorized—i.e. on the cultural and political beliefs of the subject) and a stable core concept, which is not contingent on the influence of inferential information, leads us to think that the mastery of this word requires both inferential and referential competence. More specifically, the common core of the concept might be composed of referential information (which is not influenced by the different, ideologically coined definitions of the term and does not depend on what the subjects have learned about freedom and believe freedom to be), while the 'periphery' (so to speak) of the concept is inferentially determined and therefore subject to the influence of social and cultural factors. In this case the emotional attachment everyone exhibits with respect to freedom could be (*somehow*—in a way that still needs to be explained) traced back to the referential information everyone's concept of freedom is composed of. A full mastery will therefore require the availability of both kinds of information.

The analysis we carried out on the referential dimension of the concepts that are characterized by an emotional dimension—one should think for example of 'friendship', which we discussed previously (Chap. 5, Sect. 10)—led us to think that theoretical concepts like 'freedom' (as distinguished from intratheoretical ones) are also characterized by a referential component analogous to that of emotional concepts which *can be traced back to information driven by the body of the subject, i.e. to 'proprioceptive' or propriosensitive information* (on the definition or proprioceptive/propriosensitive see Chap. 2). From this point of view, 'freedom' can be considered analogous to 'cell' or 'electron' in the sense that it is characterized by an inferential meaning which is defined in relation to a specific (political/commonsensical) theory of freedom and a referential component which stabilizes its semantics. There is of course an important difference: theoretical terms coined in a scientific field have referents in the external world that are accessible through specific instruments. Instruments are clearly not part of the picture when we are speaking of theoretical terms in ordinary usage. Freedom, friendship or devotion surely cannot be observed with a microscope or with a telescope. *The specificity of 'freedom' in comparison to 'cell' or 'electron' is that the internal referential core of the concept relies on 'proprioceptive' (especially emotional) information rather than external, perceptual information.*

Because of this characteristic, the core concept of freedom is not intersubjectively observable like that of 'cell' or 'electron'. However, even though it is composed of subjective information, which is available to the first person only, it can also be considered at least to some extent intersubjective by virtue of the common bodily constitution of human beings, which assures that different people will nevertheless experience analogous 'proprioceptive' states. It is possible that the differentiation between perceptual and propriosensitive information is not absolute and that the core of a theoretical concept might consist of a mixture of both kinds of information. However, the example we are considering also clearly shows that the one or the other type of information will be preponderant for specific kinds of concepts. The core of concepts like 'cell' or 'electron' will e.g. consist predominantly of perceptual information, while the core of concepts like freedom will chiefly entail propriosensitive information. *Thus, we will have two kinds of theoretical concepts which are different from each other in an extremely important aspect:*

- *some theoretical concepts will have a core based (mainly) on external perceptual information—we will indicate these with the label 'E-theoretical';*
- *others will be based (mainly) on internal propriosensitive information—these will be indicated with the label 'I-theoretical'.*

E-theoretical and I-theoretical concepts are both theoretical in the sense that they are characterized by a strong inferential component whose content largely depends on relationships with other concepts. However, while intratheoretical concepts are entirely composed of inferential information, in the case of both I- and E-theoretical concepts this information is anchored by a referential core.

As a consequence of the double composition—inferential and referential—of (I- and E-) theoretical concepts, the mastery of the corresponding words will be twofold too. On the one hand, to master these words people will need to have some knowledge of their definition, i.e. they will need to know how they relate to other words. On the other hand, people who master these words will also have a referential knowledge of what they refer to: they will have some 'intuitive' knowledge about what kind of experience is linked to these concepts. In other words, in order to master an E- or I-theoretical word a person needs to be able to connect it with the perceptual/propriosensitive information, which the corresponding core concept is based on. People will master e.g. words like 'cell' or 'freedom' if they know what a cell looks like and what freedom feels like.

If our hypothesis is correct, only the mastery of theoretical words is directly related to a perceptual/'proprioceptive' referential component, while the mastery of intratheoretical words is primarily linguistic. As a consequence, we should *feel*—in a perceptually salient sense—'freedom', while we shouldn't *feel* 'democracy' in the same sense. This does not mean that we cannot imagine democracy in any perceptually relevant sense since we hypothesize that the knowledge of intratheoretical words depends on the knowledge of other words, some of which are observational and have therefore a referential component. Thus, intratheoretical concepts might also have some *indirect* perceptual anchoring and recall concrete objects like e.g. a ballot box, voting booth, voting station, the act of voting, the television broadcast transmitting the results, people waiting in a queue to vote (this is suggested e.g. also by Prinz 2002: 180). In addition, we might also hypothesize that democracy might also be related (at least according to some definitions) to freedom, and in this case the word would have also an indirect 'proprioceptive' anchoring. However, this association of democracy with concrete or with propriosensitive information is only indirect and is not an essential part of the concept of democracy.

As, for example, Ryle ([1949] 2009: Chap. 1, Sect. 2) suggests, the association of an ideal entity with concrete entities is a categorical error. As Ryle argues, the university as an institution that organizes a number of research and teaching activities is not the building in which these activities take place. If our understanding of 'university' relies on reference to a building, this shows that we do not understand the meaning of this word in its abstract sense. In the same way, the relationships between 'democracy' and concrete objects like ballot boxes, voting booths, voting stations, or freedom are based merely on associations; they have nothing to do with the reference of the word. As a consequence, the concept does not have any direct referential component; but is characterized only by inferential relationships with other concepts which might have referents. The case of 'democracy' appears radically different from that of 'quark', since a 'democracy' seems much more concrete than 'quark'. However, this difference is only an impression which stems from the fact that we are quite familiar with 'democracy' but unfamiliar with 'quarks'. In fact, the two cases are entirely comparable. Both democracy and quark cannot be perceived in any direct way, but can only be defined in the scope of a theory via other words/concepts.

However, if our hypothesis is correct, these observations do not apply to 'freedom' because this concept also includes a referential core of an internal—i.e. 'proprioceptive' or propriosensitive—kind. In the next sections, we will try to clarify this hypothesis. Starting with the example of 'freedom' we will try to show precisely what the analogy between *I*-theoretical and emotional concepts consists in. Secondly, we will put forward some (rough) criteria to identify *I*-theoretical concepts and address the question of the composition and origin of *I*-theoretical concepts in general.

3 The Deep Sense of 'Freedom' and the Issue of What 'Freedom' Basically Refers to: An Example of an I-theoretical Concept

Deep down all human beings unanimously agree to consider some specific extreme situations as indubitable violations of freedom; therefore, our concept of freedom is not *entirely* variable and dependent on what we believe in. This is the idea we suggested in the previous section. To account for this view, we need to show that the concept of freedom does not consist only of inferential information, but also has a referential component. A proposal as to what this referential component might be can be developed starting from Lakoff's hypothesis about how the concept of freedom is originally formed.

According to Lakoff, small children develop a primigenial concept of freedom starting from empirical and perceptual experience: the experience of the freedom to move their own body. "The most basic idea of freedom is freedom of motion" (Lakoff 2008: 180; see also Lakoff 2006: 29ff). He suggests that the intuitive concept of freedom small children form is based on "three fundamental types of bodily movement: locomotion, movement from one place to another; movement of the arms to grasp objects; and movement of the body to perform actions" (Lakoff 2008: 180). The freedom to move in these three senses is what constitutes the central core of our concept of freedom. On this central core is grounded also the commonsensical intuition that everyone's concept of freedom shares some common elements in spite of wide variability. These common elements are built on the perceptual experience of the freedom to move in contrast to the perceptual experience of being prevented from moving. In other words, Lakoff suggests that people's understanding of the concept of freedom lies ultimately in the bodily experience of being able to move versus being kept from moving in the sense of locomotion, grasping, and the performance of action. In this sense, the perceptual (in our view it would be more appropriate to say 'proprioceptive' or propriosensitive) experience of 'being able to move' versus 'being kept from moving' is the referential information everyone's concept of freedom is based on *at its core*.

The composition of this core can also explain why freedom is a 'visceral concept', which everybody is emotionally strongly attached to. In fact, in Lakoff's view, the common emotional attachment to freedom everyone experiences in everyday life can be traced back to the empirical experience this concept is built on. At a very basic level 'freedom' is understood only in terms of the experience of feeling free to move as opposed to the experience of being kept from moving. It is this aspect of the hindrance of bodily movement that makes freedom something so emotionally pregnant, because the inhibition of physical movement is indeed odious and insufferable. As Lakoff states: "We can now see why freedom is a visceral concept. It is tied […] to our ability to move and to interference with moving. There is little that is more infuriating than interference with our everyday bodily movements. It is the embodiment of freedom […] that makes it such an important and emotionally powerful concept" (Lakoff 2006: 31). Even though the emotional attachment to freedom concerns only the core of the concept, which relies to the motoric aspects of freedom *only*, the odiousness and the suffering that people associate with the idea of a forced restriction of their own bodily movements is not bound to the core but spreads across the concept, including to its peripheral inferential complex.

This core of freedom as freedom to move the body is only a small part of the complex concept of freedom people have, which stretches over a much larger territory than the one occupied by the original (i.e. embryonic) core. As a matter of fact, the core of the concept describes only a concrete concept, based entirely on empirical experience. To become the theoretical concept of freedom in its complex and complete form—i.e. to become the concept we rely on in political and social discussions—the core must be transformed and extended.

The transformation the concept of freedom goes through when it changes from its initial concrete forms to its final theoretical characterization cannot be explained using a simplistic view of abstraction, according to which the theoretic, abstract characterization is obtained staring from the concrete features through the mere addition and/or subtraction of properties. Instead, it must rather be explained as following a more complex genetic path which starts from an empirical invariant—the core, formed on the basis of experience—and proceeds by extension of this core according to specific ways that depend both on further cognitive processes and on the epistemological context external to the subject. In fact, the solution that Lakoff proposes—even though only through fragmentary suggestions scattered in various works—is quite interesting and offers a highly plausible general answer to this problem. According to Lakoff, from a cognitive point of view, the concept of freedom in its complex and complete form, as a theoretical, political concept, is developed through the extension of the core and is carried out on the basis of 'frames' or 'cognitive models' and 'metaphors'.

Metaphors, in general, are means that "allow conventional mental imagery from sensorimotor domains to be used for domains of subjective experience" (Lakoff and Johnson 1999: 45). According to Lakoff, some topics—even familiar ones—are too abstract to be "comprehended on their own terms. Instead, we must understand them in terms of other entities and experiences" (Lakoff and Johnson 1980: 177) which are more embodied, concrete and accessible. These last constitute a source domain that can be used to determine the topic domain: to define the elements it is composed of and the relationships among them. Through the use of metaphors, correspondences between the two conceptual domains are set up: the source domain is mapped into the topic domain and the conceptual structure of the target domain is built up and understood as partially isomorphic with the source domain.[14]

A first extension of the core is due to the application of a so-called primary metaphor, i.e. of a simple (i.e. not compositional) metaphor that must be based directly on experience. A primary metaphor "has a minimal structure and arises naturally, automatically, and unconsciously through everyday experience by means of conflation, during which cross domain associations are formed" (Lakoff and Johnson 1999: 46). Because these metaphors are not only naturally and automatically produced by the human cognitive system, but are also universally applied to early experience, which is relatively independent of cultural variables, they tend to be universal conceptual means that emerge and begin to be used quite early in cognitive development. In Lakoff's view, a first extension of the primigenial concept of freedom through primary metaphor occur when we start conceiving locomotion, grasping and bodily movements as having an aim, i.e. the aim of reaching a desired destination; of getting some desired object and of performing the desired action (Lakoff 2006: 29). This leads to the more abstract concept of freedom as *freedom to achieve a goal*.[15] As a result of the way in which this concept has been

---

[14]To get a sense of the debate on metaphor and its cognitive function see e.g. Gentner and Bowdle (2002: 18–21, 2008: 109–128) Johnson (2010: 401–414), Gentner et al. (2001: 199–253), Fauconnier and Turner (2008: 53–66) and Wolff and Gentner (2011: 1456–1488).

[15]Lakoff distinguishes between two senses of freedom, which are respectively the idea of "freedom from" and the idea of "freedom to": see e.g. Lakoff (2008: 180). However, this distinction is not relevant here. For a further development of this point, which explores the possibility of distinguishing three main ideas of freedom (and not only two, as Lakoff does) see: Dellantonio and Pastore (2011, 2014).

extended: "Whenever there is an issue of setting and achieving goals, there is an issue of freedom—freedom thought of metaphorically, viscerally, in terms of functioning physically with your body in space to carry out some purposeful action" (Lakoff 2006: 29).

Already this first extension of the concept leaves unspecified many aspects of prime importance and opens some semantic blanks that need to be filled in through further specifications: Where does the freedom of someone to pursue his/her own goals finish and where does the freedom of someone else start? What are the goals that a person must or can be free to pursue? These blanks can be filled in only through a further expansion of the concept that allows it to establish inferential links with other concepts—which might themselves be contested, ambiguous and problematic—such as e.g. JUSTICE, EQUITY, HUMAN NATURE, DAMAGE, OPPORTUNITY, CAPACITY and so on (Lakoff 2006: 39ff). According to Lakoff, these concepts reciprocally fill in each other's blanks, that is, determine and specify each other when they are inserted and integrated in the complex world-view each adult has (and has inherited in an individual way from its culture). Lakoff calls this world-view a 'cognitive model' (Lakoff 1987: 68) or 'frame' (Lakoff 2006: 10–14, 2008: 250ff) to indicate that it consists of a structured set of knowledge, i.e. of a non-rigorously defined theory of limited extent (Lakoff 2006: 14). If the specification of the core-concept of freedom is carried out on the basis of different world-views, we will obtain two different and differently determined concepts of freedom. For this reason, Lakoff comes to the conclusion that: "When the blanks are filled in by progressives and conservatives, what results are two radically different ideas expressed by the same word, 'freedom'" (Lakoff 2006: 15). And further: "The idea of freedom has different interpretations, depending on your moral and political worldview" (Lakoff 2006: 22).

Freedom has been described in very differing terms by the various political, social and religious traditions; thus, depending on the description people embrace, they categorize 'freedom' differently and have a different understanding of what freedom (in a theoretical, political sense) is about. As Lakoff also suggests, for e.g. conservatives, 'freedom' is primarily about the free market, about property rights (and therefore also about the protection of one's property from taxation) and about the protection of traditional values. On the contrary, for progressives, to be free means to have the possibility 'to have' and 'to choose a life', so to speak, and therefore to get a free education, enjoy access to free healthcare, to earn a fair wage and to live according to your personal values (see Lakoff 2006). This means that, the situations or states of affairs a person identifies as cases of realized freedom depend on what concept of freedom s/he has developed: i.e. depend on which inferential connections his/her concept of freedom entertains with other relevant conceptual elements in his/her frame/cognitive model.

However, beyond the differences in the definition of freedom and in its inferential characterization, since freedom is a theoretical (instead of an intratheoretical) concept, some kind of stability and intersubjective agreement/understanding of what freedom consists of is granted by its referential, perceptual core. Indeed, it is because such a core exists that all people—regardless of the ideology they believe in—share some minimal aspects of their referential competence with respect to the word 'freedom'. If Lakoff's hypothesis is correct, this minimal sharing of referential competence is due to and can be traced back to our bodily dimension and to the experience of motor constraints.

From this point of view, the stability and the intersubjectivity of the meaning of an intratheoretical term depend solely on the stability of the intersubjective agreement on its definition (i.e. its inferential characterization). By contrast, the stability of the meaning of a theoretical term does not depend uniquely on intersubjective agreement about its definition, but relies primarily on the referential, perceptual core of the concept. Nevertheless, this core is exclusively made up of perceptual information and it supports only a concrete understanding of freedom in terms of 'being able to move' versus 'being kept from moving'. Therefore, the stability of the core does not result in the theoretical (political, high-level) concept being intersubjectively stable, instead of contested, even though people will always have the impression that it is intersubjectively stable; they will feel that in spite of all the different interpretations, everybody understands the concept in the same way. In fact, the intersubjective agreement of what freedom is takes place only at a very basic level and uniquely concerns a set of central cases.

The general idea beyond these observations is that the more a concept is composed of referential information, the less the content of this concept is determined by inferential knowledge and therefore by the specific theory/view this knowledge comes from. Intratheoretical concepts are the most variable, since the information they consist in is primarily cultural and 'theory-laden'. Theoretical concepts are less variable and less dependent on inferential knowledge, since they have a referential core that keeps them stable, *at least to a minimal extent*. Still, also theoretical concepts undergo a certain (even high) variability when we move away from the core concepts and from the cases that are more close to it.

With respect to the problem of establishing to what extent a specific core can stabilize a concept, it is not irrelevant to determine what kind of core we are speaking about. In a case like e.g. 'cell', the referential core of the concept can only be accessed through the mediation of instruments; however, since this method is external and intersubjective, it can be publicly observed and it can constitute an allocentric verification criterion for the concept itself. In the case of concepts like 'freedom', the referential core is internal and it stems from bodily experience. This makes the concept of freedom, at least in some respects, more similar to 'fear' or 'jealousy' than to 'cell' or 'electron'. The affinity between 'freedom' and 'cell' or 'electron' is only due to its inferential structure which is driven by a theory—and more specifically by social, political, economic and, more often, commonsensical views (theories and structured belief systems) on what freedom should be. However, as for the kind of information its core is based on, 'freedom' is more similar to emotional concepts (see Chap. 5).

In fact, considering the importance that the component of the emotional attachment assumes with respect to the concept of freedom (freedom is primarily something everybody strives for and nobody is willing to sacrifice), it is plausible that the emotional component of this concept is even stronger than Lakoff's theory suggests. According to Lakoff, the core experience the concept of freedom is based on consists in *voluntary* movements of our body (i.e. of the possibility to move freely as contrasted with being hindered from moving). However, the body is also characterized by spontaneous and involuntary motility that constitutes something like a tacit experiential continuum, or a psychic *basso continuo* which permanently accompanies the subject. In other words, everyone always experiences not only his/her voluntary movements, but also his/her involuntary motility connected with vital functions (e.g. the micro-movements of respiration and of the blood flow, the micro-movements of the eye, the change in the tension of the facial muscles when a facial expression changes, the change of blood pressure with the corresponding sensation of ease and/or unease, but also sensations related to the interaction and integration of body parts, to posture, to muscle tome or to the change of the motoric balance of the body parts, etc.).

In the previous chapters, we discussed the plausibility of the hypothesis that in its complexity this basic bodily information can be considered the experiential grounds on which the cognitive system forms concepts relative to its internal, physical and emotional states (on this specific point see Chap. 5, but see also Chap. 2 for a related issue concerning how this kind of information might be the starting point for forming a sense of self as animate being). Since 'freedom' is characterized by a very strong emotional component that seems to be excessive if compared with the degree of pleasure/frustration we experience when we are able to carry out voluntary movements or when we are prevented from doing so, we could hypothesize that the information our core concept of freedom consists in does not include only or primarily voluntary movements, but also *the experience of the spontaneous motility of the bodily which is connected with the vital functions*.

According to this hypothesis, the child develops his/her core concept of freedom starting from the tacit experience of his involuntary motility, which becomes conscious and salient when this motility is somehow momentarily inhibited. When something impedes the normal involuntary motility of the body, what the child perceives is a momentary inhibition of his/her vital functions, which is surely accompanied by an emotion of a different quality and intensity in comparison to that associated with the unease of being kept from carrying out a voluntary movement. In this sense, the core concept of freedom is not formed using the experience of 'feeling free', but is built starting from the opposite experience of being kept from moving. Therefore, what we concretely experience is the lack of freedom, while we form our positive concept of freedom in opposition to that experience: we are free when we are *not* kept (even very briefly) from carrying out our involuntary movements. In the same way, the positive attachment we experience towards freedom is formed in opposition to the direct experience of the suffering due to the (even very brief and partial) inhibition of spontaneous motility (i.e. of the vital functions).

If the concept of 'freedom' is originally formed starting from our subjective experience of being free to move in the sense of not being prevented from carrying out the spontaneous motility of the body, and the difference between a condition of freedom and one of constraint is perceived in terms of the difference between the experience of the normal functioning of our spontaneous motility and the experience of a momentary inhibition of this motility, then the experience of freedom corresponds to that of feeling alive while the deprivation of freedom is related to the feeling of losing our life. If this hypothesis is correct, then the central cases of freedom that depend more directly on the core concept and that are therefore less contested will be those connected with the free course of spontaneous motility, while the situations in which people will more unanimously agree that an infringement of freedom occurred will concern cases in which vital functions are forcedly inhibited. This would explain, among other things, why so many metaphors we use to express the restriction of freedom rely on notions that recall the forced and violent inhibition of the involuntary movements that allow us to stay alive: e.g. *suffocate freedom, strangle freedom, suppress freedom.*

To sum up, we agree with Lakoff that the concept of freedom is build starting from the somatic perceptual experience subjects have of their own movements. However, we rely on a different notion of movement, which does not refer primarily to voluntary actions, but rather to the tacit experience subjects have of their spontaneous and involuntary motility. In this way, what we get is an even deeper and emotionally more meaningful embodiment of this concept. If the experience of our spontaneous and involuntary motility is indeed at the base of our sensation of being alive, then the hindrance of it would become an attack on life itself, since it would imply an inhibition of vital functions. Understood in this way, the core concept of freedom is related to life itself: to the sensation of being alive we experience through our involuntary movements and to the freedom to stay alive—therefore also to our survival instinct which is the instinct to perpetuate at any cost the motility which is at the base of life. According to this interpretation, the emotional attachment to freedom becomes far more meaningful and visceral since it is directly connected with life as a physical state opposed to death. The hindrance of a voluntary movement is bothersome, while the hindrance of an involuntary movement leads to death. This view establishes a somatic connection between the idea of inhibition of freedom and loss of life. Therefore, the terror of death and the sensations connected with the survival instinct and the desire to preserve life become part of the core concept of freedom, and establish an emotional equivalence between freedom and life.

Independently of whether one prefers Lakoff's view on the genesis of freedom or the stronger interpretation we put forward here, the main idea remains the same. The concept of freedom is characterized by a perceptual core, which is not primarily based on information derived from the external world and from the 'external' senses, but from information of a *'proprioceptive' and qualitative kind.* Our representation of 'freedom' is formed starting from information of a bodily nature which is analogous to that which emotional core concepts are made of (see Chap. 5). 'Proprioceptive' (i.e. propriosensitive) and qualitative information contribute to the constitution of the core concept: but while this core is only a small part of the theoretical, adult concept of freedom, it is still the original basis for the concept (starting from which the theoretical concept has been developed) and the permanent referential element of the concept.

If this hypothesis is correct, the word 'freedom' is not entirely an inferential construct, and its understanding does *not only* depend on the inferential competence one has developed with respect to it, but is also characterized by an independent referential dimension. From this point of view, the capacity to *fully* master the word 'freedom' will also require that the subject is able to perceive and to process 'proprioceptive' and qualitative information. To be more precise, from a structural point of view our semantic competence with respect to 'freedom' is analogous to that for complex emotional terms, whose mastery requires broad inferential competence, but also relies in part on competence of a referential kind. Even though this referential element is only a secondary component of the concept in comparison to its inferential dimension, this is the only stable part of the concept and is therefore what assures some intercultural and intersubjective stability in the understanding of freedom.

4 Abstract Concepts: Standard Approaches and New Insights

For the most part, 'classical' theories of concepts (i.e. the theories that prevail in the literature and experimental studies mostly rely on) consider abstract concepts to be purely linguistic representations, which are inferentially determined through the relationships they entertain with other linguistic representations. This is possibly also the reason why the issue of how abstract concepts are constituted and how they support linguistic mastery has only sporadically been addressed by the literature. Two prominent approaches in cognitive (especially psycholinguistic) research which address the processing and comprehension of abstract words since the 1970s/1980s stand out. They are Allan Paivio's *Dual Coding Theory* (see e.g. Paivio 1971a, b, [1986] 1990, 2007) and the *Context Availability Theory* developed by Schwanenflugel and Shoben (1983), Schwanenflugel and Stowe (1989) and Schwanenflugel (1991). These are usually considered to be alternative hypotheses on how abstract concepts are represented which give rise to different accounts of our capacity to master abstract words. Let us examine them more closely.

According to Paivio, the differentiation between abstract and concrete terms is "psychologically real" in the sense that concrete and abstract terms are *not* mentally represented in the same way. Paivio's theory suggests that the cognitive system is composed of two different symbolic subsystems connected to each other: a verbal one, specialized for the processing of language, and a nonverbal one, also called the *imagery* system, whose function is mainly to perceptually analyze the external world and generate mental images of it. The word 'imagery', as Paivio uses it, describes both this system and the capacity this system enables to dynamically form and recall mental images. Mental images are not meant as visual pictures only, but basically describe traces stored in memory of all kind of sensations—acoustic, olfactory, haptic and gustatory. In this sense, an 'acoustic image' would be the nonverbal representation of a sound, an olfactory image would be the nonverbal representation of a smell, etc. As Paivio specifies: "Our minds 'contain' memory isomorphs of how entities and events look, sound, feel" (Paivio 2007: 25).

> The most general assumption in dual coding theory is that there are two classes of phenomena handled cognitively by separate subsystems, one specialized for the representation and processing of information concerning nonverbal objects and events, the other specialized for dealing with language. In keeping with my earlier usage, I will often refer to the nonverbal (symbolic) subsystem as the imagery system because its critical functions include the analysis of scenes and the generation of mental images (both functions encompassing other sensory modalities in addition to vision). The language-specialized system will be referred to as the verbal system (Paivio [1986] 1990: 53–54; see also Paivio 2007: 33–34).

According to Paivio, these two coding systems are structurally distinct and can therefore be independently active; however, since they are also functionally interconnected, the activity in one system can trigger the activity of the other.

In this model, abstract terms are represented in the verbal system only, while concrete terms are represented both in the verbal and in the imaginal system. The imaginal code is perceptual and "gives rise to conscious (reportable) imagery when activated" (Paivio 2007: 39). From this point of view, our knowledge of abstract words is linguistic only, while our understanding of concrete words also relies on the capacity to form an image of what they denote. "Thus, if I am asked to describe my dining room table, I first experience an image of that table, which I can then describe by naming its components and adjectival attributes" (Paivio [1986] 1990: 62). Since concrete linguistic inputs directly activate both the verbal and the imaginal system, while abstract linguistic inputs only directly activate the verbal system, concrete linguistic inputs will be easier and faster to understand than abstract linguistic inputs. According to Paivio words evoke images and images evoke words: Paivio explains referential meaning as an "associative connection between imaginal and verbal representations, such that an object or picture can be named, and the name can evoke an image"(Paivio 1971a: 11).

The view that is usually contrasted with Paivio's *Dual Coding Theory* is Schwanenflugel and colleagues' *Context Availability Theory*. This theory was developed starting from a supposed flaw in Paivio's theory, which hypothesized that abstract terms are processed more slowly than concrete terms because they are not directly connected to an image that would make the word easier to understand and thus aid (i.e. also speed up) processing. Schwanenflugel and colleagues observe that concrete words are indeed sometimes understood faster, but note that this facilitation effect takes place only in certain cases and specifically *only* when linguistic inputs are presented without an appropriate context: "In isolation, it took longer to comprehend abstract sentences than concrete sentences. However, when context was provided, this comprehension time difference disappeared" (Schwanenflugel and Shoben 1983: 86). The fact that this effect disappears when a context is provided challenges—according to Schwanenflugel and colleagues—the *Dual Coding Theory* and the idea that concrete linguistic inputs activate two different systems, a verbal and an imaginal one. Schwanenflugel and colleagues put forward an alternative general model, according to which comprehension consists of the capacity to make interconnections between ideas and more specifically to "relate the to-be-comprehended materials to one's prior contextual knowledge" (Schwanenflugel and Stowe 1989: 116). This theory gives an alternative explanation of the reasons why abstract linguistic contents are processed more slowly than concrete linguistic contents when they occur in absence of a context, while they are processed equally fast when they are presented in a paragraph. The general idea is that "Comprehension difficulties emerge when the reader is unable or slow to relate the incoming message to his or her prior knowledge. These difficulties can be overcome through presentation of such hard-to-understand materials in a supportive stimulus context (Bransford and McCarrell 1974), because such contexts make the necessary prior knowledge readily available for the reader." (Schwanenflugel and Stowe 1989: 116).

The *Context Availability Theory* rules out the notion that comprehension is supported—at least in the case of concrete terms—by imaginal representations of some kind (i.e. by the retrieval of information of a sensory nature). On the contrary, it suggests that understanding depends in any case on grasping the inferential relationships among words. From this point of view, the fact that abstract words are processed more slowly than concrete ones when they are presented in isolation must be explained uniquely on the basis of the greater/smaller difficulty of retrieving the relevant prior knowledge necessary for comprehension in these different cases. Schwanenflugel and colleagues suggest that "where concrete concepts possess both context-independent and context-dependent properties, abstract concepts possess mainly context-dependent information (possibly because these concepts appear in so many contexts). Thus, without context, concrete concepts have easy accessibility to certain properties whereas abstract concepts will have very little information available" (Schwanenflugel and Stowe 1989: 95). So, concrete concepts are processed faster than abstract ones when they are presented in isolation because people can easily retrieve context-independent properties of concrete concepts but not those of abstract ones. However, this difference disappears when the respective words are inserted in a paragraph, because the additional verbal information facilitates the retrieval of context-dependent information.

Context Availability Theory is usually considered an alternative to *Dual Coding Theory* for explaining how concrete and abstract words are understood. However, according to some authors the findings reported by Schwanenflugel and colleagues are in fact compatible with *Dual-Coding Theory*. *Dual Coding Theory* hypothesizes that the understanding of abstract words is almost entirely dependent on inferential information (since the connection of abstract words with the imaginal system is only indirect, i.e. mediated by other words), while the comprehension of concrete words also relies on sensory information. This can be considered in line with the result that, when more verbal information is available, abstract words are processed more easily and therefore faster. As e.g. Prinz suggests:

> The dual-code interpretation is consistent with the context-availability model because it says that abstract words are processed faster when more information is available. It predicts that abstract words need a richer verbal context for rapid retrieval. The dual-code model also predicts that there will be no comparable improvement in performance on concrete words because performance is already near ceiling levels when concrete words are presented in isolation. The referent of a concrete word can be readily imagined even when no verbal context is provided (Prinz 2002: 132).

Independently of the fact that the evidence provided to support the *Context Availability Theory* can or cannot (at least under certain conditions) also be explained on the basis of the *Dual Coding Theory*, what really matters with respect to the aims of this work is that the *Context Availability Theory* proposes an explanation that goes—more than Paivio's *Dual Coding Theory*—in the direction of a radically inferentialistic view of linguistic understanding. Understanding is interpreted as a recovery of prior information of a linguistic kind; the sensory information coming from the external world is assigned only a marginal role as the source of context-independent information regarding the properties of the perceivable objects denoted by the words (or of the observable circumstances described by the sentences). From this point of view, *Context Availability Theory* gives a purely inferential explanation not only of abstract concepts but also of concrete concepts, implicitly suggesting that the only difference between them is that in the case of concrete concepts there is more linguistic information immediately available for their determination.

Because of this aspect, *Context Availability Theory* does not account for the distinction between referential and inferential competence and it reduces our semantic competence to a capacity to link some words to others. In this respect, Paivio's theory has greater explanatory power since it does not explain linguistic understanding on the basis of the verbal symbolic system only, but assigns a central role to reference at least as far as concrete words are concerned. Indeed, according to Paivio the comprehension of concrete words relies primarily on the non-verbal system through which imaginal representations are processed; a system which codifies information about external objects in an analogue, iconic and isomorphic way. This idea of imaginal representations is de facto[16] largely compatible with Marconi's thesis that our competence with respect to concrete words is not only of a verbal kind, but also includes a non-verbal capacity for identifying the perceptual (analogical)[17] features of the objects denoted by the words. As Paivio points out:

> Both classes of words [abstract and concrete] have interconnections with the representations of other words in the verbal system. The specific nature and structure of the verbal-associative networks for concrete and abstract words presumably differ in systematic ways that reflect differences in the contexts in which they have been acquired and used, but in general it can be said that concrete and abstract words are semantically differentiated by the degree of availability of referential interconnections. Concrete words have both

[16]In this respect, it should be noted that Marconi might not agree with Paivio, since in some places he argues against representationalism in general.

[17]For a clarification of the notion of 'analogic' see Chap. 3, Sect. 5.

> referential and verbal-associative meaning, whereas abstract words depend relatively more on verbal-associative interconnections for their meaning (Paivio [1986] 1990: 123).

And this is the reason why "comprehension is more dependent on imagery in the case of concrete than abstract sentences" (Paivio [1986] 1990: 219).

The dual coding approach was—like all theories supporting the existence of mental imagery developed in the same period (for an overview see e.g. Nigel 2013)—a reaction against the classical computational view according to which representations are abstract, amodal and language-like symbolic structures unrelated to the physical and functional features of their referents and not bound to the perceptual system. As Paivio points out:

> From its inception, the dual coding approach was a reaction against that singular view in that it emphasized the importance of nonverbal imagery as a mode of thought that was distinct from, though coordinated with, verbal processes. However, dual coding and other imagery-based approaches promptly found themselves in competition with a computational analogue of verbal mediation [...]. Purportedly amodal and abstract, propositions were nonetheless described as if they were mental sentences that served as basic elements in cognitive models much as verbal mediators had done earlier. Prepositional theorists were, if anything, even more single minded in insisting that all information, whether verbal or nonverbal, was transformed into this common descriptive format and then operated on by processes of a similar form (Paivio [1986] 1990: vii).

Computational views explain conceptual thought and linguistic capacities on the basis of internal relations between symbols, without adequately accounting for the problem of how these are connected to the external world. Imagery is introduced to explain how such amodal symbols are anchored to nonlinguistic perception so that people can understand in a referentially salient sense the meaning of words. In this respect, Paivio's approach is entirely consistent with—and in fact even anticipates—the perspective of "grounded cognition", which is focused on the so-called "symbol grounding problem", i.e. the problem of explaining "How is symbol meaning to be grounded in something other than just more meaningless symbols?" (Harnad 1990: 340—on this see Chap. 2, Sect. 1) In this sense, even though Paivio's hypothesis of a dual coding system might be dated, the general issue he tries to resolve through the notion of imagery is still debated and can be interpreted in terms of the general question of the link between language (verbal representations) and perception and thus also of the distinction between referential and inferential competence.

Paivio's theory not only gives a psychological explanation of the difference between referential and inferential competence we introduced by presenting Marconi's views; it goes a step further in the direction of the thesis we are suggesting here. First of all, Paivio embraces the *empiricist principle*[18] according to

[18]Paivio describes his as an empiricist approach. The form of empiricism he puts forward is, however, in certain respects anomalous in comparison to other theories that go under the same name. Besides the empiricist principle mentioned in the text, Paivio's empiricism is characterized primarily (a) by the idea that linguistic representations are multimodal, instead of amodal; (b) by an opposition to computational approaches; (c) and by the adoption of a radically inductive method that aims to develop all hypotheses starting from empirical data (see Paivio [1986] 1990: Chap. 1).

which "all knowledge derives form perceptual, behavioral and affective experiences with the world" (Paivio 2007: 25). Coherently with this principle, he also maintains that no representation can be entirely abstract, but all representations maintain a link to experience. With respect to this point, in an attempt to clarify an aspect of his theory which he thinks is often misunderstood, Paivio specifies: "Abstract nouns are not assumed to be represented *only* by the verbal system if this is intended to mean that they cannot arouse images. The correct statement is that they are less likely to arouse images or do so with greater difficulty than concrete pairs because of less direct access to the imagery system" (Paivio [1986] 1990: 170). Secondly, he embraces as a "guiding theoretical assumption" the idea that "internal (mental) representations have their developmental origin in perceptual, motor, and affective experience" (Paivio [1986] 1990: 55). The fact that he mentions "motor" and above all "affective experience" shows that he agrees to consider internal experience as a possible source of perceptual information we can represent in the imagery system and we can rely on to perceptually ground our linguistic knowledge.[19]

Indeed, in his recent book Paivio (2007: 39) maintains that his view is analogous to that of Lawrence Barsalou which suggest that all kind of representations must be perceptually grounded and that internal information is central to the representation of abstract concepts: "[…] the same symbol formation process that operates on the physical world can also operate on introspective and proprioceptive events. As a result, the introspective symbols essential to many abstract concepts can be represented in a perceptual symbol system" (Barsalou 1999: 600). Especially in his later writings, Paivio also maintains that the non-verbal symbols that allow our understanding of words include *both external sensory information produced by the sensory-motor system and internal information of a proprioceptive and affective kind* (see also Paivio 2007: 33ff).

---

[19]In the philosophical and in the classical psychological tradition, terms like "affect", "affective" or "affection" assume a different meaning. In fact, the word "affection" is derived from Latin terms like affectus and afficere used to translate the Greek term "pathos" which indicates the experience of any kind of event or modification caused by the interaction with an entity other than myself. The verb "to affect" in English preserves in part this original meaning. The word is used in this way e.g. by Aristotle in his doctrine of categories (Aristotle 1963). However, in his On the Soul Aristotle also used the word "affection" in another more restricted and specific sense to indicate only the passive modifications of the psyche that occur without the active and voluntary participation of the subject. In this sense "affections" are identified with and described as passions. In the modern era the word "affection" was used in this restricted sense by Descartes (Brown 2006) and Spinoza (Lebuffe 2010). This interpretation of "affection" in the sense of passion is also that inherited by psychological research. Here "affection" indicates not only the passive modifications of the psyche, but also the effects of this alteration (Dixon 2003). Thus, the word implicitly recalls the idea of an effect caused by some modifications. In a non-metaphysical context, these modifications must concern primarily the body, thus we should use "affection" to indicate any state which is the effect of some bodily modification. In this sense, interoception and proprioception are univocally particular kinds of affective states while emotions belong to the class of affections are affective states only insofar as they are caused by bodily modifications. However, this sense of the word has been lost in the contemporary usage of the term used in the psychological research in which affective and emotional have become synonyms.

Even though Paivio is surely more inclined than Barsalou to consider abstract concepts as different from concrete concepts and to think that understanding them requires a primarily inferential process, both Barsalou's and Paivio's[20] theories constitute attempts to show that external and internal sensory experience are essential for the constitution of abstract concepts. However, neither specifies what precise function this internal information plays nor how people come to develop abstract concepts from sensory information.

Some suggestions as to how to deal with these issues come instead from another family of theories about concepts that do embrace an embodied perspective: the so called conceptual metaphor theories developed primarily by George Lakoff, Mark Johnson and Mark Turner that we already mentioned when discussing Lakoff's hypothesis about 'freedom' (see Sect. 3 of this chapter as well as Johnson 1987; Lakoff 1987; Lakoff and Johnson 1980; Lakoff and Turner 1989; Turner 1987).[21]

According to metaphor theories, concrete conceptual domains of knowledge are used to describe abstract conceptual domains. More specifically, abstract concepts (or concepts which are more abstract than others) are understood on the basis of their mapping to (more) concrete ones. The (more) concrete concept serves as a source domain for the comprehension of an abstract (a more abstract) target concept. By linking the structural elements of the concrete concept—which are well-known and familiar—to other analogous elements characterizing the (more) abstract concept, comprehension of the abstract concept is facilitated. The abstract concept LIFE can e.g. be understood in terms of the much more concrete and familiar concept of JOURNEY: this is the origin of the metaphor of life as a journey shared by many cultures (Lakoff and Turner 1989: Chap. 1). Using this metaphor, the structural elements of life can be understood analogically in terms of the structural elements of a journey—as the elements constituting the concept of journey are mapped onto the concept of life. According to Lakoff's and Turner's description, birth can be understood in term of an arrival (indeed, we usually say that a baby is on the way, and when it has been born we announce his/her arrival); while death can be conceptualized as a departure (indeed, we describe death as 'passing away', 'being gone', 'departing' etc.). In addition, purposes can be seen as destinations and the means for achieving purposes as routes to reach these destinations. "Difficulties in life are impediments to travel. Counselors are guides. Progress is the distance travelled. Things you gauge your progress by are landmarks. Choices in life are crossroads", etc. (Lakoff and Turner 1989: 3). The understanding of the structure of the concrete and familiar concept of journey allows us to confer an analogous

[20]For a systematic discussion of the analogies and of the differences between the two see e.g. Paivio (2007: Chap. 5).

[21]The idea that abstract concepts are grounded metaphorically in embodied and situated knowledge is also shared by other models developed in the field of cognitive linguistics: see e.g. Gibbs (1994), Coulson (2000) and Fauconnier and Turner (2002).

structure on to the concept of life which is—*in its abstract dimension*—more difficult to grasp.[22]

As e.g. Barsalou (1999: 600) also underlines, in order for an abstract concept to be understood on the basis of a concrete one through mappings between respective constituents, both concepts must already be available to the cognitive system; therefore, some sort of target concept must already be formed prior to the metaphorical determination: a person must e.g. already have some concept of life in order to build an analogy between it and a journey. In this sense, metaphor can only be used to further elaborate on and to extend the content of abstract concepts in order to enrich and to determine them more precisely. However, we always also need to represent abstract concepts somehow directly (i.e. in a nonmetaphorical fashion). The notion of a 'core concept' which we also examined in relation to Lakoff's analysis of 'freedom' might be the answer to this problem pointed out by Barsalou. The formation of abstract concepts starts from a core concept made of sensory information that is progressively enlarged using metaphorical procedures. This idea is congruent with Lakoff and Johnson's suggestion that "abstract concepts have a literal core but are extended by metaphors" (Lakoff and Johnson 1980: 272).

Let us for example consider the case of LIFE on the basis of Lakoff's and Johnson's model, but taking into consideration the hypothesis we put forward in Chap. 4 on the possible origin of the categories of animacy meant as aliveness. If that analysis was correct, then the capacity to classify things as alive/not-alive (animate vs. inanimate) is very basic: it is prelinguistic and relies on perceptual/ 'proprioceptive' (propriosensitive) information only. In this sense, according to our hypothesis we don't need an abstract (complex and linguistic) concept of life in order to discriminate between animate and inanimate instances. However, as adults we do in fact also have an abstract, fully developed and at least in part culturally determined concept of life: this might differ depending on how we conceive of life; depending on what we have learned about life: on the model of life we have developed on the basis of our experience and our culture. Using Lakoff's and Johnson's model, it is plausible to assume that in this case the category of animate constitutes the core concept of life (in Chap. 4 we suggested that the categorical information becomes part of the adult concepts it generates), while the abstract (adult, fully developed and culturally determined) concept of 'life' is formed through an extension of this core realized on the basis of a metaphorical procedure.

This same idea applies to all abstract—or, more precisely, I-theoretical concepts —which, in our view, are characterized by a 'proprioceptive' core concept developed prior to any further inferential specification. The case of LIFE might be special

[22]The idea of animacy and of being alive has already been discussed in Chap. 4 of this work. If that analysis is correct, the capacity to classify things as alive/not-alive is very basic and doesn't require a metaphorical process to be realized. The need for metaphorical extension relates rather to the abstract (adult, fully developed and culturally determined) concept of 'life'. This same conclusion applies to all abstract—or, more precisely, I-theoretical concepts—since they have a 'proprioceptive' core which is developed prior to any further inferential specification of the concept.

since we suggest that its core consists in the information we use to operate an early prelinguistic categorical distinction. Indeed, in the perspective we put forward the core concept of I-theoretical concepts does not need to be related to categories, but, might be constituted of any kind of propriosensitive information. The hypothesis of a 'proprioceptive' core concept would explain how a target concept is originally formed and what it consists of before the metaphorical specification occurs. The function of the metaphorical mapping is to extend the core concept and to lead to the development of the adult, fully developed and culturally determined abstract concepts we use in daily life. In this sense, it can be hypothesized that every abstract (I-theoretical) concept can be metaphorically understood or specified as far as its periphery is concerned by identifying systematic correspondences between it and a more concrete source (Lakoff and Johnson 1980 mention e.g. CAUSATION, SUBSTANCE, ARGUMENT, TIME, IDEAS, MIND etc.). In addition, every abstract concept can be simultaneously understood in terms of different other more concrete concepts which highlight different aspects of it or point to various ways in which it can be structured. Consider the example of the abstract concept 'argument'; this can be seen from different perspectives: e.g.:

- *as war* ("Your claims are indefensible." "He attacked every weak point in my argument." "His criticisms were right on target." "I demolished his argument." Lakoff and Johnson 1980: 4ff):
- *as a building* ("We need to construct strong argument for that." "I haven't figured out yet what the form of the argument will be." "We need to buttress the theory with solid arguments." "The argument collapsed." Lakoff and Johnson 1980: 46)
- *as a journey* ("When we get to the next point, we shall see that philosophy is dead." "*So far*, we've seen that no current theories will work." "We will proceed in a step-by-step fashion." Lakoff and Johnson 1980: 90); or
- *as a container* ("Your argument doesn't have much content." "That argument has holes in it." Lakoff and Johnson 1980: 92–93).

Broadly speaking, metaphors have the function bringing a concept into line with general models meant as structured sets of knowledge or, in other terms, as non-rigorously defined theories of limited extent that make sense of the elements they are composed of because they establish inferential relations among them (e.g. if life is a journey, life has a beginning, an end, it lasts for a limited amount of time, thus, we should try to make good use of that time, etc.). In this sense, the hypothesis that the concrete core concepts are extended through the use of metaphors offers an explanation of how the inferential relations might be set up that contribute to the constitution of an abstract concept. This conclusion is perfectly coherent with the hypothesis we embraced about abstract concepts and with the terminology we adopted, according to which so called abstract concepts are and should be called *theoretical* concepts (we distinguished between E-theoretical and I-theoretical concepts depending on whether their core is built on the basis of information driven by the external senses or by the internal monitoring of bodily states).

Since we acknowledge the possibility that not all abstract concepts have a concrete core, but that at least some of them are purely inferential constructs built up from other concepts, we next introduced the notion of intratheoretical concepts as opposed to both I- and E-theoretical concepts. Also in the case of intratheoretical concepts, the general idea is that they consist of an inferential network which is structured as a small theory in which each element contributes to determining the others. If Barsalou is right, and the application of metaphors requires that not only the source but also the target concept already exist in some form before the application of any mapping procedure, then intratheoretical concepts that do not have a perceptual core, but need to be built from scratch cannot originate in this way. They might be formed using more direct definitional procedures. In fact, we don't need to assume that the only way to build theoretical concepts is through the application of metaphorical procedures either. I- and E-theoretical concepts could also be built through the direct identification and determination of an inferential network. The fundamental element of our hypothesis is not the use of metaphors for the enlargement of core concepts, but the idea that this extension occurs through the building of a model that links a concept in a relationship of reciprocal determination. Metaphorical mapping is only one of the possible (indirect) ways in which these models might be formed.

To sum up, the idea we started with in this chapter is that all concepts have a perceptual basis. In the case of intratheoretical concepts this base is only indirect and derived from the other concepts that are used for their definition. According to our hypothesis, a large number of so-called abstract concepts are however not intratheoretical, but rather theoretical and are formed starting with a perceptual or 'proprioceptive' core. The perceptual or 'proprioceptive' base on which theoretical concepts are built is not homologous for all concepts. We cannot exclude e.g. that the core of I-theoretical concepts also includes, at least sometimes, perceptual information from the external senses. In the same way, the core of E-theoretical concepts might also include, at least sometimes, 'proprioceptive' (propriosensitive) information derived from internal self-monitoring. However, we hypothesize that E-theoretical concepts are composed mainly of perceptual information, while I-theoretical concepts mainly consist in 'proprioceptive' information.

Most of the abstract concepts we use in our everyday life are different from 'cell' or 'electron' in the sense that they are not theoretical constructs with an external referential component we can *perceptually* access in an indirect way (e.g. through the use of some device and/or technique). In fact, only scientific concepts can be E-theoretical in the sense of the definition we gave previously. Therefore, the abstract concepts we generally use must be either I-theoretical or intratheoretical (*in the last section of this chapter we will put forward a criterion in order to discriminate between the two*). If the view we are presenting is correct and it is indeed possible to identify a 'proprioceptive' core for most of the abstract concepts we use, then our semantic competence with respect to the majority of the commonsensical 'abstract' words we master relies on I-theoretical concepts.

5 The Structural Analogy Between Emotion and Abstract Concepts

The hypothesis discussed in this chapter suggests that not all concepts that do not have an observable referent in the external world are inferential constructs related to experience only in an indirect manner through the mediation of other related terms. On the contrary, at least some concepts that do not have a concrete referent which can be perceived though the senses do have a referential core based on 'proprioceptive' (propriosensitive) information. We have called the concepts that have a core of this kind 'I-theoretical'. We distinguished these concepts from what we call from 'E-theoretical' concepts which have a referential core based on perceptual information and from 'intratheoretical' concepts which do not have a such a core. But both theoretical and intratheoretical concepts are characterized by a broad inferential structure which plays an essential role in the mastery of the corresponding word meanings. Since the core of I-theoretical concepts consists of information derived from the internal monitoring of the body, I-theoretical concepts turn out to be similar to emotional concepts which are also characterized by a core of 'proprioceptive' information (the issue of emotional concepts has been discussed extensively in Chap. 5).

In the analysis we carried out in Chap. 5, we were able to distinguish different kinds of emotional concepts that varied along a spectrum that ran from basic emotions to highly complex emotions. Basic emotions, we said, share an important aspect with concrete concepts in the sense that the referential component plays the major role with respect to their mastery. The analysis we carried out previously (Chap. 5, Sects. 1–3) shows that the same principle also applies to words referring to internal states of a physical kind (i.e. to bodily experiences like e.g. pain, hunger, thirst etc.). Indeed, according to our hypothesis the structure of concepts referring to internal states (bodily experiences) and basic emotions is analogous to that of concrete concepts with the exception that their referents are internal instead of external. The main difference between terms denoting concrete objects and words referring to bodily experiences and basic emotions is that the former rely on a perceptual schema of information driven by the external senses, while the latter are based on a perceptual schema composed of 'proprioceptive' information coming from the system of internal bodily self-monitoring (for a definition of the notion of perceptual schema see Chap. 3, Sects. 5–7). By contrast, concepts referring to complex emotions consist largely of inferential information and are therefore much more unstable and dependent on the specific definitions/characterizations they acquire in various cultures (for the difference between physical states, basic and complex emotional concepts see Chap. 5). While in the case of concepts denoting bodily experiences, basic emotions and concrete objects the referential component plays the dominant role in the mastery of the corresponding word, in the case of I-theoretical concepts and non-basic emotional concepts the inferential component takes the lead.

I-theoretical concepts are therefore structurally analogous to concepts referring to complex emotions. I-theoretical concepts are similarly characterized by a 'proprioceptive' core, and by a large periphery made of inferential information which depends on specific cultural factors. Since inferential information has a major role with respect to the composition of the concept, people's understanding of the respective words will be culturally and possibly subjectively variable and depend mainly on the definitions they have learned and models they have acquired. Nevertheless, even in these cases inferential information alone is not sufficient. To fully master these words, subjects need to also have referential competence and therefore they need to develop a corresponding concept which also includes an experiential core consisting (mostly) of 'proprioceptive' information.

This structural analogy between I-theoretical and emotional concepts can be sustained using different lines of evidence. In the following part of this section we will introduce two of these which are particularly relevant and at least apparently unrelated to each other. (a) The first argument is driven by cognitive research on concepts and particularly psycholinguistic studies; while (b) the second is built on several observations on the deficits of semantic competence exhibited by people with Autism Spectrum Disorder that further expand on the analysis developed in the previous chapter (see Chap. 5, Sect. 10). In this respect, a preliminary specification is needed. We previously introduced the term (I and E) theoretical versus intertheoretical as substitutes for the word 'abstract', which we considered to be too ambiguous. However, in order to discuss these two lines of evidence, we must return to the opposition of concrete and abstract, since these are the notions the debates we are going to consider rely on. If needed, we will use our terminological distinctions to specify what kind of abstract concepts are meant in the various contexts.

(a) As far as the cognitive research on concepts is concerned, the existence of a structural analogy between emotional concepts and certain abstract—in our vocabulary theoretical and more specifically *I-theoretical*—concepts is sustained by psychological studies concerning processing differences for concrete and abstract words and so-called 'concreteness effects'. These studies show that concrete words have some processing advantages with respect to abstract words; on the whole, they are processed faster and more accurately than abstract words; in addition, memory is better for concrete than for abstract words.[23] These studies also focus on emotional concepts showing that they exhibit complex, controversial, but relevant similarities to abstract concepts. We can roughly distinguish between three main lines of research in the last decades with respect to patterns of difference between concrete, emotional and abstract concepts (corresponding more or less to three different temporal 'stages'). (i) According to the first line of research, emotion terms can be incorporated with abstract terms; (ii) according to the second, emotion terms are distinct from both abstract and concrete terms; while (iii) according to the third

[23]For a brief review of the early and more recent studies on so-called 'concreteness effects' see e.g. Kousta et al. (2011: 15).

line of research, even though emotion terms can be differentiated from abstract terms in some respects, emotional and abstract terms are intimately related to each other.

(i) In the first stage of the concreteness effects literature, emotion words are often included in the same class as abstract terms (see e.g. Reed and Dick 1968; Chiarello et al. 1987; Eviator et al. 1990; Nelson and Schreiber 1992). Even though usually no justification for this classification is given, the reason why emotional terms are included in the class of abstract term can be traced back to the way in which the abstract words for most studies were chosen. These studies generally consider abstractness to be opposed to concreteness and apply the equation that low concreteness is the same as abstractness: abstract words are those that do not refer to (or only indirectly refer to) something observable (perceivable) in the external world. For the most part, the concrete and non-concrete words used in these studies are taken from the MRC psycholinguistic database (Coltheart 1981; Wilson 1988)[24] which was assembled by merging a number of smaller previous databases[25]—and which provides the concreteness rating for 150,837 words (These ratings were assigned by asking a number of people to rate the degree of concreteness of a list of words. In the next section, we will come back to the issue of the instructions given to people in order to accomplish this task). If one considers concreteness ratings only, emotion words are indistinguishable from abstract words since both have low concreteness ratings. For this reason, the studies that selected abstract words from this database by considering only their concreteness ratings automatically included emotion words with abstract words.

(ii) In the second stage, researchers began to criticize the previous incorporation of emotion terms in the class of abstract terms and they hypothesized that emotion terms constituted a distinct group from both concrete and abstract words (Altarriba et al. 1999). These authors maintained that low concreteness ratings could not be the only criterion used to select abstract words. They showed that emotion words differ from non-emotional abstract words in terms of other relevant semantic properties for which the database provided information, specifically their imageability and context availability ratings. Roughly speaking (we will discuss this aspect in more detail in the next section), imageability is defined in terms of the capacity to arouse mental images of things or events. These images are not meant in a strict sense as visual mental pictures only, but in a broad sense as the content of some sensory experience. The context availability ratings of a term depend on how many other words people can recall in a short time that are related to the target term, i.e. on the ability to easily and quickly recall the possible contexts of occurrence for a given word. As e.g. Altarriba and colleagues note:

[24]The database is available online at http://websites.psychology.uwa.edu.au/school/MRCDatabase/uwa_mrc.htm. The database provides information about 26 different linguistic properties. The ones we will consider in this word are concreteness and imagery ratings.

[25]In the studies before 1981 other databases were used.

> On both the concreteness and the context availability scales, emotion words were rated lower than abstract and concrete words. However, on the imagery scale, emotion words were rated higher than abstract words but lower than concrete words. These results indicate that emotion words possess characteristics different from those of either abstract or concrete words. Therefore, the inclusion of emotion words in an abstract category may have biased the results of previous research on concreteness effects (Altarriba et al. 1999: 583).

As they specify in a later work: "[...] the results indicate that concepts represented by emotion words are more imageable and are easier to think of a context for than abstract words but are less concrete than abstract words" (Altarriba and Bauer 2004: 397).

The authors who follow this line of research hypothesize that the reason why emotion words have higher imageability and context availability ratings (i.e. the reason why they seem easier to image and to connect to other related terms than abstract words) is that—even though emotions are not concrete objects that can be experienced through the senses—*they are perceptually connected to specific images*. Referring to the previous research on the universality of facial expressions of basic emotions (Ekman et al. 1972; Izard 1971), they speculate that each emotion is connected with a specific facial expression. Because of this connection, they suggest that "it is plausible that an image may be more accessible for emotion words than for abstract words. For example, the concrete word *tiger* may elicit an image of a tiger, and the emotion word *scared* may elicit an image of a scared facial expression" (Altarriba and Bauer 2004: 394–395). Regardless of the explanation they put forward for the differences in the various ratings (which is actually quite far from the one we will suggest in the next section), the conclusion they draw from their findings is that "the concepts represented by emotion words are characteristically different from and activate different relationships than abstract words" (Altarriba and Bauer 2004: 407).

(iii) In the third stage, the literature on concreteness effects again began to consider the similarities between abstract and emotion words as far as their very low concreteness rating is concerned and tried to establish new theoretical connections between them putting forward the hypothesis that both were grounded in internal information. For example, Lawrence Barsalou and Katja Wiemer-Hastings argue that abstract concepts are grounded in "introspective properties" which are meant to derive both from affective states and mental states (i.e. states related to beliefs, opinions, etc.; see Barsalou and Wiemer-Hastings 2005). A similar position is also maintained in different studies by Wiemer-Hastings and colleagues, whose aim is to show how "introspective features, such as emotions, are likely a central property of abstract concepts" (Wiemer-Hastings and Xu 2005: 721). According to these authors, abstract as well as emotional concepts are qualitatively different from concrete concepts, since only concrete concepts are characterized by perceptual features. By contrast, the more a concept is related to introspective properties the more it is perceived as abstract, i.e. as not-concrete (Wiemer-Hastings et al. 2001).

This view is taken further by other authors such as Stavroula-Thaleia Kousta and Gabriella Vigliocco who propose that abstract words are different from concrete ones because they are characterized by a statistical preponderance of affective information with respect to concrete words (Vigliocco et al. 2009; Kousta et al. 2011). According to their view, both concrete and abstract concepts consist of two different types of information: "experiential information" and "linguistic information". The "linguistic information" is a language based kind of information analogous to the inferential information that has been discussed in this work. The "experiential information" is of a twofold kind, or more specifically is drawn from two different sources depending on whether we are considering concrete or abstract concepts. The experiential information concrete concepts rely on is qualified as sensory-motor information, meant as information derived from sensory-motor interactions with the outside world. In contrast, the experiential information abstract concepts are made of is of an affective kind, i.e. it consists of the experience people have of their inner states and particularly of their emotions. As Vigliocco and colleagues maintain:

> The apparent dichotomy between concrete and abstract word meanings arises because of a statistical preponderance for sensory-motor information to underlie concrete word meanings and a preponderance for affective and linguistic information to underlie abstract word meanings. While sensory-motor information is statistically more preponderant for concrete word meanings, affective and linguistic information is statistically more important for abstract word meanings, both for their acquisition and their subsequent representation in the adult system (Vigliocco et al. 2009: 223).

Independently of the specific position they argue for, their studies show that there is a relevant relationship between emotional and abstract concepts that can be observed in particular in their double line of opposition to concrete concepts. First of all, emotional terms have—like abstract words—very low concreteness ratings (the concreteness ratings for emotional terms are even lower than those for abstract terms that are more neutral from the point of view of their emotional valence). Secondly, unlike concrete concepts, abstract concepts are characterized—as emotional concepts—by a preponderance of affective information. This affective information constitutes the direct experiential component of the abstract concept. Like concrete terms, abstract terms consist of different proportions of experiential and linguistic information. Concrete concepts are more or less concrete depending on the extent to which sensory information driven by the external world dominates linguistic information, while abstract concepts are more or less abstract depending on the extent to which affective information dominates linguistic information. Thus, according to this last line of research, abstract concepts are similar to emotion concepts since both rely on a mixture of experiential and linguistic information, while this experiential information is identified with affective (internal) information.

(b) The structural similarity between emotional and abstract concepts (the name we favor is I-theoretical concepts) can also be supported on the basis of clinical evidence related to deficits in semantic competence exhibited by subjects affected by Autism spectrum disorder also. Even when they are highly intelligent, these subjects have serious difficulties mastering not only emotional terms, but also abstract terms.[26]

Abstract terms are characterized by the fact that unlike concrete terms they do not have an observable referent in the world. Some clues to the reasons why people with autism spectrum disorder have difficulties understanding these words might come from the already mentioned (Chap. 5, Sect. 10) bibliography of Temple Grandin, an animal psychology professor who suffers from autism spectrum disorder and who gives a detailed description of her insurmountable difficulties in understanding all kinds of words whose referent is not an object, a process or an event directly observable in the external world. Writing of her childhood, she reports that it was extremely problematic for her to learn words that were not connected to visual pictures, or verbs or indexical forms: "As a child, I left out words such as 'is,' 'the,' and 'it,' because they had no meaning by themselves. Similarly, words like 'of' and 'an' made no sense. Eventually I learned how to use them properly, because my parents always spoke correct English and I mimicked their speech patterns. To this day certain verb conjugations, such as 'to be,' are absolutely meaningless to me" (Grandin 2006: 15).

The problems in understanding abstract word meanings Grandin ascribes to herself are related to the lack of a concrete referent for these words which would allow her to construct a mental picture that could serve as surrogate for the word meaning. In fact, to understand a word that does not have an external observable referent she uses a strategy of 'externalization' of meanings based on the conversion of words into symbolic pictures built up in her imagination using mainly sensory (visuospatial) information.

> Growing up, I learned to convert abstract ideas into pictures as a way to understand them. I visualized concepts such as peace or honesty with symbolic images. I thought of peace as a dove, an Indian peace pipe, or TV or newsreel footage of the signing of a peace agreement. Honesty was represented by an image of placing one's hand on the Bible in court. A news report describing a person returning a wallet with all the money in it provided a picture of honest behavior (Grandin 2006: 17).

[26]Several empirical research studies have shown that people suffering from Autism Spectrum Disorder exhibit difficulties in understanding and in mastering both emotional and abstract terms (see, e.g., Hermelin and O'Connor 1967; Toichi and Kamio 2001; Ropar and Peebles 2007; on this see also Chap. 5, Sect. 10). On the contrary, they exhibit a communicative and cognitive style oriented towards 'concreteness', i.e. they seem to better understand terms that are related to external sensory experience, especially visual experience, or whose meaning can be directly or indirectly conveyed in visual images (Kamio and Toichi 2000; Mitchell and Ropar 2004; Gaigg et al. 2008; Alderson-Day 2011; Kunda and Goel 2011; Muth et al. 2014; see also Lohmar 2016: 265 and 273; Esposito et al. 2016). This cognitive style of people suffering from Autism Spectrum Disorder was the object of different interpretations. A recently proposal by Tatyana Glezerman suggests that this is due to a dominance of the left-hemisphere which processes mainly sensorimotor stimuli and visual images (Glezerman 2013: 26ff and 234ff). By contrast, Alexithymic subjects exhibit no particular deficits in mastering abstract words (Wotschak and Klann-Delius 2013).

What Temple Grandin describes is a form of 'thinking in pictures', in which the picture serves as a stable and handy element which can give concreteness and provide something like a referent to the words of a language. As she writes:

> I think in pictures. Words are like a second language to me. I translate both spoken and written words into full-color movies, complete with sound, which run like a VCR tape in my head. When somebody speaks to me, his words are instantly translated into pictures. [...] One of the most profound mysteries of autism has been the remarkable ability of most autistic people to excel at visual spatial skills while performing so poorly at verbal skills. When I was a child and a teenager, I thought everybody thought in pictures. I had no idea that my thought processes were different. [...] Unlike those of most people, my thoughts move from video-like, specific images to generalization and concepts. For example, my concept of dogs is inextricably linked to every dog I've ever known. It's as if I have a card catalogue of dogs I have seen, complete with pictures, which continually grows as I add more examples to my video library (Grandin 2006: 3; 3–4; 12).

To understand the meaning of a word Temple Grandin must grasp a referent consisting of an object, event or behavior which can be represented through a picture that confers concreteness on it, i.e. tangibleness. If a word does not have a directly observable reference, Grandin must build some surrogate of it to use as a stable element for the semantic mastery of the word. This surrogate makes explicit some essential characteristics/relationships expressed by the abstract word as in the examples of honesty and peace.

Using this strategy, Temple Grandin was able to develop a capacity to master abstract words—as well as emotion words (see Chap. 5, Sect. 10)—which appeared almost normal in the sense that it was well suited to the relevant circumstances and in compliance with the linguistic behavior of normal people. However, she exhibits only an abnormal, 'impersonal appropriation' of the semantics of abstract words which suggests that the conceptual representations she developed to support word mastery do not include the corresponding appropriate internal feelings: if we consider e.g. her mastery of 'honesty' and 'peace', she does not seem to know what honesty and peace are (*feel like*). In the case of abstract words as in the case of emotional terms the semantic competence of Temple Grandin turns out to be impaired and different from that exhibited by normal subjects; and this becomes even more clear if we consider that she reports that she is never really sure that she has understood an abstract or emotional word properly.

The case of Temple Grandin shows that semantic competence with respect to abstract words is not—as one might think (e.g. on the basis of Marconi)—entirely inferential, but is also characterized by a different component of a referential kind. Grandin has no difficulty in learning inferential information. On the contrary, as already pointed out (see Chap. 5, Sect. 10), people with Autism Spectrum Disorder use inferential definitions and descriptions to learn word meanings much more than others (an iconic example is Hobson's patient who tries to understand the meaning of 'friendship' by learning all of its possible definitions). Grandin also has no difficulty in mastering words that have an observable referent in the external world; in fact, the alternative strategies she develops to overcome her deficits in language mastery are based on the attempt to picture some external, perceptual referent for the words she has difficulties understanding.

The fact that people with Autism Spectrum Disorder exhibit analogous difficulties for mastering emotional and abstract/I-theoretical words—such as 'peace', 'honesty' or, going back to the example discussed in the previous chapter, 'friendship' (Chap. 5, Sect. 10)—offers important support for the hypothesis that these two word classes are structured in the same way, in the sense that mastering them relies on concepts which are characterized by a referential core made of 'proprioceptive' (propriosensitive) information. This leads us to think that the linguistic deficit people with Autism Spectrum Disorder exhibit for abstract/I-theoretical words has an origin and a cause analogous to that hypothesized for emotional words—one that can be traced back to a difficulty in getting or —more plausibly—processing 'proprioceptive' information to form an integrated and global representation.

6 Abstraction (in Summary) and the Issue of Establishing Which Concepts Do Have an Internal Grounding

In this chapter, we tried to argue that "the standard picture of meaning" (Sect. 1 of this chapter) according to which abstract words are entirely inferential constructs is wrong or at least highly imprecise in an essential respect: at least some of the concepts that support our mastery of so-called abstract words are grounded in sensory information, i.e. they are composed of a core concept consisting of sensory information. Thus, we preferred to avoid as much as possible the word 'abstract' and we choose to speak of theoretical and intratheoretical concepts: we called those concepts that retain an empirical grounding theoretical, in spite of their broadly inferential composition, while we called concepts that are entirely constituted by inferential relationships intratheoretical. In the field of science, theoretical concepts might be grounded in external sensory information since the use of instruments allows us to observe phenomena which are impossible to experience with the senses only. However, grounding for the concepts we use in everyday life, is almost exclusively internal. This internal grounding has been described as '*proprioceptive*' (or *propriosensitive*, for a definition see Chap. 2), and more specifically analogous to that we identified and explained in Chap. 5 with respect to terms referring to internal states and to emotions.

The joint conclusion we can draw from the analyses in this and the previous chapter may be summarized as follows.

- Concepts/words referring to bodily experiences and basic emotions are strongly internally grounded; from a semantic point of view they are therefore analogous to concrete concepts in the sense that referential competence plays a central role with respect to semantic mastery and even prevails over inferential competence.

- Concepts/words referring to non-basic emotions and a number of abstract terms in ordinary usage like 'friendship' or 'freedom' are on the contrary I-theoretical in the sense that they are for the most part inferential constructs, but also have an internal grounding consisting of 'proprioceptive' (propriosensitive) information. In the case of theoretical concepts, both components of semantic competence—the referential and the inferential—are essential to achieve full semantic mastery, even though the inferential component might be more important to ensure an adequate use of the word. Going back to the more classical term 'abstract', one could state that these concepts are at an intermediate level of abstraction in the sense that they are inferential constructs with a referential core concept.
- Finally, there are concepts/words that are indeed entirely inferential constructs—that are properly abstract according to the standard definition. These have been defined as intratheoretical and have been described as purely linguistic constructs developed in the context of a theory. Their content/meaning depends entirely on their inferential relationships with other words used to define them. The mastery of this word class is primarily if not entirely due to inferential competence, while the link to experience is only indirect and the meaning of such words relies on other words for definition.

This classification presupposes something like *different kinds of concreteness* and *different degrees of abstraction*: concepts/words referring to bodily experiences and basic emotions are described as analogous to concrete concepts/words, while abstract concepts/words can be more or less abstract depending on whether they are intratheoretical (and rely entirely on inferential information) or theoretical (and are partially grounded in perceptual—or more often in 'proprioceptive'—information).

This idea of a different form of concreteness manifestly contradicts the classical opposition between abstract and concrete according to which we can consider as concrete only terms/concepts whose meaning/content is fixed by their relations with observable properties—i.e. whose referents are objects in the external world that can be perceived through the senses. As a matter of fact, from this perspective, words referring to bodily experiences and emotions (basic as well as non-basic) are univocally abstract (on this see also above Sect. 5). At the same time, the standard picture of abstraction is also not compatible with the view that some abstract concepts/words retain a direct grounding in sensory information of an internal kind and it offers a different account of the reason why we can distinguish between different degrees of abstraction. Indeed, according to the standard picture of language, concepts/words can be more or less abstract depending on *how 'far' they are from the observational vocabulary*: i.e. depending on whether the inferential relationships that define their content/meaning include (more or less) elements of the observational vocabulary. If a word/concept is defined using inferential elements that belong to the observational vocabulary, this will be less abstract than a word/a concept whose meaning/content is defined by inferential relationships that do not have any direct connection to the observational vocabulary.

The view we propose does not deny that it is possible to distinguish different degrees of abstractness on these grounds. However, we propose also a different—more salient and 'macroscopic'—way to differentiate among words/concepts with regard to their degree of abstractness based on the idea that many words/concepts that are usually considered abstract do have *a direct, partial grounding in sensory experience and this experience is mainly of a 'proprioceptive' kind.* If this is the case, the most relevant way to establish the degree of abstractness of specific concepts/words consists in establishing the proportions of *internal or external* sensory information and inferential information that characterize them.

While this classification of concrete and abstract might provide a clear result in line of principle, it might also appear quite speculative from a practical perspective since we haven't offered any criterion to distinguish among concept/word classes that are more or less abstract (intratheoretical or I-theoretical). In fact, other than the differentiation among words referring to concrete objects or properties, to internal states, to basic and non-basic emotions, the distinction between I-theoretical an intratheoretical words/concepts turns out to be particularly problematic, since we don't have any clear *semantic* parameter to establish which words/concepts belong to the one or the other class (i.e. which words/concepts have an internal grounding and which do not). In the following part of this section we will try to show that two measures developed by psycholonguistic research can help to address this issue.

The first measure which is partially—or rather, as we will see, indirectly—relevant with respect to this aim is the so-called concreteness measure that assesses the degree of concreteness of words. If abstractness is considered to be the opposite of concreteness—as it usually is—then low concreteness is equivalent to/accounts for high abstractness. These measures are obtained asking a number of people to assign ratings of concreteness to a number of words on the basis of specific norming instructions that specify what kinds of words must be considered more or less concrete. Thus, the idea of concreteness these ratings are based on depends directly on these norming instructions; to understand this process we will need to analyze—at least briefly—these instructions and their variations from the time they were first introduced in the sixties until the most recent collection of ratings in 2013.

Concreteness ratings were first introduced by Spreen and Schultz (1966), and then proposed again in the same form by Paivio et al. (1968): while Spreen and Schultz (1966) defined the ratings in terms of 'low concreteness' and 'high concreteness', Paivio et al. (1968) labeled the ends of the scale as 'high concreteness' versus 'high abstractness'. Later collections until very recent times continued using these latter instructions. The most important source of the concreteness ratings collected with these norming instructions is the (already mentioned) MRC Psycholinguistic Database (Coltheart 1981; Wilson 1988).[27] Spreen and Schultz's instructions for concreteness read as follows:

[27]The database is available online at http://websites.psychology.uwa.edu.au/school/MRCDatabase/uwa_mrc.htm. The database provides information about 26 different linguistic properties. The ones we will consider in this work are concreteness and imagery ratings.

> Nouns may refer to persons, places and things that can be seen, heard, felt, smelled or tasted or to more abstract concepts that cannot be experienced by our senses. The purpose of this experiment is to rate a list of words with respect to 'concreteness' in terms of sense-experience. Any word that refers to objects, material or persons should receive a high concreteness rating; any word that refers to an abstract concept that cannot be experienced by the senses should receive a low concreteness rating. Think of the words 'chair' and 'independence'. 'Chair' can be experienced by our senses and therefore should be rated as high concrete; 'independence' cannot be experienced by the senses as such and therefore should be rated as low concrete (abstract) (Spreen and Schultz 1966: 460).

Paivio, Yuille and Madigan describe their instructions for the concreteness ratings as follows (the second part is identical to Spreen and Schultz's instructions for concreteness):

> The printed instructions were essentially those used by Spreen and Schulz (1966) with the modification that the ends of the scale were defined in terms of abstractness-concreteness rather than low concreteness-high concreteness, as was done by Spreen and Schulz. For example, the present instructions stated that, 'Any word that refers to objects, materials, or persons should receive a *high concreteness* rating; any word that refers to an abstract concept that cannot be experienced by the senses should receive a high abstractness rating, [...] (Paivio et al. 1968: 5).

These instructions were intended to convey a notion of concreteness according to which a word is concrete if it refers to something that can be perceived through the senses—that can be seen, heard, felt, smelled or tasted; it is abstract (or at least not-concrete) otherwise. However, as pointed out e.g. also by Connell and Lynott (2012: 461) the example mentioned in the instructions—"persons, places and things", and above all "objects, materials or persons" as well as "chair" versus "independence"—bias people to rely on a different definition of concreteness for their ratings, a definition which is closer to our everyday understanding of this notion, according to which *only material objects are concrete*. From this perspective, 'chair' is surely concrete, but 'smell', 'sound' or 'flavor' are not. More generally, from the point of view of common sense, exclusively *visual stimuli* (like light, shadow, colors—everything that can only be seen, but not also touched or perceived by other senses), as well as exclusively *acoustic, gustatory, olfactory or tactile stimuli* that are not caused by something that can also be experienced visually or by another additional sense *do not belong to the range of concrete things*. The association between concreteness and materiality leads to a multisensory view of concreteness whereby an object is considered concrete only if it is perceived through more than one sense simultaneously.[28]

[28]Connell and Lynott's (2012) position concerning what idea of concreteness the norming instructions of Spreen and Schulz (1966) and of Paivio et al. (1968) give rise to differs to some extent from ours. Connell and Lynott are persuaded that the original norming instructions given to the participants to assign concreteness ratings to the words were misleading because they induced people to focus primarily on visual perception and to overlook the other external sensory modalities (the other four senses). We agree with Connell and Lynott that the original instructions were misleading, but we think that they led people to consider only things that can be perceived through multiple senses at once as concrete.

This way of understanding concreteness in terms of materiality is not in accordance with the idea of concreteness—or the opposite notion of abstractness—that psycholinguist studies rely on (i.e. words are concrete if their referents can be perceptually experienced). The latter depend on a further basic assumption: words/concepts that rely on sensory information are processed differently from words/concepts that are not directly linked to sensory information. This difference reflects the (already mentioned) "concreteness effect" by which: "Abstract concepts are traditionally thought to differ from concrete concepts by their lack of perceptual information, which causes them to be processed more slowly and less accurately than perceptually-based concrete concepts" (Connell and Lynott 2012: 452).

Since concreteness ratings are used to investigate the concreteness effect, it is clear that they need to be assigned on the basis of an idea of concreteness compatible with this effect, an idea which classifies as concrete all and only those things that can be experienced through at least one of the external senses. To achieve this aim, a very recent collection of ratings for 40 thousand common English words acknowledges the criticisms and suggestions proposed by Connell and Lynott (2012) and Lynott and Cornell (2009) and formulates norming instructions that make explicit reference to each of the five senses separately (Brysbaert et al. 2014).[29]

Going back to the issues brought up in Sect. 4 of this chapter concerning the criteria we may use to qualify a concept/word as abstract, it is important to point out that both the original norming instructions as well as the revised ones used by the rating collection of 2013 produce the same outcomes for words referring to emotions or internal states: since emotions or internal states are not perceived by the external senses, terms referring to them must necessarily be considered as abstract.

However, the concreteness construct is not the only measure used in the psycholinguistic literature to assess the degree of concreteness/abstraction of words. There is indeed also another construct—the so-called measure of imageability—that is usually considered as completely equivalent to and therefore also interchangeable

[29]They used the following instructions: "Some words refer to things or actions in reality, which you can experience directly through one of the five senses. We call these words concrete words. Other words refer to meanings that cannot be experienced directly but which we know because the meanings can be defined by other words. These are abstract words. Still other words fall in-between the two extremes, because we can experience them to some extent and in addition we rely on language to understand them. We want you to indicate how concrete the meaning of each word is for you by using a 5-point rating scale going from abstract to concrete. A concrete word comes with a higher rating and refers to something that exists in reality; you can have immediate experience of it through your senses (smelling, tasting, touching, hearing, seeing) and the actions you do. The easiest way to explain a word is by pointing to it or by demonstrating it (e.g. To explain 'sweet' you could have someone eat sugar; To explain 'jump' you could simply jump up and down or show people a movie clip about someone jumping up and down; To explain 'couch', you could point to a couch or show a picture of a couch). An abstract word comes with a lower rating and refers to something you cannot experience directly through your senses or actions. Its meaning depends on language. The easiest way to explain it is by using other words (e.g. There is no simple way to demonstrate 'justice'; but we can explain the meaning of the word by using other words that capture parts of its meaning) […]" (Brysbaert et al. 2014: 906).

with that of concreteness. The persuasion that the imageability and concreteness measures are equivalent stems from the way in which imageability was originally conceived.

The imageability measure was introduced for the first time by Paivio et al. (1968) in order to support the view that developed into the (already mentioned) *Dual-Code Theory*. The measure was based on Paivio's general perspective on cognitive processing and language understanding which was developed during the 60s and 70s (mainly in Paivio 1965, 1971b and Paivio et al. 1968) and which assumed its current and final configuration in the 80s (in particular with Paivio [1986] 1990). More recent works (especially Paivio 2007) reaffirm the original theory and discuss it in the light of more general hypotheses on how mind and brain developed from an evolutionary point of view. We have already discussed some fundamentals of Paivio's *Dual Coding Theory* (Sect. 4 of this chapter). In summary, the theory suggests that the cognitive system consists of two structurally distinct symbolic subsystems: a verbal one, specialized for language processing only and a nonverbal one, the so-called imagery system, that generates 'mental images' of the events and objects perceived in the external world and allows for referential acts (Paivio [1986] 1990: 63ff). Mental images are memory traces of all kinds of perceptions: they are not merely mental pictures, but also nonverbal representation of sounds, of olfactory, tactile and gustatory sensations.

According to the *Dual Coding Theory*, the imaginal and verbal systems are functionally interconnected and activity in one initiates activity in the other. Concrete words represented in the verbal system are directly connected with the corresponding mental images stored in the nonverbal system; abstract words are linked to the imagery system only indirectly through other words that define them and have a corresponding sensory image. "Both classes of words [abstract and concrete] have interconnections with the representations of other words in the verbal system. […] Concrete words have both referential and verbal-associative meaning, whereas abstract words depend relatively more on verbal-associative interconnections for their meaning" (Paivio [1986] 1990: 123). For this reason "comprehension is more dependent on imagery in the case of concrete than abstract sentences" (Paivio [1986] 1990: 219). While concrete words directly evoke the mental images they are connected to in the other system easily/quickly (and this makes—among other things—the referential act easier/faster), abstract words evoke images only indirectly and thus more slowly and with difficulty. According to Paivio, the ease or difficulty with which words evoke mental images can be measured and the corresponding ratings can be used as an indirect criterion to determine the degree of concreteness of the corresponding words. Paivio defines imageability (or imagery) ratings as a measure of the image-evoking capacity of words, i.e. of their capacity to arouse sensory (i.e. perceptual) experiences stored in memory (Paivio 1963, 1965; Paivio et al. 1968). Since concrete words are hypothesized to have a higher image-evoking capacity than abstract words, greater imageability ratings must be associated with higher concreteness ratings. For this reason, both constructs are hypothesized to measure the same underlying variable.

Paivio's hypothesis suggests that imageability and concreteness must be perfectly correlated. In fact, a very high correlation was confirmed—*at least at a general level*—by the data he collected together with his collaborators (Paivio et al. 1968) as well as by many other later studies by other authors (see e.g. Benjafield and Muckenheim 1989; Christian et al. 1978; Friendly et al. 1982; Gilhooly and Logie 1980; Rubin and Friendly 1986; Schwanenflugel et al. 1988; Toglia and Battig 1978).

Because of this high correlation, these two measures began to be considered as equivalent and perfectly interchangeable in the literature. As e.g. Jamie Reilly and Jacob Kean maintain: "Although imageability and concreteness are technically different psycholinguistic constructs, the correlation between these variables is so strong that many authors use the terms interchangeably. Here we make the same assumption of synonymy between imageability and concreteness in terms of theory (i.e., concreteness effects—imageability effects)" (Reilly and Kean 2007: 158). The same point has been made more recently by Cornell and Lynott (2012: 453): "Imageability ratings are frequently used interchangeably with concreteness ratings in the experimental literature […] because of their high correlation and theoretical relationship in dual coding theory".

However, already in their seminal article in which the imageability measure was introduced for the first time Paivio and collaborators pointed out some anomalies with these constructs. First of all, they noted that "the correlation between I [Imageability] and C [Concreteness], although substantial, is not as high as one might expect if it is assumed that both scales measure the same underlying variable" (Paivio et al. 1968: 7). Secondly, they observed that the reason why the correlation is less perfect than expected is due to a number of words that exhibit a contrasting pattern, since their imageability ratings are significantly higher than their concreteness ratings. According to Paivio's analysis:

> the nouns for which I and C are most discrepant have common characteristics. […] Most of these are words with strong emotional and evaluative connotations. The largest group consists of terms referring to affective reactions or affective attitudes: affection, agony, amazement, amour, anger, anxiety, devotion, fun, gaiety, gratitude, grief, happiness, hatred, hope, hostility, humor, insolence, jealousy, joviality, joy, kindness, love, loyalty, misery, mood, panic, passion, pleasure, pride, sadness, and shame. Others in this category are labels for attitudes and emotional situations, or are generally evaluative in meaning: blessing, bravery, chaos, charm, Christmas, courtship, death, glory, obedience, obsession, safety, tragedy and vanity (Paivio et al. 1968: 7).[30]

These anomalies were confirmed more recently by other studies. Altarriba and colleagues show e.g. that abstract, concrete and emotion words are reliably different

[30]As Paivio and collaborators point out, only a few words exhibit the contrasting pattern in which C exceeds I. However, this anomaly could easily be explained showing that they were all very unusual words (e.g. armadillo, chloride, surtax). Since these words are extremely infrequent, it is plausible that people know that these are concrete words and assign therefore high concreteness ratings, but cannot associate them with specific sensory information stored in memory and therefore give them low imageability ratings (Paivio et al. 1968: 7).

from each other when rated on concreteness, and imageability: "Analyses revealed that, for concreteness, concrete words received the highest ratings, followed by abstract, and then by emotion words. For [...] the imageability scale, concrete words received the highest ratings, followed by emotion, and then by abstract words" (Altarriba et al. 1999: 579). The paradoxical situation pointed out first by Altarriba and colleague as well as by Kousta and colleagues is that: "[...] Although emotion words were rated as more abstract than other abstract words, they were higher in imageability [...] than other abstract words (and lower than concrete words)" (Kousta et al. 2011: 20).

The explanation Paivio and colleagues have offered for these contrasting patterns is that in some cases non-concrete words easily evoke some sort of sensory information and for this reason they get high imageability ratings, even though they have low concreteness ratings. More specifically: "These words appear to have the common property of having been associated with sensory experience (usually affective in nature) but not specific things or classes of things, hence their relatively high imagery and lower concreteness ratings" (Paivio et al. 1968: 7). *Thus, external sensory information (on which concreteness is measured) is not the only kind people rely on when they assign imageability ratings.*

Paivio's hypothesis is that imageability ratings are assigned also on the basis of affective information. This is in line with the thesis we argued for in Chap. 5 according to which affective (i.e. emotional) information is a kind of sensory information which is not driven by the external world, but that is experienced internally and depends ultimately on the body and on bodily states. If this is the case, emotion words have high imageability ratings in spite of their low concreteness because—even though we don't experience emotions through the external senses—emotion words easily evoke internal sensory experience.

This explanation of the discrepancies reported in the literature between imagery and concreteness suggests that imageability and concreteness do not measure one and the same thing, but imageability also tracks something other than the link between words and external sensory information. In particular, it tracks the link between words and any kind of sensory information—be it external or internal. This conclusion implies that, *for some reason*, when people assigned imageability ratings to the words, they considered not only the ease/difficulty with which a word evoked external sensory experience stored in memory, *but also the ease/difficulty with which a word evoked internal sensory experience stored in memory.* This reason must be traced back first of all to the norming instructions given to the participants to assign their ratings. The norming instructions given for the assignment of imageability ratings were first introduced by Paivio et al. 1968:

> Nouns differ in their capacity to arouse mental images of things or events. Some words arouse a sensory experience, such as a mental picture or sound, very quickly and easily, whereas others may do so only with difficulty (i.e., after a long delay) or not at all. The purpose of this experiment is to rate a list of words as to the ease or difficulty with which they arouse mental images. Any word which, in your estimation, arouses a mental image (i.e., a mental picture, or sound, or other sensory experience) very quickly and easily should be given a high imagery rating: any word that arouses a mental image with difficulty or not

> at all should be given a low imagery rating. Think of the words 'apple' or 'fact'. Apple would probably arouse an image relatively easily and would be rated as high imagery; fact would probably do so with difficulty and would be rated as low imagery. […] Make your rating by putting a circle around the number from 1 to 7 that best indicates your judgment of the ease or difficulty with which the word arouses imagery. The words that arouse mental images most readily for you should be given a rating of 7; words that arouse images with the greatest difficulty or not at all should be rated 1; words that are intermediate in ease or difficulty of imagery, of course, should be rated appropriately between the two extremes (Paivio et al. 1968: 4).

Unlike the instructions devised for the collection of concreteness ratings which explicitly mention the external senses, the instructions given for the assignment of imageability ratings are more ambiguous: external senses are not explicitly mentioned, while the stress is placed generically on 'sensory experience' ("Any word which, in your estimation, arouses a mental image (i.e., a mental picture, or sound, or other sensory experience) very quickly and easily should be given a high imagery rating"). This leaves open the possibility that participants assign high imageability ratings not only to words denoting perceivable entities or properties in the external world, but also to words referring to things that can be experienced directly through 'proprioception'—such as e.g. emotions and internal states. Happiness, pain, fear, love, hunger are not experienced through the external senses and do not evoke any visual or acoustic image; however, people have a direct internal experience of happiness, pain, fear, love, hunger etc. and they can quite easily evoke an 'image' of what it is like to experience happiness, pain, fear, love, hunger, etc.

This interpretation is compatible with Paivio's position that the notion of 'image' must be understood in a wide sense as the memory trace of any kind of external sensory experience as well as affective experience. However, it also extends the association between images and affective experience suggesting that people can imagine (i.e. recall the memory trace of) any kind of internal sensory experience, i.e. a 'proprioceptive' one. In fact, affective experience can be considered as a particular kind of 'proprioception'.

The concreteness ratings of words denoting emotions are even lower than the ratings people assigned to abstract words. This means that the norming instructions for concreteness did indeed convey an interpretation of concreteness according to which everything that cannot be perceived through the external senses is not concrete. On the contrary, the fact that emotion words had much higher imageability ratings than abstract words suggests that the norming instructions for imageability were interpreted differently and as a consequence imageability ratings were assigned also taking into account whether a word evoked not only external, but also internal sensory information, i.e. 'proprioception'.

According to this interpretation, imeageability is a twofold construct: on the one hand, imeageability is indeed a direct measure of concreteness, since imageability measures the ease/difficulty with which words evoke external (perceptive) sensory experience stored in memory, where for an object to be concrete it needs to be strongly linked to external (perceptive) sensory experience. On the other hand, according to our hypothesis, imageability also measures the ease/difficulty with which a word evokes internal (i.e. 'proprioceptive') sensory information; in this case concreteness and imageability are dissociated measures. Thus, in order to use imageability as a measure of the internal sensory experience evoked by a word we need to consider imageability ratings in relation to concreteness ratings. More specifically, since imageability ratings are a joint measure of the link of words with both external and internal sensory experience, the extent of a word's connection with internal sensory information can be obtained by subtracting its concreteness rating from its imageability rating. The higher the difference between the two ratings, the higher the 'weight' of internal sensory experience for the imageability rating. If the interpretation of imageability we put forward is correct, then the difference between imageability and concreteness can indeed give us a tangible (even though not completely accurate) measure of the internal sensory information aroused by a word.

We tested this hypothesis in an empirical study (Dellantonio et al. 2014) analyzing the concreteness and the imageability ratings included in the MRC database which is the most widely-used source for both these measures. On the basis of the database we showed that imageability and concreteness ratings are uncorrelated not only in the case of emotion words, but in general for all words referring to internal states, i.e. to bodily experiences such as balance, relaxation, movement, tremor, sit, rest, jump, run, walk, ache, sick, hunger, thirsty, warmth, itch, pain, cold, etc. Our statistical analysis showed that both for emotion words and words referring to bodily experiences the difference between imageability and concreteness ratings was significantly higher than that of a randomly chosen control group of words. Since both words denoting bodily experiences and emotions rely to a large degree on internal information, we consider this difference as evidence in support of the hypothesis that imageability measures not only the link between words and external sensory experience, but also the degree of the internal grounding of words.

Our analysis showed that the concreteness ratings of words denoting bodily experiences are significantly higher than the concreteness ratings of words denoting emotions, but it also revealed that that the imageability ratings of words denoting bodily experiences are significantly higher than the imageability ratings of words denoting emotions. This is compatible with the idea that bodily experiences are considered to be more perceptual than emotional states and therefore more concrete. As we maintained in the previous chapter, emotional (i.e. affective) information is a kind of 'proprioceptive' information. However, the 'proprioception' of emotions is more diffuse, less precisely localized in specific body parts and more complex to process than the 'proprioception' of other bodily experiences (on this see Chap. 5): this explains why emotions are felt to be less concrete than other internal bodily experiences like hunger, warmth, relaxation etc. At the same time, however, 'proprioception' is more directly/strongly connected with internal sensory experience and this is the reason why the imageability ratings of words denoting bodily experiences are higher than those of emotion words.

In this study, we also tested a number of technical words corresponding to what we previously called intratheoretical terms (e.g. DEMOCRACY, FACTOR, FUNCTION,

HYGIENE, INSTANCE, OBLIVION, OUTCOME, PRECURSOR, REGENCY, SPAN, SUBTRACTION, VETO etc.) for which understanding relies entirely on their definition. If imageability measures the degree of the internal grounding of a word, then the imageability ratings for these words should be very low and should correlate with (similarly low) concreteness ratings. The analysis confirmed this point since it revealed that for this word class imageability and concreteness ratings perfectly correlate and are equally extremely low.

If this interpretation of imeageability is correct and imageability—when it diverges from concreteness—is a measure of the degree to which words are grounded in internal sensory experience (i.e. on 'proprioception' in a broad sense), then the imageability construct might be used to solve the problem we started with in this section. Specifically, the difference of imageability and concreteness may give a measure of the internal grounding of a word and indicate whether a word/concept has a referential component/an experiential core concept (and is therefore *I*-theoretical) or does not have such a component/core, is an entirely inferential construct and therefore intratheoretical. More specifically, according to the definition we proposed, I-theoretical and intratheoretical terms/concepts are different from each other because while the former have a direct link to *internal* sensory experience, the latter are linked to sensory experience only indirectly. On the basis of this difference a plausible hypothesis is that I-theoretical terms more easily evoke internal ('proprioceptive') sensory information than intratheoretical terms, even though they do not evoke any *external* sensory experience because they are not concrete. Thus, words exhibiting a high difference between imageability and concreteness (i.e. high imageability but low concreteness) might be considered to be *I*-theoretical, while words exhibiting equally low imageability and concreteness might be considered to be *intra*theoretical.

This is an entirely new perspective on imageability, which—if further confirmed—could offer strong evidence in favor of an 'embodied view' of language according to which words are grounded not only on external, but also on internal, body-related sensory experience. Moreover, taken together the imageability and concreteness constructs could even provide a measure of both internal and external grounding to use in research (for a closer analysis of this method based on the comparison of imageability and concreteness ratings see Pastore et al. 2015).

Comparing imageability and concreteness ratings to establish whether a concept/word is I-theoretical has apparently anomalous implications that need to be discussed more specifically. If we qualify all the concepts/words that have low concreteness and high imageability ratings as I-theoretical, words/concepts referring to emotions and bodily experiences become I-theoretical too. To a certain extent this is a natural consequence of our hypothesis, since we suggest that *all* the non-concrete and non-intra theoretical concepts include both a referential component of a 'proprioceptive' (propriosensitive) nature and an inferential component consisting in linguistic information. In this sense, there is a commonality or, more specifically, a continuum from I-theoretical words/concepts and words/concepts referring to emotions, and those referring to bodily states. On the other hand, according to our hypothesis the referential and the inferential components have different 'weights' in these various cases. In other words, *all* non-concrete and non-intra theoretical concepts include these two components, but in different proportions. These differences have relevant consequences for linguistic mastery, since the mastery of words that rely on concepts characterized by a predominance of referential information depends primarily on capacities of a referential kind, while the mastery of words that rely on concepts characterized by a predominance of inferential information depends primarily on competence of an inferential kind. This is the reason why we described and classified word/concepts referring to internal states and basic emotions (that are characterized by a predominance of referential information) as more similar to concrete concepts than to *I*-theroretical concepts, the latter consisting predominantly of inferential information.

The imageability construct is not accurate enough to help us establish whether a concept is characterized by a predominance of 'proprioceptive' information or inferential information. However, according to the interpretation we offer, the imageability scale does offer a tangible measure of the link specific words enjoy with internal information and it therefore gives us a means for distinguishing at least between non-concrete concepts/words that do rely on internal ('proprioceptive') information and intratheoretical words/concepts. Assuming that we can identify words referring to basic emotions and internal states a priori on a semantic basis, the imageability measure offers a way to establish the problematic distinction we mentioned at the beginning of this section between I-theoretical and intratheoretical concepts/words.

Using the data included in the MRC database it is possible to identify a posteriori some words that rely heavily on internal information because they exhibit a particularly high difference between imageability and concreteness. We calculated the means of this difference for the words included in the database (7684 words) and choose only the ones that significantly exceeded that mean (standard deviation >2.00). Among these were words denoting emotions, bodily experiences or states closely related to emotions and to bodily experiences such as ROMANCE, HORROR, PARADISE, INTOXICATION, RELAXATION, DISTRESS, PLUNGE, WARMTH, but also other apparently different kinds of words. In fact, a particularly high discrepancy between imageability and concreteness is also exhibited by specific religious words like DEVIL, GODDESS, DEMON, SIN, CHRISTMAS, ETERNAL, DEVOTION, HELL, PARADISE, SPIRIT, ANGEL and by a number of words that are difficult to classify from a semantic point of view, even though they can surely be considered to be abstract, according to the classical definition, including not only FREEDOM and FRIENDSHIP (which we discussed in detail previously), but also others like ADOLESCENCE, ADULTERY, AUTHENTIC, AUTUMN, CHAOS, CHILDHOOD, CONFUSION, DANGER, DARKNESS, EMERGENCY, FAILURE, FANTASY, GHOST, GRADUATION, HOLIDAY, HORROR, IMPATIENCE, INDEPENDENCE, INFINITY, INTIMATE, MAGIC, MARRIAGE, MATERNITY, MATURITY, MIDNIGHT, MYSTERY, OBEDIENCE, PAST, POVERTY, REFLECTION, SUMMER, VANITY, WEIRD (for a closer analysis of these words and an attempt to classify them see Pastore et al. 2015).

If we are right, then these are all I-theoretical words/concepts characterized not only by an inferential component, but also by an internal grounding in 'proprioceptive' information. Thus, the concepts that support our competence with these words must also include a core of 'proprioceptive' origin. If the analysis we conducted for the example of 'freedom' is correct, then this core concept would be essential for the acquisition of the corresponding word. In addition, if the analysis of the example of friendship discussed here (Sect. 5) as well as previously (see Chap. 5, Sect. 10) can indeed be extended to all I-theoretical concepts, a 'proprioceptive' core concept is also essential to achieve a full mastery of all the corresponding word, since these words require not only inferential capacity, but also specific referential competence that relies on the perceptual (i.e. 'proprioceptive' or propriosensitive) information the core concept is based on.

References

Alderson-Day, B. (2011). Verbal problem-solving difficulties in autistic spectrum disorder and atypical language development. *Autism Research, 7*(6), 720–730.

Altarriba, J., & Bauer, L. M. (2004). The distinctiveness of emotion concepts: A comparison between emotion, abstract, and concrete words. *The American Journal of Psychology, 117*(3), 389–410.

Altarriba, J., Bauer, L. M., & Benvenuto, C. (1999). Concreteness, context availability, and imageability ratings and word associations for abstract, concrete, and emotion words. *Behavior Research Methods, Instruments and Computers, 31*(4), 578–602.

Aristotle. (1963). Categories. In J. L. Ackrill (Ed.), *Aristotle, categories and de interpretatione* (pp. 3–42). Oxford: Clarendon Press.

Barsalou, L. W. (1999). Perceptual symbol systems. *Behavioral and Brain Sciences, 22*(4), 577–660.

Barsalou, L. W. (2003). Abstraction in perceptual symbol systems. *Philosophical Transactions of the Royal Society of London: Biological Sciences, 358,* 1177–1187.

Barsalou, L. W., & Wiemer-Hastings, K. (2005). Situating abstract concepts. In D. Pecher & R. Zwaan (Eds.), *Grounding cognition: The role of perception and action in memory, language, and thought* (pp. 129–163). New York: Cambridge University Press.

Benjafield, J., & Muckenheim, R. (1989). Dates of entry and measures of imagery, concreteness, goodness, and familiarity for 1,046 words sampled from the Oxford English Dictionary. *Behavior Research Methods, Instruments & Computers, 21*(1), 31–52.

Berkeley, G. ([1710] 1999). *Principles of human knowledge and three dialogues.* Oxford: Oxford University Press.

Bransford, J. D., & McCarrell, N. S. (1974). A sketch of a cognitive approach to comprehension: Some thoughts on what it means to comprehend. In W. Weimer & D. Palermo (Eds.), *Cognition and the symbolic processes* (pp. 189–229). Hillsdale, NJ: Erlbaum.

Brown, D. J. (2006). *Descartes and the passionate mind.* Cambridge: Cambridge University Press.

Brysbaert, M., Warriner, A. B., & Kuperman, V. (2014). Concreteness ratings for 40 thousand generally known English word lemmas. *Behavior Research Methods, 46*(3), 904–911.

Canfora, L. (2006). *Democracy in Europe: A history of an ideology* (S. Jones, Trans.). Oxford: Blackwell.

Carey, S. (2009). *The origin of concepts.* Oxford: Oxford University Press.

Carnap, R. ([1928] 2003). *The logical structure of the world and pseudoproblems in philosophy* (R. A. George, Trans.). Chicago: Open Court.

Carnap, R. (1956). The methodological character of theoretical concepts. In H. Feigl & M. Scriven (Eds.), *Minnesota studies in the philosophy of science. Volume 1: Foundations of science and the concepts of psychology and psychoanalysis* (pp. 38–76). Minneapolis: Minnesota University Press.

Chiarello, C., Senehi, J., & Nuding, S. (1987). Semantic priming with abstract and concrete words: Differential asymmetry may be postlexical. *Brain and Language, 31*(1), 43–60.

Christian, J., Bickley, W., Tarka, M., & Clayton, K. (1978). Measures of free recall of 900 English nouns: Correlations with imagery, concreteness, meaningfulness, and frequency. *Memory & Cognition, 6*(4), 379–390.

Clark, J. M., & Paivio, A. (1989). Observational and theoretical terms in psychology. *American Psychologist, 44*(3), 500–512.

Coltheart, M. (1981). The MRC psycholinguistic database. *The Quarterly Journal of Experimental Psychology Section A: Human Experimental Psychology, 33*(4), 497–505.

Connell, L., & Lynott, D. (2012). Strength of perceptual experience predicts word processing performance better than concreteness or imageability. *Cognition, 125*(3), 452–465.

Coulson, S. (2000). *Semantic leaps: Frame-shifting and conceptual blending in meaning construction.* Cambridge, New York: Cambridge University Press.

Cruse, A. (2000). *Meaning in language. An introduction to semantics and pragmatics.* Oxford: Oxford University Press.

Crystal, D. (1995). *The Cambridge encyclopedia of the English language.* Cambridge: Cambridge University Press.

Dellantonio, S., & Pastore, L. (2011). 'Libertà' senza significato? Concetti astratti, cognizione e determinismo linguistico. *Rivista Internazionale di Filosofia e Psicologia, 2*(2), 164–186.

Dellantonio, S., & Pastore, L. (2014). Freedom and moral judgment. A cognitive model of permissibility. In L. Magnani (Ed.), *Model-based reasoning in science and technology. Theoretical and cognitive issues* (pp. 339–361). Berlin, Heidelberg: Springer.

Dellantonio, S., Mulatti, C., Pastore, L., & Job, R. (2014). Measuring inconsistencies can lead you forward: Imageability and the X-ception theory. *Frontiers in Psychology, 5,* 708. doi:10.3389/fpsyg.2014.00708.

Dixon, T. (2003). *From passion to emotions. The creation of a secular psychological category.* Cambridge: Cambridge University Press.

Ekman, P., Friesen, W. V., & Ellsworth, P. (1972). *Emotion in the human face.* Elmsford, NY: Pergamon.

Esposito, G., Dellantonio, S., Mulatti, C., & Job, R. (2016). Axiom, anguish, and amazement: How aitistic traits modulate emotional mental imagery. *Frontiers in Psychology, 7,* 757. doi:10.3389/fpsyg.2016.00757.

Eviator, Z., Menn, L., & Zaidel, E. (1990). Concreteness: Nouns, verbs, and hemi-spheres. *Cortex, 26*(4), 611–624.

Fauconnier, G., & Turner, M. (2002). *The way we think: Conceptual blending and the mind's hidden complexities.* New York: Basic Books.

Fauconnier, G., & Turner, M. (2008). Rethinking metaphor. In R. Gibbs (Ed.), *The Cambridge handbook of metaphor and thought* (pp. 53–66). New York: Cambridge University Press.

Friendly, M., Franklin, P. E., Hoffman, D., & Rubin, D. C. (1982). The Toronto Word Pool: Norms for imagery, concreteness, orthographic variables, and grammatical usage for 1,080 words. *Behavior Research Methods & Instrumentation, 14*(4), 375–399.

Gaigg, S., Gardiner, J. M., & Bowler, D. M. (2008). Free recall in Autism spectrum disorder: The role of relational and item-specific encoding. *Neuropsychologia, 46*(4), 983–992.

Gentner, D., & Bowdle, B. (2002). *Metaphor processing, psychology of. Encyclopedia of cognitive science* (pp. 18–21). London: Nature Publishing Group.

Gentner, D., & Bowdle, B. (2008). Metaphor as structure-mapping. In R. Gibbs (Ed.), *The Cambridge handbook of metaphor and thought* (pp. 109–128). New York: Cambridge University Press.

Gentner, D., Bowdle, B., Wolff, P., & Boronat, C. (2001). Metaphor is like analogy. In D. Gentner, K. J. Holyoak, & B. N. Kokinov (Eds.), *The analogical mind: Perspectives from cognitive science* (pp. 199–253). Cambridge, MA: MIT Press.

Gibbs, R. W. (1994). *The poetics of mind: Figurative thought, language, and understanding.* Cambridge: Cambridge University Press.

Gilhooly, K. J., & Logie, R. H. (1980). Age-of-acquisition, imagery, concreteness, familiarity, and ambiguity measures for 1,944 words. *Behavior Research Methods & Instrumentation, 12*(4), 395–427.

Glezerman, T. B. (2013). *Autism and the brain: A neurophenomenological interpretation*. Berlin: Springer.

Grandin, T. (2006). *Thinking in pictures and other reports from my life with Autism*. New York: Vintage Books.

Harman, G. (1987). (Nonsolipsistic) conceptual role semantics. In E. Lepore & B. Loewer (Eds.), *New directions in semantics* (pp. 55–81). London: Academic Press.

Harnad, S. (1990). The symbol grounding problem. *Physica D: Nonlinear Phenomena, 42,* 335–346.

Hermelin, B., & O'Connor, M. N. (1967). Remembering words by psychotic and subnormal children. *British Journal of Psychology, 58*(3–4), 213–218.

Hume, D. ([1740] 2009). *A treatise of human nature*. London: The Floating Press.

Izard, C. E. (1971). *The face of emotion*. New York: Appleton-Century-Crofts.

Jackendoff, R. (1985). *Semantics and cognition*. Cambridge, MA: MIT Press.

Johnson, M. (1987). *The body in the mind: The bodily basis of meaning, imagination, and reason*. Chicago: Chicago University Press.

Johnson, M. (2010). Metaphor and cognition. In D. Schmicking & S. Gallagher (Eds.), *Handbook of phenomenology and cognitive science* (pp. 401–414). Berlin, New York: Springer.

Kamio, Y., & Toichi, M. (2000). Dual access to semantics in Autism: Is pictorial access superior to verbal access? *Child Psychology & Psychiatry & Allied Disciplines, 41*(7), 859–867.

Kousta, S.-T., Vigliocco, G., Vinson, D. P., Andrews, M., & Del Campo, E. (2011). The representation of abstract words: Why emotion matters. *Journal of Experimental Psychology: General, 140*(1), 14–34.

Kunda, M., & Goel, A. K. (2011). Thinking in pictures as a cognitive account of Autism. *Journal of Autism and Developmental Disorders, 41*(9), 1157–1177.

Lakoff, G., & Johnson, M. (1980). *Metaphors we live by*. Chicago: Chicago University Press.

Lakoff, G. (1987). *Women, fire, and dangerous things*. Chicago: University of Chicago Press.

Lakoff, G. (2006). *Whose freedom? The battle over America's most important idea*. USA: Picador.

Lakoff, G. (2008). *The political mind. Why you can't understand a 21st-century politics with an 18th-century brain*. Greenwood: Viking.

Lakoff, G., & Johnson, M. (1999). *Philosophy in the flesh: The embodied mind and its challenge to Western thought*. New York: Basic Books.

Lakoff, G., & Turner, M. (1989). *More than cool reason: A field guide to poetic metaphor*. Chicago: Chicago University Press.

Lebuffe, M. (2010). The anatomy of the passions. In O. Koistinen (Ed.), *The Cambridge companion to Spinoza's Ethics* (pp. 188–222). Cambridge: Cambridge University Press.

Locke, J. ([1690] 1999). *An essay concerning human understanding*. Electronic classic series. The Pennsylvania State University. ftp://ftp.dca.fee.unicamp.br/pub/docs/ia005/humanund.pdf.

Lohmar, D. (2016). *Denken ohne Sprache. Phänomenologie der nicht-sprachliches Denken bei Mensch und Tier im Licht der Evolutionsforschung, Primotologie und Neurologie*. Berlin: Springer.

Lynott, D., & Connell, L. (2009). Modality exclusivity norms for 423 object properties. *Behavior Research Methods, 41*(2), 558–564.

Marconi, D. (1997). *Lexical competence*. Cambridge, MA: MIT Press.

Mitchell, P., & Ropar, D. (2004). Visuo-spatial abilities in Autism: A review. *Infant and Child Development, 13*(3), 185–198.

Murphy, G. L. (2004). *The big book of concepts*. Cambridge, MA: MIT Press.

Muth, A., Hönekopp, J., & Falter, C. M. (2014). Visuo-spatial performance in Autism: A meta-analysis. *Journal of Autism and Developmental Disorders, 44*(12), 3245–3263.

Nelson, D. L., & Schreiber, T. A. (1992). Word concreteness and word structure as independent determinants of recall. *Journal of Memory and Language, 31*(2), 237–260.

Nigel, T. (2013). Mental imagery. In E. N. Zalta (Ed.), *The stanford encyclopedia of philosophy*. https://plato.stanford.edu/archives/spr2017/entries/mental-imagery/.

Paivio, A. (1963). Learning of adjective-noun paired-associates as a function of adjective-noun word order and noun abstractness. *Canadian Journal of Psychology, 17*(4), 370–378.

Paivio, A. (1965). Abstractness, imagery, and meaningfulness in paired-associate learning. *Journal of Verbal Learning and Verbal Behaviour, 4*(1), 32–38.

Paivio, A. (1971a). Imagery and language. In S. J. Segal (Ed.), *Imagery current. Cognitive approaches* (pp. 7–32). New York, London: Academic Press.

Paivio, A. (1971b). *Imagery and verbal processes*. New York: Holt, Rinehart & Winston.

Paivio, A. ([1986] 1990). *Mental representations: A dual coding approach*. New York: Oxford University Press.

Paivio, A. (2007). *Mind and its evolution: A dual coding theoretical approach*. Mahwah, NJ: Erlbaum.

Paivio, A., Yullie, J., & Madigan, S. (1968). Concreteness, imagery, and meaningfulness values for 925 nouns. *Journal of Experimental Psychology. Monograph Supplement, 76*(1), 1–25.

Pastore, L., Dellantonio, S., Mulatti, C., & Job, R. (2015). On the nature and composition of abstract (theoretical) concepts: The x-ception theory and methods for its assessment. In L. Magnani, P. Li, & W. Park (Eds.), *Philosophy and cognitive science. Volume 2. Western and eastern studies* (pp. 35–58). Berlin: Springer.

Prinz, J. J. (2002). *Furnishing the mind. Concepts and their perceptual basis*. Cambridge, MA: MIT Press.

Putnam, H. (1975). The meaning of 'meaning'. In H. Putnam (Ed.), *Philosophical papers. Volume 2: Mind, language and reality* (pp. 215–271). Cambridge: Cambridge University Press.

Quine, W. V. O. ([1960] 2001). *Word and object*. Cambridge, MA: MIT Press.

Reed, H. B., & Dick, R. D. (1968). The learning and generalization of abstract and concrete concepts. *Journal of Verbal Learning and Verbal Behavior, 7*(2), 486–490.

Reilly, J., & Kean, J. (2007). Formal distinctiveness of high- and low imageability nouns: Analyses and theoretical implications. *Cognitive Science, 31*(1), 1–12.

Robinson, H. (1999). Editor's introduction. In G. Berkeley (Ed.), *Principles of human knowledge and three dialogues* (pp. ix–xxxvii). Oxford: Oxford University Press.

Ropar, D., & Peebles, D. (2007). Sorting preference in children with Autism: The dominance of concrete features. *Journal of Autism and Developmental Disorders, 37*(2), 270–280.

Rosen, G. (2012). Abstract objects. In E. N. Zalta (Ed.), *The stanford encyclopedia of philosophy*. http://plato.stanford.edu/archives/spr2012/entries/abstract-objects/.

Rubin, D. C., & Friendly, M. (1986). Predicting which words get recalled: Measures of free recall, availability, goodness, emotionality, and pronunciability for 925 nouns. *Memory & Cognition, 14*(1), 7994.

Ryle, G. ([1949] 2009). *The concept of mind* (60th anniversary Ed.). London, New York: Routledge.

Schwanenflugel, P. J. (1991). Why are abstract concepts hard to understand? In P. J. Schwanenflugel (Ed.), *The psychology of word meanings* (pp. 223–250). Hillsdale, NJ: Lawrence Erlbaum Associates.

Schwanenflugel, P. J., Harnishfeger, K. K., & Stowe, R. W. (1988). Context availability and lexical decisions for abstract and concrete words. *Journal of Memory and Language, 27*(5), 499–520.

Schwanenflugel, P. J., & Shoben, E. J. (1983). Differential context effects in the comprehension of abstract and concrete verbal materials. *Journal of Experimental Psychology. Learning, Memory, and Cognition, 9*(1), 82–102.

Schwanenflugel, P. J., & Stowe, R. W. (1989). Context availability and the processing of abstract and concrete words in sentences. *Reading Research Quarterly, 24*(1), 114–126.

Spreen, O., & Schulz, R. W. (1966). Parameters of abstractions, meaningfulness and pronunciability. *Journal of Verbal Learning and Verbal Behaviour, 5*(5), 459–468.

Toglia, M. P., & Battig, W. F. (1978). *Handbook of semantic word norms*. Hillsdale, NJ: Erlbaum.

Toichi, M., & Kamio, Y. (2001). Verbal associations for simple and common words in highfunctioning Autism. *Journal of Autism and Developmental Disorders, 31*(5), 483–489.

Turner, M. (1987). *Death is the mother of beauty: Mind, metaphor, criticism*. Chicago: Chicago University Press.

Van Fraassen, B. C. (1980). *The scientific image*. New York: Oxford University Press.

Verein Ernst Mach. ([1929] 2012). In F. Stadtler & E. T. Uebel (Eds.), *Wissenschaftliche Weltauffassung. Der Wiener Kreis*, Herausgegeben vom Verein Ernst Mach. Reprint on the first edition on behalf of the Institute Vienna Circle on the Occasion of ist 20th anniversary, with translations into English, French, Spanish and Italian. Wien, New York: Springer.

Vigliocco, G., Meteyard, L., Andrews, M., & Kousta, S. (2009). Toward a theory of semantic representation. *Language & Cognition, 1*(2), 219–247.

Wiemer-Hastings, K., & Xu, X. (2005). Content differences for abstract and concrete concepts. *Cognitive Science, 29*(5), 719–773.

Wiemer-Hastings, K., Krug, J., & Xu, X. (2001). Imagery, context availability, contextual constraint, and abstractness. In *Proceedings of the 23th Annual Conference of the Cognitive Science Society* (pp. 1134–1139). Mahwah, NJ: Lawrence Erlbaum.

Wilson, M. D. (1988). The MRC Psycholinguistic Database: Machine Readable Dictionary, Version 2. *Behavioural Research Methods, Instruments and Computers, 20*(1), 6–11.

Wolff, P., & Gentner, D. (2011). Structure-mapping in metaphor comprehension. *Cognitive Science, 35*(8), 1456–1488.

Wotschack, C., & Klann-Delius, G. (2013). Alexithymia and the conceptualization of emotions: A study of language use and semantic knowledge. *Journal of Research in Personality, 47*(5), 514–523.

Concluding Remarks

This book began by considering the scientific status of psychology and more generally the antimentalist stance of contemporary research on mind. If a discipline must fulfill the criteria established by the natural sciences and originally described by logical positivists and behaviorists in order to be considered scientific, then psychology can only become a science by excluding those objects that cannot be observed or verified experimentally from a third person perspective from its domain of study. This would require a major change from its traditional focus: not only would psychology need to overcome mind-body dualism, it would also need to reject the mentalist idea that mental phenomena, by virtue of being internal, conscious and introspectively accessible only to the first person, are absolutely different in kind from physical phenomena.

This antimentalistic stance is not only at the basis of philosophical and psychological behaviorism (which tried to get rid of internal states *tout court*) but also remains a hidden leitmotif in the later functionalist and cognitivist traditions. In fact, even though these traditions acknowledge that mental states are indeed the internal causes of observable behaviors, they continue to pursue the methodological precept that mental states must be described according to an extrinsic, third person perspective: they must be identified uniquely starting from external and publicly observable criteria, excluding any reference to introspection and to the qualitative dimension of the state itself, i.e. to the way it is consciously perceived by the subject who is in it. The so-called phenomenal character of experience remains excluded from investigations that follow these lines of research. In fact, cognitive approaches focus on a notion of representation that is defined in terms of information (representations store information) where information is interpreted as something that can be defined in objective and observable terms on the basis of its external sources without appealing to any subjective aspects of experience accessible only to the first person (i.e. by introspection).

The issue we raise in this work is whether this notion of representation and especially the idea of information it relies on can indeed explain our cognitive capacities or whether we need to rely on a more complex idea of information that also includes subjective factors. We support the latter argument and do so by choosing a rather specific perspective which allows us to investigate in detail *how a*

S. Dellantonio and L. Pastore, *Internal Perception*, Studies in Applied Philosophy, Epistemology and Rational Ethics 40, DOI 10.1007/978-3-662-55763-1

particular type of subjective experience influences our cognition (i.e. informs our representations) at least with respect to some cognitive capacities. Indeed, we do not address subjective experience in general terms but rather concentrate on a specific kind, viz. bodily experience. Furthermore, the cognitive capacity we consider most closely is categorization, i.e. the ability to identify and classify instances which is, among other things, the precondition for acquiring and mastering the meaning of words in natural languages.

Our analysis starts with a brief historical and theoretical discussion of the notions of representation and information that are at the basis of the explanatory apparatus in current cognitive research. We show that and how they have been derived from and built on the mind/computer analogy and examine why the computer metaphor leads to an idea of representation and information that is partially misleading and, in any case, inappropriate for describing human cognition. Two main issues are identified in this respect:

(i) First, we discuss the so-called 'grounding problem'. We need to explain how it is possible for humans, unlike computers, to acquire information on the basis of experience. Moreover, if we think of the cognitive system as a set of mechanisms operating on representations conceived as abstract and arbitrary symbols like the ones used in computer programs, we need to explain how these symbols can be meaningful, i.e. related to and grounded in perceptual information.

(ii) However, our analysis of the notion of information also demonstrates that ontologically, computers and living creatures are certainly not one and the same thing, since the latter are biological organisms equipped with a living body. Further, this body is not merely a neutral medium that transmits information from the outside world to the mind, but is itself a primary source of information which is available for cognitive processing and which also deeply 'contaminates' the experience we obtain through perception. Thus, we suggest that any account of cognition must not only better explain the relationship between representations and perception; it must also clarify how our first-person, direct, and aware experience of bodily states enters into our cognitive processes and how it influences them.

To address the issue of how conceptual representations might be linked to perception (i), we consider more closely the main families of theories that try to account for how concepts are mentally represented. We show that—in spite of their differences—they share the very same explanatory principle: they conceive of concepts as being represented in the form of feature lists which are linguistically described. This view is presented as extremely problematic. On the one hand, a list of features *per se* does not reveal how conceptual objects are actually structured; recalling a well-known remark of Gregory Murphy: "A pile of bird features does not make a bird—the parts need to be tied together in just the right way" (Murphy 2004: 48). On the other hand, if we assume that we can identify and classify some objects because we have learned a list of linguistically described features that belong to them (e.g. BIRDS have feathers, wings, head, legs, eyes, lays eggs, flies, etc.), we need to explain how we came to apprehend what these features concretely correspond to (e.g. what FEATHERS, WINGS, HEAD etc. are). Because of these problems, we suggest that the gap between perceptual experience and conceptual information must be bridged by assuming that concepts must include both a perceptual component (which we call a 'perceptual core concept') and linguistic information.

Only if we assume that concepts include both these kinds of information—this is the thesis we argue for—can we explain the capacity exhibited by people to use the corresponding words in their language. Borrowing a distinction introduced by Diego Marconi, we suggest that human semantic competence is indeed characterized by two main components: inferential competence (we are able to connect the words we know with others in our language and say e.g. that birds have feathers, wings, heads, legs, eyes, lay eggs, fly, etc.) and referential competence (we can say what objects or circumstances in the world correspond to a word). While the linguistic (feature-based) information accounts for the inferential component of our competence, the perceptual core concept explains its referential component.

To define exactly what this perceptual core concept might look like, we first analyze Prinz's notion of 'proxytype' which describes a visual schema of a class of objects that allows us to classify new observable instances as belonging or not belonging to it. This notion turns out to be extremely useful for explaining categorization as far as concepts corresponding to concrete, basic terms are concerned. However, it cannot explain how other conceptual classes are perceptually grounded and therefore needs to be extended. We do so by referring to Mark Johnson and George Lakoff's notion of 'schematic image'. A schematic image is conceived as composed of information coming from different modalities (i.e. bodily, kinesthetic, and perceptual information) and serves the aim of classifying multiple concrete phenomena which differ in form and content. We argue that the notions of 'proxytype' and 'schematic image' offer insight into how perceptual information must be organized to serve as a foundation for our conceptualizing capacity. Taken together, they can be described as 'perceptual schemata', i.e. as patterns or models of all kinds of perceptual information available to the cognitive system, *including bodily experience*, that accomplish a quasi-conceptual function in the sense that they are used to recognize and to classify different instances through perception. In fact, we suggest the label '*perceptual schema*' which indicates both how the perceptual core concept is actually structured and what information it consists of.

As for the problem of human versus artificial cognition (ii), we consider a line of criticism of the tendency of cognitive studies to conflate the ways in which living and artificial systems convey and process information that goes back to Ulrich Neisser and continues in contemporary embodied cognition. To address this, we develop an alternative view. Building on a tradition that goes from Charles Sherrington to James Gibson to more recent perspectives like those of Antonio Damasio and Bud Craig, we suggest that humans are equipped with a *propriosensitive perceptual system* that keeps track of all the information produced by the body and that the experience produced by this system influences external perception as well as cognition. We define this experience as 'proprioceptive' in a broad sense and use 'proprioceptive' (in inverted commas) as a synonym for propriosensitive. Moving beyond the current technical use of the term *proprioception* (which denotes only the information provided by the muscular and skeletal systems), we use it in the Gibsonian sense as a more embracing notion that includes all the forms of experience people have of their bodily states, i.e. kinesthetic, sensorimotor and postural information as well as sensations generated by and related to the general state of the body and its parts and organs (usually called 'interoception'). Starting with Damasio and Craig's views, we suggest that emotional experience is also—at least in part—an effect of propriosensitive perceptual states. We rely on Peter Goldie and Matthew Ratcliffe's analysis of emotional states to explain the relationship between emotions and the body as well as the way in which this affects our external perception and cognition.

While the first two chapters of the book provide an overview of the role that propriosensitive information plays in cognition, in later chapters this issue is addressed in relation to specific cognitive activities and, more specifically, in relation to categorization. We do not consider categorization in general, but enter into detail regarding specific cases of categorization that seem to be more intimately related to 'proprioception' and appear to be paradigmatic for discussing what exactly 'proprioceptive' (propriosensitive) information is and how it is involved in cognitive processing.

(1) First, we consider the case of *categories as different from, even though related to, that of concepts*. (2) Second, we investigate several specific conceptual classes that depend more heavily or directly than others on 'proprioception', i.e. *concepts denoting bodily experiences* (such as 'pain', 'thirst', 'hunger' etc.) and *emotions*. (3) Finally, we consider whether concepts describing unobservable entities or abstract conditions such as 'quark', 'friendship', 'democracy' and 'freedom' rely only on linguistic information or whether it is possible to differentiate between them by considering whether they are based, at least in part, on propriosensitive information.

(1) As for the issue of categories, we suggest that the synonymic use of 'category' and 'concept' in the cognitive sciences in misleading. Concepts—so we argue—are neither the first nor the only means through which our experience is organized. As has been maintained by a long-standing philosophical tradition and confirmed by some recent cognitive and developmental studies, in order for our experience to be organized according to the fine-grained and culturally driven criteria implicit in our conceptual apparatus, it must first be differentiated into broad categorical partitions. This is the precondition for the further and more specific organization of our experience carried out by concepts. In the hypothesis we discuss, categories are defined as macro-classes of highly general kinds that are produced spontaneously from the mind in a universal manner before any conceptual or linguistic organization of the experience occurs. The specific problem we investigate with respect to categories concerns their genesis: do we need to assume that they are innate or is it possible to explain how they might be formed from experience? Here is where the 'propriosensitive system' comes into play. As a matter of fact, our aim is to show that propriosensitive information might be an essential component of the information we use to form our categories.

We do not discuss this question in general. We instead analyze a particular case of categorical partition that has been the subject of a lively debate in the recent literature—i.e. the ANIMATE/INANIMATE dichotomy—since there is wide agreement on the fact that it develops extremely early in cognitive development, that it is cross-cultural and that it leads to the acquisition of our folk-biological concepts. The literature points out that infants can identify biological entities by relying on movement indicators: they classify together all the objects that exhibit a self-propelled, non-mechanical and goal-directed movement and differentiate them from other objects that do not move in this way. The point on which nativist and non-nativist (empiricist) views diverge concerns the origin of infants' knowledge of movement: how do they know that animate objects move in a certain way, while inanimate ones don't? Or, to put it in other terms, how do they know that certain movement characteristics are relevant for classifying objects? The most common answer to these questions is that infants rely on innate knowledge.

The solution we propose here takes a different tack: the cognitive system does not rely exclusively on information of exteroceptive origins, but also and simultaneously works on propriosensitive experience. Thanks to this additional information, humans (and other living creatures) are capable of 'sensing' the movements of their own body. And this means they might also be able—on the basis of "apperceptive transfer" (in the phenomenological sense proposed by Edmund Husserl)—to project this information onto the external world and to identify (single out) those objects which exhibit the same sort of movement that they perceive in themselves. Thus, the original categorical dichotomy that is usually described as ANIMATE versus INANIMATE would be more accurately described in term of the opposition 'IT MOVES LIKE ME' versus 'IT DOES NOT MOVE LIKE ME'. This process of analogizing apperception 'couples' and 'analogizes' propriosensitive information available in the first person with information extracted from perceived objects. The availability of propriosensitive information along with the idea that we can project this onto the external world might explain why the distinction between ANIMATE/INANIMATE arises so early in our cognitive development without recourse to any assumption of innate knowledge.

(2) To analyze specific conceptual classes that appear to depend more heavily or directly than others on propriosensitive information, we return to our earlier discussion on how categorization might work for words like e.g. 'cat' or 'vase' that refer to perceivable objects in the external world. Regarding this word class, we follow Marconi's view and suggest that we need to distinguish between referential and inferential competence. Moreover, to account for referential competence from a cognitive point of view, we put forward the hypothesis that human concepts include a perceptual component which describes the observable characteristics of the objects in the form of a *perceptual schema*. The issue we raise here concerns the concepts/words that *do not* denote entities that we can observe through the external senses, but rather states which we experience internally such us hunger, pain or emotions. Can we speak of referential competence in such cases? Does 'proprioception' allow us to identify and classify internal states on a referential basis—i.e. perceptively (propriosensitively)? Could we also speak of a perceptual schema in

relation to internal, propriosensitive information? We propose that the answer to this question is positive: even though words denoting the internal states of subjects such as bodily experiences and emotions do not refer to intersubjectively identifiable entities in the external world, they have some kind of referent that can be accessed through direct perception ('proprioception').

Our thesis is that the categorization of bodily experiences such us 'pain' or 'hunger' is based solely on propriosensitive information and in particular on the intersection of four main characteristics of this information: the intensity of such feelings, their localization in the body, their co-occurrence with other signals and above all their specific quality. The dimensions determined by these characteristics form the perceptual schema that we use to identify and classify these states from a referential point of view.

The case of emotions is certainly more complex than that of bodily experiences. On the basis of a discussion of several perceptual and cognitive theories of emotions, we suggest that emotions are not all alike, and argue that we must distinguish at least between basic emotions which appear to be universal and more directly related to bodily feelings and non-basic emotions which depend more heavily on cognitive factors. To discuss how emotions might be identified and classified on the basis of propriosensitive information, we rely on John Lambie and Anthony Marcel's differentiation between 'emotion states' and 'emotion experiences' to describe first-order and second-order emotion awareness. An emotion state is the feeling we have of the bodily states and changes that occur when we experience an emotion, while the emotion experience is the fully developed and integrated emotion we have and are aware of when we reflect on what we are experiencing. Also, taking this distinction further, we show that the same characteristics we suggested for the identification and classification of bodily experiences—specific qualitative sensations; somatic localization; specific intensity; presence/absence of specific concomitant sensations—can be used for the identification and classification of, at least, basic emotions. As for non-basic emotions, we conjecture that they cannot be identified and classified uniquely on the basis of propriosensitive information, but require further cognitive information for categorization.

These reflections are followed by the presentation of clinical evidence on the semantic competence of people suffering from Alexithymia and Autism Spectrum Disorder. This evidence corroborates both the idea that the referential identification and classification of bodily states and emotions falls on a continuum and the hypothesis that people are capable of identifying and classifying their emotions only if they have a truly, fully integrated emotional experience—i.e. including second order emotion awareness in the sense indicated by Marcel and Lambie.

(3) Concepts/words denoting external observable objects have a referential component which is accessible through perception. Concepts/words denoting internal states like bodily experiences and emotions have a referential component which is accessible through 'proprioception'. Starting from these conclusions, one could hypothesize that there is a third class of concepts/words denoting unobservable entities or abstract conditions that has neither a perceptual nor a 'proprioceptive' referential component, i.e. that is not perceptually grounded, either

externally or internally and fully relies on inferential information. This might well be the case—so we argue—but the class of concepts/words that are purely inferential constructs is much smaller and more internally differentiated than one might think.

First, if we accept the idea that concepts/words can have either external or internal perceptual grounding, then we must conclude that there are two kinds of 'concreteness'. On the one hand, concepts/words might be concrete in the classical sense, because they have an observable referent in the external world; on the other hand, they might be concrete because they have an internal referent that is detected by the propriosensitive system. Second, words/concepts might be more or less abstract depending on whether they entertain a weaker or stronger link to sensory experience. This link might be either to the perceptual or to the propriosensitive system.

To explain and to justify the introduction of these distinctions, we first rely on the discussion of theoretical/intratheoretical concepts carried out in the field of philosophy of science: while theoretical concepts are—at least indirectly—related to experience, intratheoretical concepts are merely inferential constructs. We illustrate these two different kinds of concepts in science—arguing that 'electron' and 'cell' are theoretical concepts while 'quark' is an intratheoretical concept—and we discuss how the same distinction can be applied to concepts/terms in ordinary usage which relate to propriosensitive information. The case we focus on most closely is that of 'freedom' (which, in line with Lakoff's view, we consider to be linked to propriosensitive information and therefore more like a theoretical concept) versus 'democracy' (which we consider to be mainly inferential and therefore more like an intertheoretical concept). Secondly, to sustain this position we present some evidence from studies on deficits in the semantic mastery of abstract words exhibited by people suffering from Autism Spectrum Disorder. Finally, we show that one of the classical views put forward in psycholinguistic research—i.e. Alan Paivio's *Dual Coding Theory*—can explain how abstract concepts are mentally represented in line with our analysis. Indeed, Paivio, also in agreement with Lawrence Barsalou's conception, suggests that internal (introspective and affective) information might be involved in the way we understand abstract words. Finally, we consider two scales that have been used in psycholinguistic research to measure the degree of concreteness versus abstractness of words and we show that—used jointly—these scales can provide a measure of the internal versus external grounding of specific words.

Our analysis of the function played by propriosensitive information with respect to cognition and more specifically with respect to categorization in specific cases is far from conclusive and the hypotheses we put forward in this regard do not constitute, unfortunately, any fully developed theory on the matter. However, our inquiry points in a very clear direction, indicating that propriosensitive information is an essential component of the information humans use for their cognitive processes. If the perspective we suggest is correct, then research on the nature and the role of internal information must return to the center of psychological research. Since internal information is primarily of a qualitative character, subjective qualities, in spite of their largely ineffable character, can no longer be excluded from psychological approaches to cognition.

Subjective qualities might need to be excluded from investigation in the physical sciences, but they are an intrinsic part of the object of psychology. This makes psychology a unique discipline that cannot be—at least not fully—assimilated by the natural sciences. As Thomas Nagel so clearly stated: "What has made modern physical science possible is the method of investigating the observable physical world not with respect to the way it appears to our senses [...]. The whole idea of objective physical reality depends on excluding the subjective appearances from the external world and consigning them to the mind instead. But it follows inexorably from this strategy that the same methods of objective physical understanding cannot be successfully applied to the subjective contents of the mind themselves" (Nagel 1994: 65–66). If we exclude these subjective qualities from psychological investigation, our explanation of how cognition works—of how we develop our knowledge—will be limited, unsatisfactory, and most probably even seriously misleading. Indeed, if our analysis is correct, not only are the states through which we experience the world intrinsically qualitative, but qualities are part of the information that forms the buildings blocks of our knowledge. Thus, we cannot address the issue of how our cognitive processes operate without considering the role that qualities and qualitative states play in them. The development of new ways for investigating these states as rigorously as possible is certainly one of the frontiers, if not the frontier for cognitive research.

References

Murphy, G. (2004). *The big book of concepts*. Cambridge, MA: MIT Press.

Nagel, T. (1994). Consciousness and objective reality. In R. Warner & T. Szubka (Eds.), *The mind-body problem: A guide to the current debate* (pp. 63–68). London, New York: Blackwell.